The Logica Yearbook 2020

The Logica Yearbook 2020

Edited by

Martin Blicha
and
Igor Sedlár

© Individual authors and College Publications 2021
All rights reserved.

ISBN 978-1-84890-376-0

College Publications
Scientific Director: Dov Gabbay
Managing Director: Jane Spurr

www.collegepublications.co.uk

Original cover design by Laraine Welch

All rights reserved. No part of this publication may be reproduced, stored in a retrieval system or transmitted in any form, or by any means, electronic, mechanical, photocopying, recording or otherwise without prior permission, in writing, from the publisher.

Preface

This volume is special, or at least we hope it to be. Logica 2020 was supposed to be held, as usual, in Hejnice, Czech Republic, in late June 2020. However, as so much else over the last year, the normal course of our working lives was disrupted by the Covid-19 pandemic. We felt, together with the rest of the Logica organizing team, that it makes little sense to organize Logica virtually; simply too much from what makes Logica what it is for so many people would be lost if we could not meet in person.

However, as we already had a number of high quality submissions in our hands before March 2020, we decided to continue in the Logica tradition at least by publishing a volume of papers based on the submitted abstracts.

We would like to thank the Institute of Philosophy of the Czech Academy of Sciences for its continuous support of the Logica series and the management of Hejnice Monastery for their understanding of the reasons that led to cancellation of Logica 2020. We thank the Czech Science Foundation for financial support (grant no. 20-18675S). We are grateful to College Publications and its managing director, Jane Spurr, for our consistently pleasant cooperation during the preparation of each Logica Yearbook Series volume. We are very grateful to the members of the Logica 2020 Programme Committee and the reviewers of the submissions we received for this volume for their time and valuable advice. Last but not least, we would like to thank all of the authors for their contributions and collaboration during the editorial process. We hope that future Logica volumes will not share the special status of the present one, and will be accompanied by live meetings enjoyed by everyone involved.

Prague, July 2021   Martin Blicha and Igor Sedlár

Table of Contents

Uniqueness of Logical Connectives in a Bilateralist Setting 1
  Sara Ayhan

Against the Countable Transitive Model Approach to Forcing 17
  Matteo de Ceglie

Infinity is Not a Size 33
  Matthias Eberl

Identity in Relevant Logics: A Relevant Predication Approach 49
  Nicholas Ferenz

Modelling Reflective Equilibrium with Belief Revision Theory 65
  Andreas Freivogel

A Useful Four-Valued Logic with Indefinite and Privative Negations:
Ammonius and Belnap on Term Negations 81
  José David García Cruz

Exploring a Result by Ghilardi: Projective Formulas vs. the Extension
Property 97
  Iris van der Giessen

Serious Statements and Plans 115
  John T. Kearns

The Third and Fourth Stoic Account of Conditionals 127
  Wolfgang Lenzen

Normative Parties in Subject Position and in Object Position 147
  Tereza Novotná and Matteo Pascucci

Logic and Human Practices 165
  Jaroslav Peregrin

A Note on Paradoxical Propositions from an Inferential
Point of View 183
  Ivo Pezlar

Three Conditionals: Contraposition, Difference-making and
Dependency 201
  Eric Raidl

A Logic of Affordances 219
   Sebastian Sequoiah-Grayson

Imperative Bilateralism 237
   Kai Tanter

Presupposition, Admittance and Karttunen Calculus 253
   Yoad Winter

Uniqueness of Logical Connectives in a Bilateralist Setting

SARA AYHAN[1]

Abstract: In this paper I will show the problems that are encountered when dealing with uniqueness of connectives in a bilateralist setting within the larger framework of proof-theoretic semantics and suggest a solution. Therefore, the logic 2Int is suitable, for which I introduce a sequent calculus system, displaying—just like the corresponding natural deduction system— a consequence relation for provability as well as one dual to provability. I will propose a modified characterization of uniqueness incorporating such a duality of consequence relations, with which we can maintain uniqueness in a bilateralist setting.

Keywords: uniqueness, bilateralism, proof-theoretic semantics, verification, falsification, connectives

1 Introduction

The question of uniqueness is the question whether a connective is characterized by the rules governing its use in a way that there is *at most* one connective playing its specific inferential role. The usual way to test this is to create a 'copy-cat' connective governed by the same rules and show that formulas containing these connectives are interderivable. Important work has been conducted showing the problematic features of certain logics leading to the failure of uniqueness for some connectives in these systems, along with refinements of the requirements for uniqueness (see Section 3.2). In this paper I will deal with bilateralist proof systems, more specifically with proof systems for the logic 2Int, which are bilateral in that they display two consequence relations: one for provability and one for dual provability (see Section 2.2). In such a setting, according to the common understanding of uniqueness, the question could be raised, whether this bilateralist proof-theoretic semantics (PTS) framework does not lead to different meanings

---

[1] I would like to thank Heinrich Wansing for the opportunity to discuss this topic and paper extensively and for his feedback, which is always helpful and on point.

depending on whether we prove or refute. Making this problem and my solution fully understandable requires laying some groundwork on bilateralism (see Section 2.1) and uniqueness (see Section 3.1) first. My aim is to show that the problems occurring in a bilateralist setting extend the problematic settings and solutions to ensure uniqueness that have been detected so far. Finally, I will propose a modification of our characterization of uniqueness that enables us to deal with uniqueness in bilateralism (see Section 3.3).

2 Bilateralism

2.1 Bilateralism and proof-theoretic semantics

The topic of bilateralism has received more and more attention in different areas within the past years including the area of PTS. In a nutshell, bilateralism is the view that dual concepts like truth and falsity, assertion and denial, or, in our context, proof and refutation should each be considered equally important, and not, like it is traditionally done, to concentrate solely on the former concepts. The debate started out in the context of considerations regarding an approach to the meaning of logical connectives, called "proof-theoretic semantics" (see Schroeder-Heister, 2018 for an extensive overview of this area, as well as Francez, 2015, which also covers the relation to bilateralism). In PTS, situated in the broader context of *inferentialism*, the meaning of logical connectives is determined by the rules of inference that govern their use in proofs. Bilateralism is therefore an approach to meaning which questions the established view, famously held by Frege (1919) and especially endorsed by Dummett (e.g., in (1976; 1981; 1991)), that denying a proposition A is equal to asserting the negation of A.[2] This has been opposed by several authors claiming that denial is a concept prior to negation and hence, should not be analysed in terms of it (see Martin-Löf, 1996; Restall, 2005). Thus, bilateralism demands an equal consideration of these dual concepts in that they should *both* be taken as primitive concepts, i.e., not reducible to each other.

Applying this to the proof-theoretic context, this amounts to demanding a proof system not only to characterize the proof (or verification) conditions of connectives but also their refutation (or falsification) conditions. Traditionally, in proof systems like natural deduction systems, the focus is only on the former, whereas, if we consider these notions to be on a par, we need to

---

[2] For an analysis of the established view as well as different ways to tackle it see also (Ripley, 2011).

Uniqueness of Logical Connectives in a Bilateralist Setting

extend these systems with rules that capture falsification conditions. This is what Rumfitt (2000) proposes in his seminal paper connecting bilateralism and PTS, in which he introduces a natural deduction system with signed formulas for assertion and denial. Wansing (2017) goes one step further and argues that considering the speech acts of assertion and denial as well as their internally corresponding attitudes of judgment and dual judgment on a par, gives rise to also considering a consequence relation dual to our usual consequence relation. He claims that, in order to take bilateralism seriously in the context of proof theory, we need to embed this principle of duality on a level deeper than that of formulas: Next to our usual consequence relation ($\vdash^+$), which captures the notion of verification from premises to conclusion, we also need to consider a *dual* consequence relation ($\vdash^-$) capturing the dual notion of falsification from premises to conclusion.[3]

2.2 Bilateralist calculi: N2Int and SC2Int

Therefore, Wansing (2017) devises a natural deduction system for the bi-intuitionistic logic 2Int, which comprises not only proofs (indicated by using single lines) but also *dual proofs* (indicated by using double lines). Also, a distinction is drawn in the premises between *assumptions* (taken to be verified) and *counterassumptions* (taken to be falsified). This is indicated by an ordered pair $(\Gamma; \Delta)$ (with Γ and Δ being finite, possibly empty multisets) of assumptions (Γ) and counterassumptions (Δ). Single square brackets denote a possible discharge of assumptions, while double square brackets denote a possible discharge of counterassumptions. The language $\mathcal{L}_{2Int}$ of 2Int, as given by Wansing, is defined in Backus-Naur form as follows:
$A ::= p \mid \bot \mid \top \mid (A \wedge A) \mid (A \vee A) \mid (A \to A) \mid (A \prec A)$.
I will in general use $p, q, r, \ldots$ for atomic formulas, $A, B, C, \ldots$ for arbitrary formulas, and $\Gamma, \Delta, \Gamma', \ldots$ for multisets of formulas. In a rule, the formula containing the respective connective of that rule is called the *principal formula*, while its components mentioned explicitly in the premises are called the *active formulas*.

As can be seen, we have a non-standard connective in this language, namely the operator of co-implication $\prec$,[4] which acts as a dual to implication, just like conjunction and disjunction can be seen as dual connectives. With

---

[3] In the spirit of Hacking's (1979, p. 292) conception of the sequent calculus as a metatheory, I use "$\vdash^+$" and "$\vdash^-$" both when talking about consequence relations in the metalanguage as well as for the sequent signs in the sequent calculus system which I will introduce below.

[4] Sometimes also called "pseudo-difference" (e.g., Rauszer, 1974) or "subtraction" (e.g., Restall, 1997), and used with different symbols.

that we are in the realms of so-called bi-intuitionistic logic, which is a conservative extension of intuitionistic logic by co-implication. Note that there is also a use of "bi-intuitionistic logic" in the literature to refer to a specific system, namely BiInt, also called "Heyting-Brouwer logic". Co-implication is there to be understood to internalize the preservation of non-truth from the conclusion to the premises in a valid inference. The system 2Int, which is treated here, uses the same language as BiInt, but the meaning of co-implication differs in that it internalizes the preservation of falsity from the premises to the conclusion in a dually valid inference (see Wansing, 2016a, 2016b, 2017, p. 30ff.).

From the viewpoint of bilateralism, i.e., considering falsificationism being on a par with verificationism, it is quite natural to extend our language by a connective for co-implication. The reason for this is that co-implication plays the same role in falsificationism as implication in verificationism: Both can be understood to express a concept of entailment in the object language. If we expect $\vdash^+$ to capture verification from the premises to the conclusion in a valid inference and $\vdash^-$ to capture falsification from the premises to the conclusion in a dually valid inference, then, just like implication internalizes provability in that we have in our system $(A; \emptyset) \vdash^+ B$ iff $(\emptyset; \emptyset) \vdash^+ A \to B$, likewise co-implication internalizes dual provability in that we have $(\emptyset; A) \vdash^- B$ iff $(\emptyset; \emptyset) \vdash^- B \prec A$.

With the two implication connectives also two negation connectives are defined: intuitionistic negation with $\neg A := A \to \bot$ and co-negation with $-A := \top \prec A$. Concerning switching between proofs and dual proofs, there is a division of labour between those negations in that we can move from proofs to dual proofs with intuitionistic negation and from dual proofs to proofs with co-negation: $(\Gamma; \Delta) \vdash^+ A$ iff $(\Gamma; \Delta) \vdash^- \neg A$ and $(\Gamma; \Delta) \vdash^- A$ iff $(\Gamma; \Delta) \vdash^+ -A$.[5]

Besides the usual introduction and elimination rules (henceforth: the *proof rules*) for intuitionistic logic, the natural deduction system N2Int, which is presented below, also contains rules that allow us to introduce and eliminate our connectives into and from dual proofs. These so-called *dual proof rules* are obtained by a dualization of the proof rules (see Wansing, 2017, pp. 32–34, for the description and the rules of the calculus) and having these two independent sets of rules is exactly what reflects the bilateralism of the proof system.

---

[5] I will not consider negation further in this paper, since I am concerned with connectives which are defined by their rules. See (Wansing, 2016a, 2016b, 2017) for a more detailed discussion, though.

N2Int

$$\frac{\displaystyle\genfrac{}{}{0pt}{}{(\Gamma;\Delta)}{\vdots}}{\dfrac{\bot}{A}} \bot E \qquad \frac{\displaystyle\genfrac{}{}{0pt}{}{(\Gamma;\Delta)}{\vdots}}{\dfrac{\overline{\top}}{A}} \top E^d \qquad \frac{\displaystyle\genfrac{}{}{0pt}{}{(\Gamma;\Delta)}{\vdots} \quad \genfrac{}{}{0pt}{}{(\Gamma';\Delta')}{\vdots}}{\dfrac{A \quad B}{A \wedge B}} \wedge I \qquad \dfrac{A \wedge B}{A} \wedge E_1 \qquad \dfrac{A \wedge B}{B} \wedge E_2$$

$$\dfrac{\overline{A}}{A \wedge B} \wedge I_1^d \qquad \dfrac{\overline{B}}{A \wedge B} \wedge I_2^d \qquad \dfrac{\overline{A \wedge B} \quad \overline{C} \quad \overline{C}}{C} \wedge E^d$$

$$\dfrac{A}{A \vee B} \vee I_1 \qquad \dfrac{B}{A \vee B} \vee I_2 \qquad \dfrac{A \vee B \quad C \quad C}{C} \vee E$$

$$\dfrac{\overline{A} \quad \overline{B}}{\overline{A \vee B}} \vee I^d \qquad \dfrac{\overline{A \vee B}}{\overline{A}} \vee E_1^d \qquad \dfrac{\overline{A \vee B}}{\overline{B}} \vee E_2^d$$

$$\dfrac{B}{A \to B} \to I \qquad \dfrac{A \quad A \to B}{B} \to E \qquad \dfrac{\overline{A} \quad \overline{B}}{\overline{A \to B}} \to I^d \qquad \dfrac{\overline{A \to B}}{A} \to E_1^d \qquad \dfrac{\overline{A \to B}}{\overline{B}} \to E_2^d$$

$$\dfrac{A \quad \overline{B}}{A \prec B} \prec I \qquad \dfrac{A \prec B}{A} \prec E_1 \qquad \dfrac{A \prec B}{\overline{B}} \prec E_2 \qquad \dfrac{\overline{B}}{\overline{B \prec A}} \prec I^d \qquad \dfrac{\overline{B \prec A} \quad A}{\overline{B}} \prec E^d$$

What I will present here additionally is a sequent calculus, which I will call SC2Int. SC2Int corresponds to N2Int in that we have a proof in N2Int of A from the pair $(\Gamma; \Delta)$ iff the sequent $(\Gamma; \Delta) \vdash^+ A$ is derivable in SC2Int, and we have a dual proof of A from the pair $(\Gamma; \Delta)$ iff the sequent $(\Gamma; \Delta) \vdash^- A$ is derivable in SC2Int. While Wansing (2017) proves a normal form theorem for N2Int, for SC2Int also a cut-elimination theorem can be proven (Ayhan, 2020). Since this means that our system enjoys the subformula property, this ensures the conservativeness of our system.[6]

---

[6]The exact relation between conservativeness and cut-elimination is debatable and, more specifically, depends on the system that is used (see Hacking, 1979; Kremer, 1988) but given

Sequents are of the form $(\Gamma; \Delta) \vdash^* C$ (with Γ and Δ being finite, possibly empty multisets), which are read as "From the verification of all formulas in Γ and the falsification of all formulas in Δ one can derive the verification (resp. falsification) of C for $* = +$ (resp. $* = -$)". Within the right introduction rules we need to distinguish whether the derivability relation expresses verification or falsification by using the superscripts $+$ and $-$. Within the left rules this is not necessary, but what is needed here instead is distinguishing an introduction of the principal formula into the *assumptions* (indexed by superscript a) from an introduction into the *counterassumptions* (indexed by superscript c). Thus, the set of *proof rules* in SC2Int consists of the rules marked with $+$ or with a, while the set of *dual proof rules* consists of the rules marked with $-$ or with c. When a rule contains multiple occurrences of $*$, application of this rule requires that all such occurrences are instantiated in the same way, i.e., either as $+$ or as $-$.

SC2Int

For $* \in \{+, -\}$:

$$\frac{}{(\Gamma, p; \Delta) \vdash^+ p} Rf^+ \qquad \frac{}{(\Gamma; \Delta, p) \vdash^- p} Rf^-$$

$$\frac{}{(\Gamma; \Delta) \vdash^- \bot} \bot R^- \qquad \frac{}{(\Gamma, \bot; \Delta) \vdash^* C} \bot L^a \qquad \frac{}{(\Gamma; \Delta) \vdash^+ \top} \top R^+ \qquad \frac{}{(\Gamma; \Delta, \top) \vdash^* C} \top L^c$$

$$\frac{(\Gamma; \Delta) \vdash^+ A \quad (\Gamma; \Delta) \vdash^+ B}{(\Gamma; \Delta) \vdash^+ A \wedge B} \wedge R^+ \qquad \frac{(\Gamma, A, B; \Delta) \vdash^* C}{(\Gamma, A \wedge B; \Delta) \vdash^* C} \wedge L^a$$

$$\frac{(\Gamma; \Delta) \vdash^- A}{(\Gamma; \Delta) \vdash^- A \wedge B} \wedge R_1^- \qquad \frac{(\Gamma; \Delta) \vdash^- B}{(\Gamma; \Delta) \vdash^- A \wedge B} \wedge R_2^- \qquad \frac{(\Gamma; \Delta, A) \vdash^* C \quad (\Gamma; \Delta, B) \vdash^* C}{(\Gamma; \Delta, A \wedge B) \vdash^* C} \wedge L^c$$

$$\frac{(\Gamma; \Delta) \vdash^+ A}{(\Gamma; \Delta) \vdash^+ A \vee B} \vee R_1^+ \qquad \frac{(\Gamma; \Delta) \vdash^+ B}{(\Gamma; \Delta) \vdash^+ A \vee B} \vee R_2^+ \qquad \frac{(\Gamma, A; \Delta) \vdash^* C \quad (\Gamma, B; \Delta) \vdash^* C}{(\Gamma, A \vee B; \Delta) \vdash^* C} \vee L^a$$

$$\frac{(\Gamma; \Delta) \vdash^- A \quad (\Gamma; \Delta) \vdash^- B}{(\Gamma; \Delta) \vdash^- A \vee B} \vee R^- \qquad \frac{(\Gamma; \Delta, A, B) \vdash^* C}{(\Gamma; \Delta, A \vee B) \vdash^* C} \vee L^c$$

$$\frac{(\Gamma, A; \Delta) \vdash^+ B}{(\Gamma; \Delta) \vdash^+ A \to B} \to R^+ \qquad \frac{(\Gamma, A \to B; \Delta) \vdash^+ A \quad (\Gamma, B; \Delta) \vdash^* C}{(\Gamma, A \to B; \Delta) \vdash^* C} \to L^a$$

$$\frac{(\Gamma; \Delta) \vdash^+ A \quad (\Gamma; \Delta) \vdash^- B}{(\Gamma; \Delta) \vdash^- A \to B} \to R^- \qquad \frac{(\Gamma, A; \Delta, B) \vdash^* C}{(\Gamma; \Delta, A \to B) \vdash^* C} \to L^c$$

$$\frac{(\Gamma; \Delta) \vdash^+ A \quad (\Gamma; \Delta) \vdash^- B}{(\Gamma; \Delta) \vdash^+ A \prec B} \prec R^+ \qquad \frac{(\Gamma, A; \Delta, B) \vdash^* C}{(\Gamma, A \prec B; \Delta) \vdash^* C} \prec L^a$$

that we can also prove admissibility of the other structural rules, this should be a safe assumption for our system.

Uniqueness of Logical Connectives in a Bilateralist Setting

$$\dfrac{(\Gamma; \Delta, B) \vdash^- A}{(\Gamma; \Delta) \vdash^- A \prec B} \prec R^- \qquad \dfrac{(\Gamma; \Delta, A \prec B) \vdash^- B \quad (\Gamma; \Delta, A) \vdash^* C}{(\Gamma; \Delta, A \prec B) \vdash^* C} \prec L^c$$

The following structural rules of weakening, contraction, and cut can be shown to be admissible in $\mathtt{SC2Int}$:

$$\dfrac{(\Gamma; \Delta) \vdash^* C}{(\Gamma, A; \Delta) \vdash^* C} W^a \quad \dfrac{(\Gamma; \Delta) \vdash^* C}{(\Gamma; \Delta, A) \vdash^* C} W^c \quad \dfrac{(\Gamma, A, A; \Delta) \vdash^* C}{(\Gamma, A; \Delta) \vdash^* C} C^a \quad \dfrac{(\Gamma; \Delta, A, A) \vdash^* C}{(\Gamma; \Delta, A) \vdash^* C} C^c$$

$$\dfrac{(\Gamma; \Delta) \vdash^+ D \quad (\Gamma', D; \Delta') \vdash^* C}{(\Gamma, \Gamma'; \Delta, \Delta') \vdash^* C} Cut^a \qquad \dfrac{(\Gamma; \Delta) \vdash^- D \quad (\Gamma'; \Delta', D) \vdash^* C}{(\Gamma, \Gamma'; \Delta, \Delta') \vdash^* C} Cut^c$$

3 Uniqueness

3.1 The notion of uniqueness

The issue of uniqueness has not received much attention in the literature. It was introduced more or less *en passant* in Belnap's (1962) famous response to the `tonk`-attack by Prior (1960) against an inferentialist view on the meaning of connectives.[7] Prior's intention in using `tonk` is to show that it leads the idea of PTS[8] *ad absurdum*. He argues that if the rules of inference governing the use of a connective would indeed be all there is to the meaning of it, then nothing would prevent the inclusion of a seemingly non-sensical connective, which ultimately trivializes our system, since it allows anything to be derived from everything. Belnap's proposal to solve this so-called *existence* issue of connectives was to demand extensions of a given system to be "conservative". In addition to that, he claims, one could wonder about the *uniqueness* issue of connectives. Once we have settled that it is allowed to extend our system with a certain connective, we can ask whether the rules of inference governing the connective characterize this connective *uniquely*.

Uniqueness as a requirement for a connective means that characterizing its inference rules amounts to exactly specifying its role in inference. There can be *at most* one connective playing this role; duplication of that connective with the same characterizing rules does not change its behaviour, neither in the premises nor in the conclusion. However, since Belnap's first requirement of conservativeness of the system was seen (by the responding literature and

---

[7] Belnap refers to a lecture by Hiż as being the actual origin of this idea.

[8] The term "proof-theoretic semantics" emerged much later of course but I use it whenever the idea fits to whatever terminology may be used in other places.

also by himself) to be far more important, the uniqueness requirement was more or less forgotten until it resurfaced in (Došen & Schroeder-Heister, 1985, 1988), which cover quite technical treatments of the issue as well as of connections to other proof-theoretic features. After that, the topic is absent from the debate for a long time again. A recent resuming of it can be found in (Naibo & Petrolo, 2015), which targets the question whether the uniqueness condition for connectives is the same as Hacking's "deducibility of identicals"-criterion[9]. Humberstone (2011, 2019, 2020b) is one of the few scholars who treats the topic quite extensively, dedicating one chapter of his monumental work on connectives to the question of uniqueness. His observations on the connections between (failure of) uniqueness of connectives, proof systems, and features of the consequence relation are of particular importance for the present purpose.

On the usual account of uniqueness two connectives # and #', which are defined by exactly the same set of inference rules and $\vdash$ being the consequence relation generated by the combined set of the rules, play exactly the same inferential role iff it can be shown for all A and B that $A \# B \dashv\vdash A \#' B$. Let us assume, for the moment, a common intuitionistic calculus and the example of conjunction. It can easily be shown that $\wedge$ is uniquely characterized by its usual natural deduction (resp. sequent calculus) rules (i.e., in our systems above: by its *proof rules*) governing it, since we can derive $A \wedge B$ from $A \wedge' B$ and vice versa, taking $\wedge'$ to be a connective governed by exactly the same rules as $\wedge$:

$$\frac{A \wedge' B}{A} \wedge'E \qquad \frac{A \wedge' B}{B} \wedge'E \qquad \frac{A \wedge B}{A} \wedge E \qquad \frac{A \wedge B}{B} \wedge E$$
$$\frac{}{A \wedge B} \wedge I \qquad \qquad \frac{}{A \wedge' B} \wedge I'$$

Thus, the interderivability requirement makes clear why, as I mentioned above, it is important to consider the underlying consequence relation when asking about the uniqueness of connectives.

Belnap's (1962, p. 133) original counterexample for satisfying the uniqueness condition is the connective plonk. We define plonk by the following rule: A plonk B can be derived from B. Since an extension with plonk (in the system Belnap is presupposing) is conservative, it can be stated that there is such a connective. However, it is not unique, since there can be another connective, which he calls plink defined by exactly the same rule,

---

[9]The condition that the structural rule of reflexivity for arbitrary formulas is provably admissible for every connective, i.e., each derivation using an application of it with a complex formula can be replaced by a derivation using applications of the rule with only atomic formulas (Hacking, 1979).

Uniqueness of Logical Connectives in a Bilateralist Setting

i.e., A `plink` B can be derived from B, which can otherwise play a different inferential role. The uniqueness requirement, as Belnap puts it, demands that another connective specified by exactly the same rules ought to play exactly the same role in inference, both as premise and as conclusion. In his system with reflexivity, weakening, permutation, contraction, and transitivity as structural rules, this amounts to showing that A `plonk` B and A `plink` B are interderivable. This, however, is not possible given that there is only this one rule governing the connectives and hence, `plonk` is not uniquely determined by its definition.

3.2 Problematic settings

There are several examples of connectives which are not uniquely characterized. This can be shown not only for 'ad hoc' connectives, in the sense that they are only thought of for this purpose, but also for connectives existing in calculi actually used, as, e.g., $\neg$ in FDE or $\square$ in system K.[10] Failure of uniqueness can—among other reasons—occur due to the specific formulation of the proof system, non-congruentiality of the logic or impurity of the rules. Humberstone (2011, p. 595f.) emphasizes that what does or does not uniquely characterize a given connective is the *set of rules* governing the connective, while sets of rules can be seen as a set of *conditions on consequence relations*.

The usual system Humberstone refers to when showing the non-uniqueness (e.g., of the examples mentioned in the last paragraph) is what he calls "sequent-to-sequent rules in the framework SET-FMLA", i.e., sequent rules with a set of formulas on the left side of the sequent operator and exactly one formula on the right. He also gives examples, however, where we have uniqueness in one particular formulation of the rules but not in another. Negation in Minimal Logic, for example, cannot be uniquely characterized by any collection of SET-FMLA-rules, but can be by others, which allow *at most* one formula on the right side of the sequent operator (Humberstone, 2020a, p. 186). Another example would be that disjunction is not uniquely characterized by its classical (or intuitionistic) rules when those are formulated in a zero-premise SET-FMLA system (Humberstone, 2011, p. 600).

Another important issue concerning uniqueness is the question of congruentiality, which can be a property of connectives, consequence relations, or

---

[10]Or for that matter $\square$ in every normal modal logic except for the Post-complete ones (Humberstone, 2011, pp. 601–605). Examples of failure of uniqueness are given in (Humberstone, 2011, 2019, 2020a; Naibo & Petrolo, 2015).

logics (depending on the specific understanding of those concepts). A logic is congruential, if for all formulas A, B, C, whenever A and B are equivalent insofar as they are interderivable according to a defined consequence relation of the logic, equivalence also holds when we replace A and B in a more complex formula C (Wójcicki, 1979).[11] This is closely connected to the notion of synonymy between formulas, since synonymy means that they are not only equivalent but also that replacing one by the other in any complex formula results in equivalent formulas. In view of (non-)congruentiality Humberstone (2011, p. 579f.) refines what I described as 'the usual account' (which he calls *uniqueness to within equivalence*) in that he claims that # is uniquely characterized by its set of rules iff every compound formed by that connective is *synonymous* to every compound (with the same components) formed by #' governed by exactly the same rules as #, which he calls *uniqueness to within synonymy*. This distinction coincides in the congruential case, but when the consequence relation is non-congruential, it can make a difference whether we demand the stronger or the weaker notion (2020a, pp. 183, 187).

Another terminological refinement is needed when we have systems with connectives governed by impure rules, i.e., rules which govern more than one connective. In this case, Humberstone (2011, p. 580f.) argues, we need to speak of the connective in question being uniquely characterized *in terms of* whichever connective also appears in its rules. An example would be a connective from another non-congruential logic, namely Nelson's constructive logic with strong negation, N4.[12] The rules governing strong negation, $\sim$, are impure because they also display other connectives, like conjunction and implication.[13] Thus, if we would ask for the uniqueness of $\sim$ in N4 (with impure rules), the question would always have to be "Is $\sim$ uniquely characterized by its rules in terms of $\wedge$ and $\rightarrow$?". In N4 this

---

[11] Wójcicki actually uses the term "self-extensional" instead of "congruential". The latter is used by Humberstone (2011, p. 175) for the case of connectives and consequence relations.

[12] I choose this example because N4 and 2Int are related in that strong negation $\sim$ in Nelson's logic can be read as a direct toggle between proofs and dual proofs, if it were added to 2Int, i.e., we would have $\vdash^+ A$ iff $\vdash^- \sim A$ and $\vdash^- A$ iff $\vdash^+ \sim A$.

[13] At least this is the case for the traditional (unilateral) calculi given for N4 (e.g., Prawitz, 1965, p. 97). In (Kamide & Wansing, 2012), however, there is a bilateral sequent calculus given for N4, which consists of pure rules only. This can be achieved, as in the case of SC2Int, with a system expressing different consequence relations. Likewise, Drobyshevich (2019) introduces the notion of a signed consequence relation between a set of signed formulas and a single signed formula as a bilateral variant of the notion of a Tarskian consequence relation and gives a bilateral natural deduction system for N4, which also contains pure rules only.

Uniqueness of Logical Connectives in a Bilateralist Setting

negation leads to the system's non-congruentiality, since for two formulas to be equivalently replaceable *in all contexts* it is not sufficient for the formulas to be provably equivalent, but additionally, we also need equivalence between the *negated* formulas.[14] For uniqueness this would mean that firstly, we would have to demand uniqueness to within synonymy. Secondly, it would tie uniqueness in this system to strong negation, since we would have to demand not only the interderivability of all formulas containing the connective in question with the formula containing the 'copy-cat' connective, but also the same interderivability with the strongly negated formulas. However, given that strong negation can only be uniquely characterized in terms of other connectives, this does not seem like a desirable system or a good solution to recover uniqueness.

3.3 Problems in a bilateralist system

The problem that occurs when asking about uniqueness in a bilateralist setting is closely connected to the last point addressed. However, I will show that in this case a much more intuitive solution can be given.[15] What causes trouble in the bilateralist proof systems laid out above—if we assume the common characterization of uniqueness (to within equivalence or synonymy)—is that we have two sets of rules for each connective and two consequence relations. It would make sense then to think of the proof rules as generating the consequence relation for provability and the dual proof rules as generating the dual consequence relation for dual provability. The specific consequence relation is of course important, since we usually test for uniqueness via interderivability, and in 2Int it can be shown for both relations *individually* that our connectives are uniquely characterized by only a part of the whole set of rules. Consider the case of conjunction, for example: We can show that $\wedge$ is uniquely characterized by its proof rules, since we can show (see derivations in Section 3.1) that both $(A \wedge B; \emptyset) \vdash^+ A \wedge' B$ and $(A \wedge' B; \emptyset) \vdash^+ A \wedge B$ are derivable. Likewise, taking $\wedge''$ to be a connective governed by exactly the same I^d- and E^d-rules from N2Int as $\wedge$ (resp. R^- and L^c-rules from SC2Int), we can show that it is also uniquely characterized by its dual proof rules, since $(\emptyset; A \wedge B) \vdash^- A \wedge'' B$ and $(\emptyset; A \wedge'' B) \vdash^- A \wedge B$ are derivable.

---

[14] A counterexample to congruentiality of N4 is that equivalence holds between $\sim (A \to B)$ and $(A \wedge \sim B)$ but not between $\sim\sim (A \to B)$ and $\sim (A \wedge \sim B)$ (Wansing, 2016a, p. 445).

[15] It would exceed the scope of this paper to consider all kinds of bilateral systems here, but, e.g., for Rumfitt's system with signed formulas the same problem would arise in a different guise, although the solution to maintain uniqueness might be—as in N4—not that elegant.

To show it for SC2Int:

$$\dfrac{\dfrac{\overline{(\emptyset; A) \vdash^- A}\; Rf^-}{(\emptyset; A) \vdash^- A \wedge'' B}\; \wedge'' R_1^- \quad \dfrac{\overline{(\emptyset; B) \vdash^- B}\; Rf^-}{(\emptyset; B) \vdash^- A \wedge'' B}\; \wedge'' R_2^-}{(\emptyset; A \wedge B) \vdash^- A \wedge'' B}\; \wedge L^c$$

$$\dfrac{\dfrac{\overline{(\emptyset; A) \vdash^- A}\; Rf^-}{(\emptyset; A) \vdash^- A \wedge B}\; \wedge R_1^- \quad \dfrac{\overline{(\emptyset; B) \vdash^- B}\; Rf^-}{(\emptyset; B) \vdash^- A \wedge B}\; \wedge R_2^-}{(\emptyset; A \wedge'' B) \vdash^- A \wedge B}\; \wedge'' L^c$$

However, there is no possibility to determine by this characterization that there is *only one* connective $\wedge$ because it is not possible to derive the following sequents:

$(A \wedge B; \emptyset) \vdash^+ A \wedge'' B$ $\qquad$ $(\emptyset; A \wedge B) \vdash^- A \wedge' B$
$(A \wedge'' B; \emptyset) \vdash^+ A \wedge B$ $\qquad$ $(\emptyset; A \wedge' B) \vdash^- A \wedge B$

The difference to `plonk` and `plink` is that in this case the one rule governing those connectives was 'not enough' to uniquely characterize a role in inference, while here a partial duplication of the rules (with proof rules only or dual proof rules only) is already enough for a unique characterization. So, in a way, we could say, the bilateral sets of rules overdetermine our connectives. However, since on the one hand both the proof rules as well as the dual proof rules uniquely characterize a connective, but on the other hand, there is no interderivability 'across' the consequence relations possible, how can we know that there is one conjunction with a unique meaning? Wouldn't that mean that we would be forced to say that there are actually two conjunctions, $\wedge^+$ and $\wedge^-$, one for the context of provability and one for dual provability? Thus, we could not confidently claim that our conjunction is uniquely characterized and has only one meaning in a system like N2Int or SC2Int, which would certainly have to be considered problematic.

However, let us take a look at our rules again, especially at the ones for implication and co-implication: What we can see here is that the different consequence relations are intertwined in characterizing these connectives. In N2Int this is observable by a mixture of single and double lines in the dual proof rules of implication, $\to I^d$ and $\to E_1^d$, and in the proof rules of co-implication, $\prec I$ and $\prec E_2$. In SC2Int this is indicated in the dual proof rules of implication, $\to R^-$ and $\to L^c$, as well as in the proof rules of co-implication, $\prec R^+$ and $\prec L^a$, by a mixture of $\vdash^+$ and $\vdash^-$ in the right

Uniqueness of Logical Connectives in a Bilateralist Setting

introduction rules and for the left introduction rules by the fact that active formulas are part of the assumptions *as well as* of the counterassumptions. Thus, the rules for implication as well as for co-implication need *both* consequence relations in one and the same rule application. This indicates that it would not be correct to think of the proof rules as generating the consequence relation and the dual proof rules as generating the dual consequence relation. Instead, both relations are generated by rules of both sets.[16] And this fact would support the point that we are not allowed to use different duplications of a connective when trying to show its uniqueness. Thus, when duplicating a connective, we need to use *the same duplication* for both proof rules *and* dual proof rules. By doing so, it is guaranteed that we are not talking about different connectives in different proof contexts.

So my proposal is to modify our characterization of uniqueness in a way that it also fits the context of bilateralism: In a bilateralist setting, instead of taking interderivability as a sufficient criterion for uniqueness, we also have to consider dual interderivability.

Definition 1 (uniqueness for bilaterally defined connectives) *In a bilateralist setting with consequence relations for verification as well as falsification, two n-place connectives # and #', which are defined by exactly the same set of inference rules,* play exactly the same inferential role, *i.e., are unique, iff for all $A_1, \ldots, A_n$ the formulas $\#(A_1, \ldots, A_n)$ and $\#'(A_1, \ldots, A_n)$ are interderivable as well as dually interderivable. To express this formally for the case of* 2Int:

(i) $(A \# B; \emptyset) \vdash^+ A \#' B$ *and* $(A \#' B; \emptyset) \vdash^+ A \# B$

(ii) $(\emptyset; A \# B) \vdash^- A \#' B$ *and* $(\emptyset; A \#' B) \vdash^- A \# B$.

With this definition of uniqueness we can state that all connectives of 2Int are uniquely characterized by their rules with respect to N2Int and SC2Int.[17]

A last question to consider, having Humberstone's distinction in mind, would be if this holds for *uniqueness to within equivalence* only or also for *uniqueness to within synonymy*. The question needs to be asked since 2Int is in fact also a non-congruential logic. The non-congruentiality in 2Int stems from the fact that not all formulas that are equivalent with respect to

---

[16] SC2Int shows this feature of 'mixedness' even nicer than N2Int, since in the former we have a $\vdash^*$ in *all* left rules, meaning that the rule holds for both verification and falsification.

[17] This also holds for the constants $\top$ and $\bot$, since in the case of n=0, $\#(A_1, \ldots, A_n) = \#$.

$\vdash^+$ are also equivalent with respect to $\vdash^-$. While for example $-(A \to B)$ and $A \wedge -B$ are interderivable with respect to $\vdash^+$, this does not hold for $\vdash^-$. Fortunately, the answer is that with the definition above we indeed get uniqueness to within synonymy because the following holds in 2Int: If we have equivalence, i.e., interderivability, of formulas both with respect to $\vdash^+$ as well as to $\vdash^-$, then it is guaranteed that these formulas are also replaceable in any more complex formula, i.e., then it is guaranteed that they are synonymous (for the proof see Wansing, 2016a). So the upshot of this definition is that we do not only get uniqueness to within equivalence but even uniqueness to within synonymy, without the need to consider compound formulas.

4 Conclusion

It has been made clear in other works that there are several features in logical systems which may cause problems for the claim that the connectives are uniquely characterized by the rules of that system. In this paper I examined the specific problem that occurs in a bilateralist setting in which we have two consequence relations, one for provability and one for dual provability. The refinements that are needed in such a setting differ from the ones that have been detected so far. In our specified case we also need to require that the interderivability of the formulas containing the connective is satisfied *for both consequence relations*. In other bilateral systems the specific formulation of what we require for uniqueness may differ, but in one way or another we will always need a requirement which holds not only for the context of verification (or assertion, or provability), but also for the context of falsification (or denial, or dual provability).

References

Ayhan, S. (2020). *A cut-free sequent calculus for the bi-intuitionistic logic 2Int*. https://arxiv.org/abs/2009.14787 (Unpublished Manuscript)

Belnap, N. D. (1962). Tonk, plonk and plink. *Analysis, 22(6)*, 130–134.

Došen, K., & Schroeder-Heister, P. (1985). Conservativeness and uniqueness. *Theoria, 51*, 159–173.

Došen, K., & Schroeder-Heister, P. (1988). Uniqueness, definability and interpolation. *The Journal of Symbolical Logic, 53(2)*, 554–570.

Drobyshevich, S. (2019). Tarskian consequence relations bilaterally: some familiar notions. *Synthese*.

Dummett, M. (1976). What is a theory of meaning? (II). In G. Evans & J. McDowell (Eds.), *Truth and Meaning: Essays in Semantics* (pp. 67–137). Clarendon Press.

Dummett, M. (1981). *Frege: Philosophy of Language* (2nd ed.). London: Duckworth.

Dummett, M. (1991). *The Logical Basis of Metaphysics*. London: Duckworth.

Francez, N. (2015). *Proof-theoretic Semantics*. London: College Publications.

Frege, G. (1919). Die Verneinung. Eine logische Untersuchung. *Beiträge zur Philosophie des deutschen Idealismus, 1*, 143–157.

Hacking, I. (1979). What is logic? *The Journal of Philosophy, 76(6)*, 285–319.

Humberstone, L. (2011). *The Connectives*. Cambridge, MA/London: MIT Press.

Humberstone, L. (2019). Priest on negation. In C. Başkent & T. M. Ferguson (Eds.), *Graham Priest on Dialetheism and Paraconsistency. Outstanding Contributions to Logic* (Vol. 18). Cham: Springer International Publishing.

Humberstone, L. (2020a). Explicating logical independence. *Journal of Philosophical Logic, 49*, 135–218.

Humberstone, L. (2020b). Sentence connectives in formal logic. In E. N. Zalta (Ed.), *The Stanford Encyclopedia of Philosophy* (Spring 2020 ed.). Metaphysics Research Lab, Stanford University. https://plato.stanford.edu/archives/spr2020/entries/connectives-logic/

Kamide, N., & Wansing, H. (2012). Proof theory of Nelson's paraconsistent logic: A uniform perspective. *Theoretical Computer Science, 415*, 1–38.

Kremer, M. (1988). Logic and meaning: The philosophical significance of the sequent calculus. *Mind, 97*, 50–72.

Martin-Löf, P. (1996). On the meanings of the logical constants and the justifications of the logical laws. *Nordic Journal of Philosophical Logic, 1(1)*, 11–60.

Naibo, A., & Petrolo, M. (2015). Are uniqueness and deducibility of identicals the same? *Theoria, 81*, 143–181.

Prawitz, D. (1965). *Natural Deduction: A Proof-Theoretical Study*. Stockholm: Almqvist & Wiksell.
Prior, A. N. (1960). The runabout inference-ticket. *Analysis, 21(2)*, 38–39.
Rauszer, C. (1974). A formalization of the propositional calculus of H-B logic. *Studia Logica, 33(1)*, 23–34.
Restall, G. (1997). *Extending Intuitionistic Logic with Subtraction*. https://consequently.org/papers/extendingj.pdf (Unpublished Note)
Restall, G. (2005). Multiple conclusions. In P. Hajek, L. Valdes-Villanueva, & D. Westerstahl (Eds.), *Logic, Methodology and Philosophy of Science: Proceedings of the Twelfth International Congress* (pp. 189–205). London: King's College Publications.
Ripley, D. (2011). Negation, denial, and rejection. *Philosophy Compass, 6*, 622–629.
Rumfitt, I. (2000). 'Yes' and 'No'. *Mind, 109(436)*, 781–823.
Schroeder-Heister, P. (2018). Proof-theoretic semantics. In E. N. Zalta (Ed.), *The Stanford Encyclopedia of Philosophy* (Spring 2018 ed.). Metaphysics Research Lab, Stanford University. https://plato.stanford.edu/archives/spr2018/entries/proof-theoretic-semantics/
Wansing, H. (2016a). Falsification, natural deduction and bi-intuitionistic logic. *Journal of Logic and Computation, 26(1)*, 425–450.
Wansing, H. (2016b). On split negation, strong negation, information, falsification, and verification. In K. Bimbó (Ed.), *J. Michael Dunn on Information Based Logics. Outstanding Contributions to Logic* (Vol. 8, pp. 161–189). Cham: Springer International Publishing.
Wansing, H. (2017). A more general general proof theory. *Journal of Applied Logic, 25*, 23–46.
Wójcicki, R. (1979). Referential matrix semantics for propositional calculi. *Bulletin of the Section of Logic, 8(4)*, 170–176.

Sara Ayhan
Ruhr University Bochum, Department of Philosophy I
Germany
E-mail: sara.ayhan@rub.de

Against the Countable Transitive Model Approach to Forcing

MATTEO DE CEGLIE

Abstract: In this paper, I argue that one of the arguments usually put forward in defence of universism is in tension with current set theoretic practice. According to universism, there is only one set theoretic universe, V, and when applying the method of forcing we are not producing new universes, but only simulating them inside V. Since the usual interpretation of set generic forcing is used to produce a "simulation" of an extension of V from a countable set inside V itself, the above argument is credited to be a strong defence of universism. However, I claim, such an argument does not take into account current mathematical practice. Indeed, it is possible to find theorems that are available to the multiversists but that the advocate of universism cannot prove. For example, it is possible to prove results on infinite games in non-well-founded set-theories plus the axiom of determinacy (such as $ZF + AFA + PD$) that are not available in $ZFC + PD$. These results, I contend, are philosophically problematic on a strict universist approach to forcing. I suggest that the best way to avoid the difficulty is to adopt a *pluralist* conception of set theory and embrace a set theoretic multiverse. Consequently, the current practice of set generic forcing better supports a multiverse conception of set theory.

Keywords: set theoretic multiverse, foundations of mathematics, forcing, set theory

1 Introduction

Universism is the thesis that there is only one set theoretic universe, V, that instantiates the axioms of ZFC. Such a position has been argued by several authors, for example by Martin (2001), Woodin (2011), Isaacson (2011), Shelah (2014), Barton (2020), and Livadas (2020). This universe is the so called *canonical* model of set theory, as opposed to all the others different models (such as, for instance, the constructible universe L). Although it is true that set theorists make use of all kinds of non-canonical models, universists typically insist that each of these models can be "simulated" within V

and that, in the end, they are only "simulated universes". Thus, universists typically argue against pluralist conception of set theory on the ground that the non-canonical universes that populate the so-called multiverses can be simulated within V. Before proceeding with the paper, I need to put forward a disclaimer. Throughout the paper, I will refer to "models" and "universes" interchangeably, as usually done in the multiverse literature.

Such a simulation is carried out through forcing. Forcing allows us to produce non-canonical models of V by extending its width. For example, we can use forcing to build models of $ZFC + \neg CH$, or models of $ZFC + CH$. In a very intuitive way, such an application of forcing produces an extended V that is a model of $ZFC + \neg CH$ (or $ZFC + CH$ as the case may be). Although this is the most intuitive explanation of forcing, one can understand forcing in at least three different ways:

1. the countable transitive model approach to forcing;

2. the Boolean evaluated approach to forcing;

3. the natural approach to forcing.

Among these, the first is the one that is usually accepted by set theorists. From a philosophical point of view, it is usually put forward as a defence of universism against pluralist conceptions, according to which set theory comprises a plurality of set-theoretic universes (for a defence of this approach to forcing see Barton, 2020). According to this, forcing allows us to simulate different kind of models and universes, without actually producing them. This is because forcing is applied only to a countable transitive model inside V. With the countable transitive model approach to forcing we are producing a forcing extension of that model, not of the entire set theoretic universe V. Consequently, the entirety of set theoretic practice and forcing application are actually carried out inside the single universe V. On the other hand, the natural approach to forcing can be applied to the entire set theoretic universe, and thus is the most similar and respectful of the intuition that every application of forcing produces a new set theoretic universe (for example, one in which CH is true and one in which it is instead false). It is the one defended by Hamkins and is the basis of his set theoretic multiverse (see, e.g., Hamkins, 2012), that accepts every possible set theoretic universe as legitimate and existing. Finally, the Boolean evaluated approach can be carried out looking only at the forcing relation, without appealing to models, and moreover it is needed to prove some interesting results like the Intermediate Model Theorem (this theorem says that if $V \models ZFC$ and $W \models ZFC$ is an intermediate model between V and one of its forcing extension then W is also a forcing extension of V, see Grigorieff, 1975 for details).

It is also possible to consider a purely syntactical approach to forcing, where we focus only on the partial order $(P, \leq)$. In this case we avoid the appeal to models, universes, generic filters and the like, and instead investigate only the forcing relation. In other words, we don't say that there is a model that satisfies φ, but instead only that there is a $p \in P$ which *forces* φ.

In this paper, I argue that the common interpretation of forcing is actually inadequate, since it does not align with current set theoretic practice and it restricts the range of possible results. In particular, I argue against its use in the universist's argument that the multiverse can be simulated inside the single universe V. I will do this by first arguing that, although in the single universe V we can actually simulate any non-canonical model of set theory, we cannot efficiently compare them all. In particular, we cannot resort to model-theoretic techniques, or to the tools of MAXIMIZE and restrictiveness (for example as presented by Maddy, 1996 and Incurvati & Löwe, 2016).

The paper is structured as follows. First of all I introduce the countable transitive model approach to forcing (Section 2), both from the mathematical perspective (Section 2.1) and the philosophical one (Section 2.2). After this, I present my arguments against it (Section 3), with particular attention to some examples regarding determinacy and ill-founded set theory (Section 3.1). In the end, some concluding remarks summarise the paper (Section 4).

2 The countable transitive model approach to forcing

Forcing was first developed in the 60's by Cohen as a method to produce non-canonical models of set theory (see Cohen, 2003). These models take the form of *width* extensions of V, i.e., enlargement of V "to the outside". Since its inception, forcing has been used mainly to prove the independence of certain set theoretic statements from ZFC, with the two most notable examples being the Axiom of Choice AC and the Continuum Hypothesis CH. To prove the independence of such statements, forcing is used to build two different extensions of V, one in which $ZFC + CH$ holds, and the other in which $ZFC + \neg CH$ holds (or, in the case of AC, one in which ZFC holds and one in which $ZF + \neg AC$ holds).

2.1 Mathematical details

As mentioned in the introduction, I am mainly interested in arguing against the countable transitive model approach to forcing and its use in the universist's argument that the multiverse can be simulated inside the single universe V. Before giving the details of such an argument, I need to clarify in what sense the set theoretic multiverse can be "simulated" in the single universe V. In set theory, we have the classical axiomatization ZFC and its canonical model, the cumulative hierarchy V. Through the application of set generic forcing, it is possible to produce a non-canonical model of ZFC from V. That is to say, we can "create" a new model of ZFC + some other statement: the usual example is the mutually incompatible models of $ZFC + CH$ and $ZFC + \neg CH$. In this case, we are creating two new models, V' and V^*, in which the Continuum Hypothesis is, respectively, true and false. These two models are "fatter", i.e., larger than the original V: they are usually considered *width extensions* of V, produced by the addition of new subsets to the cumulative hierarchy. There are also *height extensions* of V, produced by the additions of new ordinals, but they are not relevant for this particular argument. Crucially, set forcing cannot be applied to the whole V, but only to countable sets. Consequently, according to such approach, in order to produce a model of, for example, $ZFC + CH$, we take a countable set *in* V that "simulates" the whole universe, and apply set forcing to it to produce its width extension. But since we started with a countable set *inside* V, we are not producing a whole new universe, but only a slighter larger countable set inside the canonical universe (for a detailed account of forcing, see Nik, 2014).

The countable transitive model approach to forcing interprets the procedure just described at face value (see for example Barton, 2020). On such a view, there is nothing outside V, and when forcing a given model M we are not simulating something that could be outside V. Rather, the alternative universes that we are "seeing" are nothing but a technical artefact. This is the usual answer that universists give considering the fact that the cumulative hierarchy of sets V is usually conceived as already containing all possible subsets. Consequently, how could we add new subsets to it if it already contains all of them? The countable transitive model approach solves this impasse by appealing to the fact that all consistent first order theories have a countable model. Thus, there is a countable set $M \subset V$ that is a model of ZFC. From Cantor's diagonal argument, we know that no countable set can contain all subsets of any infinite set it contains, so our countable set M

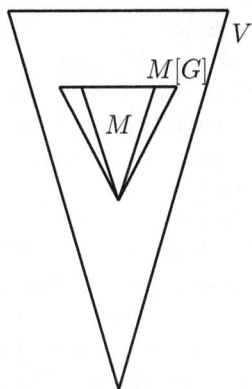

Figure 1: A representation of the countable transitive model approach to forcing

must miss some of these subsets. Forcing (following this approach) helps to "complete" the set M, by adding some of those missing sets. Moreover, we choose which subsets we want to add in such a way that the new expanded M' will be a model of the theory we are trying to study (for example $ZFC + \neg CH$).

Without delving too deeply in the details, the following is a sketch of how the forcing technique works. We start by specifying an infinite partial order $\mathbb{P} \in M$. Since M is a countable set, it follows from Cantor's diagonal argument that there is some $G \subset \mathbb{P}$ that it is not in M. We pick a G such that it is *generic filter* over M (so that the resulting operation is called *set-generic* forcing). The key properties of a generic filter are that:

1. G is a subset of $\mathbb{P}$;

2. $1 \in G$;

3. if p is a forcing condition (forcing conditions are members of $\mathbb{P}$, the smaller they are the stronger they are) weaker than q and $q \in G$, then $p \in G$;

4. if $p, q \in G$, then there exists in G a forcing condition smaller than both of them;

5. G meets all dense subsets of $\mathbb{P}$ that are in M.

The first four properties ensure that G is a filter, the last one that it is generic. We add such a G to M, and write $M[G]$ to denote the resulting extended M with the missing subset G. Let φ be the statement we want to prove. With forcing we can prove that for any such generic missing subsets of $\mathbb{P}$, $M[G] \models ZFC + \varphi$. At this point, we consider the forcing relation $\Vdash$, that connects facts about M to facts about $M[G]$. From this perspective, we are proving $ZFC + \varphi$ by proving a fact about M, i.e., that M, provided the poset $\mathbb{P}$, *forces* $ZFC + \varphi$, in symbols $M \Vdash_\mathbb{P} ZFC + \varphi$. Furthermore, note that this is only a claim about sets in M, and it does not carry any assumption about the whole V. However, it is possible to generalise such result, and prove that if there exists any M-generic G that is one of the "missing subsets" of $\mathbb{P}$, then a statement φ is forced if and only if for every such G, $M[G] \models \varphi$.

So, in other words, with forcing we can prove facts about M by adding to it some of the subsets that it is missing. Obviously such a method is more general: we are proving that any countable model $M \in V$ can be extended to the countable model $M[G] \models ZFC +$ some statement. Consequently, while the facts of the form $M \Vdash \varphi$ that we prove with forcing apply only inside V, we can interpret them as a "simulation" of an extended universe $V[G]$.

2.2 The philosophical argument in favour of universism

The appeal of this method for universists lies in the fact that it allows us to bypass the limitation of a single V that contains all possible subsets, while still being able to use forcing to produce non-canonical models of set theory. Their opposition to the multiverse conception of set theory is rooted in the fact that there are no V-generic filters, simply because, according to universists, V already has all the possible subsets, so we cannot apply our method of forcing to add some more subsets (the situation with class forcing is more complex, see for example Antos, 2018).

Consequently, the advocate of universism argues that it is not possible to apply set-generic forcing to the whole V, so that the very notion of "extension of V" is meaningless. All we can do is to use countable transitive sets inside V as simulations of the whole V, and consider the extensions of such countable models as "simulacra" of those elusive extensions of V.

Even though it may seem that the countable transitive model approach curtails the power of forcing, it is still versatile and powerful enough. For instance, in V we can construct both models of $ZFC + V = L$ and models of $ZFC + LCs$ (e.g., $ZFC +$ "there exists a measurable cardinal"), even though these two models are incompatible ($V = L$ is the Axiom of Constructability,

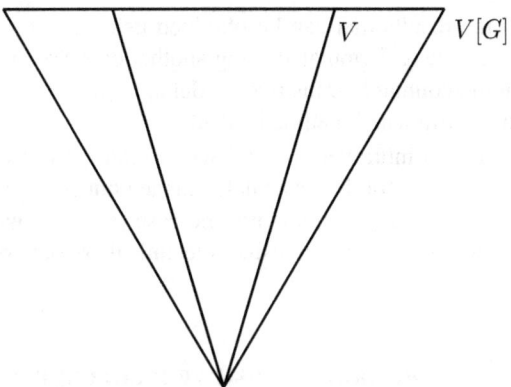

Figure 2: A representation of the natural interpretation to forcing

that says that all the sets of the universe can be built from simpler sets, and it is incompatible with the existence of most large cardinals). A point in favour of such an interpretation of forcing is the fact that every result achieved regarding forcing extensions and non-canonical models of ZFC can be achieved using such restrictive interpretation of forcing.

Moreover, even other interpretations of forcing, e.g., the Boolean evaluated method and the natural forcing interpretations, can still be used as arguments for universism. With the Boolean evaluated method, instead of assigning a truth value to the statement φ we want to force, we pick a truth value from an atomless Boolean Algebra. We then pick an ultrafilter $\mathbb{P}$ in this algebra that assigns truth values to statements in our theory T. We can finally interpret the resulting model of the theory $T + \mathbb{P}$ as extending the old one with this new ultrafilter. The natural forcing method on the other hand admits the existence of a V-generic filter missing a subset G and thus of generic extensions $V[G]$. However, both alternative interpretations of forcing can still be considered compatible with the Single Universe view:

- the Boolean evaluated method can also be seen as being carried out entirely within V, since there is no appeal to its extensions;

- the natural forcing interpretation (for details see Hamkins, 2012) can be interpreted as a technical tool that creates the illusion of real extensions of V.

Moreover, all the results that can be obtained using one of these forcing interpretations can be re-formulated using another one. So, for example, if I prove φ using the countable transitive model approach, I can also prove it using the Boolean evaluated method instead.

Thus, we have an intuitively natural way to interpret forcing from the universism perspective: forcing is actually carried out completely inside V, and even when it seems that this is not the case (with the natural forcing interpretation), we are in fact only using a technical artefact to make things easier and more intuitive.

3 Against the countable transitive model approach

The countable transitive model approach, while enticing, has some problems. First of all suppose that we accept the natural interpretation of forcing only for its instrumental value. This means not committing to the existence of the non-canonical models, however without committing to their complete non-existence either. They are useful and allows us to gain insights on the true and unique universe of sets, and as such they are admitted in our set theoretic discourse. Although this objection is compelling, I claim that a purely instrumental view of these canonical models is already a pluralist and multiversistic view of the set theoretic universe. It may well be that, if asked, a universist will profess that there exists only one set theoretic universe (albeit indeterminate and incomplete). However, if her day to day practice is actually multiversist, her argument loses strength.

Nevertheless, suppose instead that we don't want anything to do with these non-canonical extensions of V, and instead claim that we can investigate the countable models just as easily as the multiversist. In doing so, we appeal to the fact that we can use methods from model theory and methods from interpretability theory (see Visser, 2006). However, neither of these methods is ideal, and there are some problems in using them. Very briefly, interpretability theory uses the notion of isomorphism to study the *sameness* of theories (e.g., mutual interpretability, synonymy, etc.). In particular, it is possible to use it to prove that a theory is restrictive over another theory, i.e., it proves fewer isomorphisms. This approach is quite promising, but it has one important drawback: it can only be used to compare models of theories with different languages (see Visser, 2006 for details). In our example, both the theories $ZFC + V = L$ and $ZFC +$ "there exists a measurable cardinal" share the same language, so we cannot compare them using restrictiveness methods. The model-theoretic tools are also problematic. In general, it is preferable to use object-linguistic methods, i.e., methods that are carried out directly in the object language . However, model-theoretic methods are essentially metatheoretic, since they take models of a given theory to be the objects of the investigation, and we cannot do this in the theory itself. In general, a method that works directly in the theory is to be preferred, since in this way we avoid having to justify the legitimacy of our metatheory and our methods. The main consequence of this fact is that we cannot efficiently compare two non-canonical models. It would be better if we could compare them in a common context. In principle, ZFC could be such a context, but other than the technical problems just hinted there are also philosophical and interpretative problems. In particular, if we take a universist stance regarding V, we have to commit to it: is it equal to L, but without measurable cardinals, or not (but with measurable cardinals)? This forced choice will restrict the range of possible results we can prove. So the problem for the universist still stands: committing to a single universe, in which the full power of forcing techniques cannot be used, or accept a pluralist conception (even if only for instrumental reasons). Indeed, all these non-canonical models are available in the set theoretic multiverse, and we can prove all possible results comparing them, especially isomorphisms between their structures. This can be done in a better way because in the set theoretic multiverse we can define a different language for each of the universes (see Hamkins, 2012 for details). This enables us to use the methods of restrictiveness without any limitation. Moreover, we could even easily use model theoretic methods without having to retreat to the meta-theory: we can do all sorts of model theoretic constructions directly in the theory of the multiverse, in which the various universes are only objects. Clearly this cannot be done in *all* multiverses: there are multiverse conceptions perfect for this purposes and conceptions less so. However, in general, compared with a pluralistic conception of set theory, such as the ones advocated by Hamkins (2012), Steel (2014), Antos, Friedman, Honzík, and Ternullo (2018) and Gorbow and Leigh (2020), the universist conception has a lot less power in comparing these non-canonical models.

The other problem with this account is that it does not allow to compare two non-canonical models that are *mutually incompatible* in an efficient way (as already mentioned). Moreover, it is not possible to efficiently use one model in the investigation of the other one. For example, it is possible to first produce by forcing a model M of $ZFC + CH$ and then a model N of $ZFC + CH + PD$ such that $M \subseteq N$. And in this case we can use all the tools available in the model of $ZFC + CH$ in our second model of $ZFC + CH + PD$. However, consider the situation in which we first force a model M of $ZFC + \neg CH$, and then a model N of $ZFC + CH$ such that $M \subseteq N$. In this case, we cannot use any added object or method from M in the new model N (in particular, we cannot use the existence of a set x such that $|\mathbb{N}| < |x| < |\mathbb{R}|$ from the model $M \models ZFC + \neg CH$ in the model $N \models ZFC + CH$). For a less trivial example, consider $M \models ZFC +$ "there exists a measurable cardinal" and $N \models ZFC + V = L$: we cannot use the existence of a measurable cardinal in M to prove something about N, since the measurable cardinal cannot be found in N.

Another drawback of the countable transitive model approach is that it can be applied only to a very small subset of all the possible models of set theory, i.e., to only the countable transitive ones, and even this application is problematic, mainly because there are meta-mathematical problems regarding them (as Hamkins, 2012 points out). First, it follows from the Incompleteness Theorem, that if ZFC is consistent, then it cannot prove the existence of these models. A possible solution would be to assume that ZFC is consistent (so we assume $Con(ZFC)$). In this new theory, we can indeed prove that there are countable models of ZFC. However, we still cannot prove that they are *transitive*, since if they exist then $ZFC + Con(ZFC)$ is consistent (so $Con(ZFC + Con(ZFC))$). Consequently, we can either simply assume their existence (but this is not really satisfactory), or work in a stronger theory than ZFC that can serve as a suitable meta-theory. However, this move is hardly justifiable, since the same arguments will still apply to the meta-theory. Indeed, the strength of the universist's argument comes from the fact that everything is carried out inside V, and in the *theory* itself. However, the insurmountable problem of incompleteness compels the universist to carry out forcing in the meta-theory. Being a meta-theory, it has its own model, in which V is an object. Obviously this "meta-model" cannot be located inside V itself, but it has to be "outside" it. For the multiversist this is obviously not a problem (see for example Gorbow & Leigh, 2020 for an interesting construction around this problem). Even a very moderate potentialist (potentialism is one of the positions, the other being actualism, regarding the "expandability" of V: actualism states that V is not, while potentialism states that it is possible, with radical potentialism admitting expansion both in height, adding new ordinals, and in width, adding new subsets, while for moderate potentialism only one of those expansions is possible) that does not claim the existence of a full multiverse can still be able to operate "outside" V with the help of admissible sets (for details on how admissible sets work, see Barwise, 1975). On the other hand, the universist can no longer claim that everything he is doing is done inside V, and has to concede that there is actually something other than V.

Nevertheless, let's ignore this point, and concede to the universist that their countable transitive model approach to forcing can actually be carried out inside V in its entirety without any of the problems mentioned above. Even in this case, the universist argument incurs in one major problem, caused by the fact that it cannot compare two incompatible models and use them to prove something in one another. Without this possibility, the range of possible results and theorems is restricted.

3.1 Determinacy and ill-founded set theory

The inability to consider (in the sense explained above) two mutually incompatible models produced by forcing means that there are results that cannot be proved in the Single Universe context described above. In contrast, in a multiverse conception of set theory (or a "Single Universe" with an instrumental acceptance of the natural approach to forcing) we can prove a theorem in a particular model using tools and methods from an incompatible model. Obviously we should proceed with caution. Suppose we have produced two models $N \subseteq M$, such that $M \models ZFC +$ "there exists a huge cardinal" and $N \models ZFC + V = L$. In this case, we cannot use the existence of a huge cardinal in M to prove the existence of a measurable cardinal in N, even though N is contained in M and the existence of a huge cardinal implies the existence of a measurable one.

For another, more complex, example, consider the relation between the Axiom of Determinacy (AD), ill-founded set theory, and infinite games. An infinite game is a series of plays in which two players, player I and player II, alternately pick a number (this could be a natural number or, in the case of the Banach-Mazur game, a real number). After infinite many such moves, a sequence of natural numbers $(n_i)_{i \in \omega}$ (or of real numbers, but here I will stick to the simplest case) is generated: if $(n_i)_{i \in \omega} \in A$, then player I wins, otherwise if $(n_i)_{i \in \omega} \notin A$ player II wins (Figure 3 is an example of such games). The set A is the *outcome* set, or winning set. Player I has a *winning strategy* if there is a sequence of plays that he can make such that the overall sequence ends up in the outcome set, while player II has a winning strategy if he can make plays that avoids player I winning. The Axiom of Determinacy states that every infinite game is determined, i.e., one of the players has a winning strategy (for details on AD see Woodin, 1999). Note

I	r_0		r_2	...	r_n
II		r_1		r_3	...	r_{n+1}

Figure 3: An infinite game

that these does not entail that both players can have a winning strategy. We know that this axiom is incompatible with the Axiom of Choice, since using AC it is possible to build a game in which both players lose. However, it is possible to restrict ourselves to the Axiom of Projective Determinacy (PD), that instead states that the winning sets, i.e., the victory conditions, are projective sets, that is compatible with AC.

Intuitively, we can arrange and represent these infinite games as trees. All the possible sequences of the game are the various branches of the tree, and each play determines which branch we are following. The winning conditions does not change, and we can interpret the winning strategy in terms of branches: a winning strategy for player I is a branch that is in the outcome set, while a winning strategy for player II is a branch that is not in the winning set (see Figure 4).

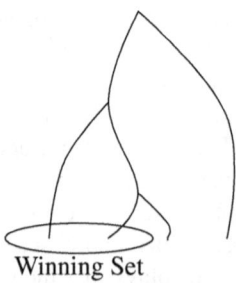
Winning Set

Figure 4: An infinite game as a tree

Now, since these infinite games are representable as trees, it is very natural to study them by means of a non-well-founded set theory (for an introduction to non-well-founded set theory, see Aczel, 1988). Non-well-founded set theory admits ill-founded sets, that is, sets that contains themselves as members. These sets are not admitted in ZFC (their existence is ruled out by the Axiom of Foundation), but they can be admitted in a non-well-founded set theory by replacing the Axiom of Foundation with one of the several anti-foundation axioms. Among them, the most important is the Anti-Foundation

Against Forcing Simulation

Axiom (*AFA*), that states that every rooted digraph (a graph in which a vertex is recognised as the root) corresponds to a unique set. For example, the loop graph (a graph with only one vertex and an edge to itself) corresponds to the set $\{x\}$. Non-well-founded set theory (for example *ZFA*, i.e., *ZFC* − Foundation + *AFA*) is very similar to classical set theory, since all the proofs and definitions in which the Axiom of Foundation is not needed still go through as normal (for example, Russell's Paradox, the Axiom of Choice, relations, pairs, natural numbers, transfinite recursion, etc.). The main difference is on how we can approach objects that could have some circularity in their definition (for example, graphs, infinite chains, etc.).

Another possible formulation of *AFA* is that every graph has a unique decoration. A *decoration* d of a graph G is a function from the vertices to natural numbers, with the following property:

$$d(g) = \{d(h) : g \to h\}.$$

In other words, the decoration function takes a vertex as an input, and gives the set corresponding to its children as the output. For example, take the leaf of a digraph: since it has no children, its decoration is the empty set. Suppose instead that we want to calculate the decoration of a vertex just before the leaf: in this case, that vertex has one child (the leaf), so its decoration is the singleton $\{\emptyset\}$. A graph can be *extended* by adding to it the decoration of its children.

It is possible to approach questions about infinite games from the perspective of non-well-founded set theory. In particular, such a perspective enables us to use tools and methods available only in non-well-founded set theory (mainly abbreviations of proofs and definitions regarding graphs and infinite trees) to the investigation of infinite games. For example, we can prove that there is a point in which player I has a winning strategy and player II no longer has one appealing to extended graphs and decorations: if that particular tree can be extended by adding a subset of the winning set, then player I has a winning strategy while player II no longer has one (since every move that she can make is already in the subset of the winning set). This is only one of the possible examples of interaction between determinacy and non-well-founded set theory, but it is possible to use the axiom that every tree corresponds to a unique set to prove more interesting results (the existence of unique winning strategies, for example). However, to get these results, we need to assume *AFA*. Consequently, if we believe only in the countable transitive model approach to forcing sketched above, we would not be able to prove these results, since in that case we cannot use objects and tools from AFA in $ZFC + PD$ (we could try to work directly in $ZFC + PD$, but this could be problematic). By contrast, even in a very simplified toy multiverse composed of only two universes, one well-founded and one non-well-founded, those results are instead available.

4 Concluding remarks

In conclusion, appealing to the fact that we can simulate any non-canonical model of set theory (and thus set theoretic universe) in the Single Universe V is in tension with current set theoretic practice. Thus, it cannot be used to defend the universist position. In particular, I claim that with the countable transitive model approach non-canonical incompatible models cannot be efficiently compared and that we cannot use them in the investigation of one another, which in turn makes it impossible to prove a number of results in set theory—results, however, that by contrast are attainable in a multiverse conception of set theory. For the sake of definiteness, I have introduced some results that cannot be proved in the Single Universe context with the appeal to ill-founded sets, but that instead can only be proved in a set theoretic multiverse. These results can be proved using the tools and methods of ill-founded set theory, and in particular the fact that every extended graph has a unique decoration. However to prove this we need both the Anti-Foundation Axiom and at least Projective Determinacy, and such a setting is not possible in the Single Universe with only the countable transitive model approach to forcing at our disposal.

References

Aczel, P. (1988). Non-well-founded sets. *CSLI Lecture Notes, 14*.
Antos, C. (2018). Class forcing in class theory. In *The Hyperuniverse Project and Maximality* (pp. 1–16). Springer.
Antos, C., Friedman, S.-D., Honzík, R., & Ternullo, C. (2018). *The Hyperuniverse Project and Maximality*. Birkhäuser, Basel.
Barton, N. (2020). Forcing and the universe of sets: Must we lose insight? *Journal of Philosophical Logic, 49*(4), 575–612.
Barwise, J. (1975). *Admissible Sets and Structures*. Springer Verlag, Berlin.
Cohen, P. J. (2003). The independence of the continuum hypothesis, II. In *Mathematical Logic In The 20th Century* (pp. 7–12). World Scientific.

Gorbow, P. K., & Leigh, G. E. (2020). The Copernican multiverse of sets. *arXiv*, 1–31.

Grigorieff, S. (1975). Intermediate submodels and generic extensions in set theory. *Annals of Mathematics*, 447–490.

Hamkins, J. D. (2012). The set-theoretic multiverse. *Review of Symbolic Logic*, *5*(3), 416-449.

Incurvati, L., & Löwe, B. (2016). Restrictiveness relative to notions of interpretations. *The Review of Symbolic Logic*, *9*(2), 238–250.

Isaacson, D. (2011). The reality of mathematics and the case of set theory. *Truth, reference, and realism*, 1–75.

Livadas, S. (2020). Abolishing platonism in multiverse theories. *Axiomathes*, 1–23.

Maddy, P. (1996). Set-theoretic naturalism. *The Journal of Symbolic Logic*, *61*(2), 490–514.

Martin, D. (2001). Multiple universes of sets and indeterminate truth values. *Topoi*, *20*, 5–16.

Nik, W. (2014). *Forcing for Mathematicians*. World Scientific.

Shelah, S. (2014). Reflecting on logical dreams. In J. Kennedy (Ed.), *Interpreting Gödel: Critical Essays* (pp. 242–255). Cambridge: Cambridge University Press.

Steel, J. (2014). Gödel's program. In J. Kennedy (Ed.), *Interpreting Gödel: Critical Essays* (pp. 153–179). Cambridge: Cambridge University Press.

Visser, A. (2006). Categories of theories and interpretations. In A. Enayat, I. Kalantari, & M. Moniri (Eds.), *Logic in Tehran* (pp. 284–341). Cambridge University Press.

Woodin, W. H. (1999). *The Axiom of Determinacy, Forcing Axioms and the Nonstationary Ideal*. De Gruyter.

Woodin, W. H. (2011). The transfinite universe. In M. Baaz, C. H. Papadimitriou, H. W. Putnam, D. S. Scott, & C. L. Harper Jr (Eds.), *Kurt Gödel and the Foundations of Mathematics: Horizons of Truth* (pp. 449–472). Cambridge: Cambridge University Press.

Matteo de Ceglie
Universität Salzburg, Fachbereich Philosophie (KGW)
Austria
E-mail: decegliematteo@gmail.com

Infinity is Not a Size

MATTHIAS EBERL

Abstract: We present a dynamic model theory that avoids the paradoxes stemming from completed infinities, but does not require any translation of formulae. The main adoption is the replacement of an actual infinite carrier set by a potential infinite one, and by a finitistic interpretation of the universal quantifier.

Keywords: potential infinite, finitism, model theory

1 Introduction

Our aim is to develop mathematics with the notion of potential infinity instead of actual infinity. Most of the technical details have been worked out and submitted partly. In this paper we give a motivation and general overview of this approach.

1.1 The idea in a nutshell

The sole and simple starting point is to regard infinity as a potential infinite. The subsequent concepts follow from that in an (almost) natural way. First, the potential infinite is basically a dynamic conception: An infinite set is a process that exhausts its elements, it is not a "flat" totality of objects. Its fundamental property is the "indefinite extensibility of the finite". The possibility to always extend a totality of objects is more fundamental than (and in contrast to) having a completed totality. Within that process, indefinitely large finite states are substitutes for the actual infinite set.

Second, thinking of sets as processes also requires to take the dependency of different process states into account. For instance, the state of a model may depend on that of the syntax, or the state of a subset depends on the state of the whole set. In this way, the paradoxes of the infinite are avoided, see Section 2.4.

Third, a potential infinite set requires a non-tautological reading of the universal quantifier (Section 3.2.1). We cannot simply interpret "for all" by referring to a completed infinite set. It is crucial that any reference to an infinite set of elements requires a reference to some specific stage; otherwise one would introduce actual infinity through the back door or already presuppose this interpretation implicitly. In order to formulate this interpretation properly we use a refinement of the Tarskian model theory.

1.2 Related work

Lavine (2009), in the Section "The Finite Mathematics of Indefinitely Large Size", introduces the concept of an *indefinitely large* size, which is nevertheless finite. His ideas are based on a work of Mycielski (1986) about locally finite theories. The central idea of Mycielski's work is that the range of the bound variables inside a single formula increases. For instance, in a formula $\forall x_0 \exists x_1 \Phi$, the variable x_0 refers to objects in some carrier set $\mathcal{M}_{i_0}$ and a further quantification, e.g., by $\exists x_1$, refers to a set $\mathcal{M}_{i_1}$, being indefinitely large relative to $\mathcal{M}_{i_0}$. Lavine uses Mycielski's results in order to argue that one is able to extrapolate results from the finite to the infinite.[1] Both Mycielski and Lavine do not investigate the notion of a potential infinite in this context.

Mycielski constructs a finite model for a finite set of formulas, implicitly leading to a dependency of the model on the syntax.[2] In order to interpret formulas of a first-order theory one has to translate them: Each bound variable is restricted by a predicate Ω_i, indicating some upper bound. Additionally one adds bookkeeping axioms in order to formulate the exchangeability of these Ω_i. Thereby Mycielski uses the common Tarskian model theory (for his purpose, a finite term model suffices).

To cope with indefinite extensibility we give a new interpretation of the universal quantifier, presented in Section 3.2, related to Mycielski's idea of the increasing ranges of quantifiers. In this way, no translation of formulas is necessary and no further axioms are required. This new model theory also allows the review of meta-mathematical properties, see Section 3.3, and proposes a way to extend the concepts to higher-order logic, see Section 3.2.4.

Lavine (2009) calls the translated, finitistic version of the axiom of infinity the "axiom of zillions". For him, the axiom of infinity is the extrapolation of the axiom of zillions from finite to infinitary set theory. In our approach ZFC is just another first-order theory. The axiom of infinity claims the existence of an infinite object, which is a representative of a potential infinite set (see

---

[1] A critical position on this extrapolation can be found in (Sereno, 1998). For a critical view on the concept of an "indefinitely large finite" see (Bremer, 2007).

[2] Note that we regard this dependency of the model on the syntax as one of the basic property of the potential infinite.

Section 3.2.4). This set is potential infinite and not actual infinite due to the non-trivial, finitistic interpretation of the universal quantifier. There are no two versions of the axioms (a finitary and an infinitary one) in our approach.

Shapiro and Wright (2006) considered the potential infinite as an indefinite extensible concept. Therein they state that "If a 'collection' is not a set, then it is nothing, has no size at all, and so can't be 'too big'" and ask: "The question, simply, is whether it is ever appropriate or intelligible to speak of all of the items that fall under a given indefinitely extensible concept". The authors come to the conclusion that there is no satisfying solution how to read such a quantification and refer to *reflection principles* as a possible answer. The interpretation that we present has an implicit reflection principle.

To sum up: If one regards the infinite as a potential infinite, one does not dispel the infinite as Mycielski does. And there is no need to use knowledge about finitely many objects in order to get knowledge about infinitely many objects, as Lavine argues. The infinite appears in a different form, it is no longer a size. An infinite set is a process and infinity creates a dependency of several process states, not of absolute cardinalities. Its new basic property is indefinite extensibility and not ultimate completion.

1.3 What it is not

Constructive logic[3] and recursion theory essentially uses the idea of a decision procedure, which plays no role in our consideration. In particular, we do not claim that there is an effective procedure to determine an indefinitely large finite state. Also, to claim the existence of an object with a specific property does not presuppose a procedure that terminates within known bounds.

We do not need any change of the axiom system as originally done by Jan Mycielski, or in Marcin Mostowski's work (see, e.g., Mostowski & Zdanowski, 2005). Mostowski uses potential infinite sets $(\mathbb{N}_i)_{i \in \mathbb{N}}$ with $\mathbb{N}_i := \{0, \ldots, i - 1\}$ as a basis of his model theory, but his approach is less dynamic and the axioms of Peano arithmetic must be adopted in a way that there exists a greatest number (see Mostowski, 2003). We also do not use paraconsistent logics to cope with the challenges of a finitistic view (cf. Bremer, 2007; Priest, 2013; Van Bendegem, 1994). Ideas based on "feasibility" are not relevant for our approach, an idea formulated, e.g., by Parikh (1971). And we do not introduce modalities: Recently Linnebo and Shapiro (2019) suggested to formalize the potential infinite using a modal reading of this notion. We also avoid vagueness or a notion of a "grey zone".

---

[3]Note that intuitionism relies on the notion of a potential infinite, but does not explain it. It refers in a naive way to infinite totalities without using some finite state.

We do not want to introduce or use a philosophical position. Our terminology and the technical treatment does not presuppose any view of mathematics such as Platonism, formalism or intuitionism. If we speak of objects that are created at some stages this should not be understood literally. We might also say that they have been conceived or revealed. Similarly, if we speak of time, later states etc., this again should not be understood literally, but figuratively.

2 The potential infinite

The locution "potential infinite" is a technical term, there is no infinity (as opposed to a finite) involved in this concept. Seen in that way, the potential infinite is a form of finitism, since at each stage there are only finitely many objects. If we refer to a potential infinite set, we necessarily have to refer to some state with finitely many elements. Often finitism is regarded as a view that postulates a *fixed* bound. With the notion of a potential infinite, the common finite vs. infinite distinction becomes a fixed vs. variable difference.

2.1 Ontological and epistemological finitism

The form of finitism that results from the potential infinite is also an *ontological* one, not an *epistemological* one. Another direction of finitism is the restriction of inferences to "finitary" ones leading to a "finitary reasoning". This is often related to Hilbert's program, which is concerned with the justification of classical mathematics by finitary reasoning. Tait (1981) claimed that finitist arithmetic coincides with primitive recursive arithmetic PRA. But his concept of finitism is based on a different idea than those of an infinite totality. As he states in (Tait, 2002): "My argument is that one can understand the idea of an arbitrary object of a given finitist type independently of that of the totality of objects of that type".

This (ontological) finitism is a justifiable position. For instance, Fletcher (1989) distinguishes between abstraction and idealization on the one hand, being basically a simplification, and extrapolation on the other hand, being an extension to a larger world. He argues that abstraction and idealization are necessary steps to obtain mathematical objects from physical objects. These procedures justify for instance the step from feasibilism to finitism. But the step to an actual infinity requires an extrapolation. Following this argument, a finitistic position is a natural result of the usual abstraction and idealization process from our experiences. The concept of an actual infinite however goes beyond these abstractions and idealizations.

2.2 Infinite entities

There are different infinite entities such as sets, spaces, lines, decimal representations, numbers etc., some numbers are also said to be infinitesimal small. Basically we distinguish between collections of objects—including structured collections, relations and functions—and single objects.

All collections of objects have some underlying set, e.g., a space has an underlying set of points or a function f an underlying set of assignments $a \mapsto f(a)$, and it suffices to explain when this set is infinite. If we call a single object infinite, e.g., a number is infinitely large or infinitesimally small, then it has an *infinite set of defining properties* such as $\omega > n$ or $\epsilon < \frac{1}{n}$ for all (infinitely many) natural numbers n. Often an infinite object has infinitely many better and better approximations. So first and primarily we have to answer the following question: *When is a set of entities called infinite?*

2.3 Paradoxes of the actual infinite

The existence of actual infinite sets are based on the "domain principle" (Hallett, 1984), that every potential infinity presupposes an actual infinity. Indefinite extensibility, as we understand it, is contrary to this principle. Nevertheless, statements still have determined truth values.

Actual infinities have several counter-intuitive properties. This starts with simple examples, e.g., there are as many natural numbers as even numbers. More complex examples are the Banach-Tarski paradox, as a consequence of the idea that a continuum is an actual infinite set of points. The deficiency of this view is not the fact that actual infinite sets have unfamiliar properties, but the fact that these "properties" stem from relations and dependencies between infinite sets which have been removed—we show this in an exemplary way in Section 2.4. Simply taking these dependencies into account prevents these paradoxes, which arise as self-made problems that have nothing to do with the mathematical content.

And even more, to introduce actual infinities does not eliminate the phenomenon of indefinite extensibility. After establishing infinite sizes in form of ordinal and cardinal numbers, the question arises naturally, what is the size of the totality of these infinite numbers. It is well known that this again leads to contradictions or paradoxes and a satisfying solution is not available[4].

---

[4]Some attempts are done with reflection principles (e.g., that from Lévy Montague) or with hierarchies of Grothendieck universes. Often a notion of small versus large is introduced, being basically a variant of the original distinction of finite versus infinite.

2.4 Removing a dependency

The cause for the paradoxes of infinity is that actual infinite sets do not allow to consider dependencies between the way its elements are exhausted. Potential infinite sets have exactly this additional structure. Each (potential) infinite set has its own way of exhausting its elements, so these processes can be related to each other. If we switch to an absolute completion of this process, not only a temporary stage, this removes the dynamic and dependency. This is best seen at one of the simplest paradoxes: An infinite set has the same size as a proper subset, whereby "same size" should mean a one-to-one correspondence of elements. For instance, the set $\mathbb{N}$ of natural numbers has the same size as the set $2\mathbb{N}$ of even numbers, given by the bijection $n \mapsto 2n$.

If we regard both sets as processes, there are different ways to relate them. The set $\mathbb{N}$ cannot be given as a whole, but solely by some state $\mathbb{N}_i = \{0, 1, \ldots, i-1\}$ and similarly $2\mathbb{N}$, say by the state $2\mathbb{N}_j = \{0, 2, \ldots, 2j-2\}$. The set $2\mathbb{N}$ is a subset of $\mathbb{N}$ only if $i \geq 2j - 1$ and there is a one-to-one correspondence only if $i = j$. Both requirements are not met at the same time if $i > 1$. But if we imagine that i and j arrive at the infinite, given by a state ω, the simultaneous satisfaction of both requirements becomes possible (intuitively since ω and $2\omega - 1$ are equal as cardinal numbers). So the paradox is removed if we do not allow this step to a completed infinite state, which moreover contains an incontinuity: $i \neq 2i - 1$ for $i > 1$ becomes $\omega = 2\omega - 1$ in the limit.

2.5 Indefinitely extensible and indefinitely large

The basic property of a (potential) infinite set $\mathcal{M}$ is its *indefinite extensibility*. But quantification requires reference to some state and the naive interpretation of "for all ..." cannot be used—it is meaningful only if there is a completed set of objects. Hence one needs a fixed finite set that is a replacement of the idealized totality of all possible elements. This is an indefinitely large stage $\mathcal{M}_i$ within $\mathcal{M}$, relative to a context of other states. An indefinitely large finite set is thus a context dependent version of the actual infinite set. Conversely, if the context can be ignored without harm, one may treat the indefinitely large set as an absolute infinite totality.

The notion of indefinite extensibility that we use includes Dummett's understanding[5]. In (Dummett, 1994) he defines: "An indefinitely extensible concept is one such that, if we can form a definite conception of a totality

---

[5]Note that whereas Dummett concludes that statements quantifying over an indefinitely extensible concept do not follow the laws of classical logic, our model-theoretic approach does not require this.

Infinity is Not a Size

all of whose members fall under that concept, we can, by reference to that totality, characterize a larger totality of all whose members fall under it." The ordinal numbers and sets are a typical example: If we refer to "all sets", this creates or reveals a new set and thus the totality of all sets has changed.

But already the natural numbers form such an indefinitely extensible concept. If we refer to the number of all numbers, then this reference creates a new number. First there is no number, hence the number of numbers is 0. So we created a first number, namely 0, and the number of numbers is 1. Henceforth there are the two numbers 0 and 1, creating number 2 and so on.

The notion of an indefinitely large finite could be seen as a *relative infinite*. If $\mathcal{I}$ denotes the set of states or indices, then a relative infinite is a relation $C \ll i$ (or $i \gg C$) between an index $i \in \mathcal{I}$ and a context $C := (i_0, \ldots, i_{n-1})$, with $i_0, \ldots, i_{n-1} \in \mathcal{I}$, stating that i is *indefinitely large* or, using a more technical notion, *sufficiently large* relative to C.

We can only investigate finitely many objects in a way that we explicitly refer to them. Say these are currently $a_0, \ldots, a_{n-1}$. Most often these objects are not fixed but variable ones, taken from some infinite sets. Assume that $a_0, \ldots, a_{n-1}$ are (variable) natural numbers, then saying that a_k is a natural number means $a_k \in \mathbb{N}_{i_k}$ for some state $i_k \in \mathbb{N}$. So the currently investigated objects, here $a_0, \ldots, a_{n-1}$, are always within a context $C = (i_0, \ldots, i_{n-1})$.

By seeing infinity as an indefinitely large finite, the infinite is not outside of an indefinitely extensible set, it is a part of it. It is only outside the region that we can reach from the current stage with our current means. The indefinitely large finite sets $\mathcal{M}_i$ with $i \gg C$ behaves exactly as actual infinite sets in the current context of investigation. But they are not completed in an absolute way, i.e., if we change the context, an extension could be necessary.

The notion of an indefinitely large finite is relative in three ways. First, it is not a single state $i \in \mathcal{I}$, but a region, e.g., $\{i \in \mathcal{I} \mid i \geq h\}$, the *indefinitely large region*. If there is a least element in this region, we call it *horizon*. Secondly, the region depends on a context $C = (i_0, \ldots, i_{n-1})$, it is thus a relation $C \ll i$. Figure 1 illustrates this situation. And thirdly, it is not a single relation $\ll$ but several ones. Their basic properties are that $C \ll i \leq i'$ implies $C \ll i'$ and additionally that $(i_0, \ldots, i_{k-1}) \ll i_k$ holds for all $k < n$. The latter expresses a dependency of the size of set $\mathcal{M}_{i_k}$ on the sizes of the sets $\mathcal{M}_{i_0}, \ldots, \mathcal{M}_{i_{k-1}}$.

If it is necessary to include the indefinitely large set $\mathcal{M}_i$ into the current context C as a further set $\mathcal{M}_{i_n}$, then the current context becomes $C' = (i_0, \ldots, i_{n-1}, i_n)$. We may then again choose an indefinitely large index $i' \gg C'$.

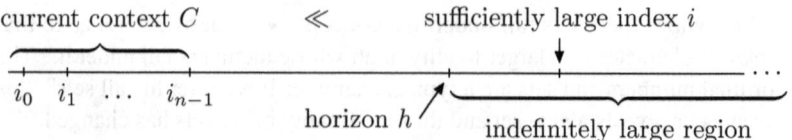

Figure 1: The structure of indefinitely extensible sets.

3 Mathematical formalization

We are now ready to present the mathematical concepts. We start with the basic concept of a system and apply it to first-order logic (FOL). In a second step we treat infinite objects when we switch to higher-order logic (HOL).

3.1 Systems: Formalizing the dynamic aspect

A dynamic object a is mathematically a function from an index set $\mathcal{I}$ to possible states a_i of a, that is, a is the family $(a_i)_{i \in \mathcal{I}}$. Similarly, a dynamic set $\mathcal{M}$ is a family of sets varying over $\mathcal{I}$, i.e., $\mathcal{M}$ is $(\mathcal{M}_i)_{i \in \mathcal{I}}$. The finitistic approach requires that $\mathcal{M}_i$ is finite, but the technical treatment makes no use of this property—we could work with other definite versus indefinite distinctions as well (these are often called "small" and "large", but we try to avoid these notions due to their connotation of size). But from our perspective we gain nothing by doing so. More relevant is the fact that we do not need any notion of finiteness of the sets $\mathcal{M}_i$.

The set $\mathcal{I}$ is equipped with a preorder $\leq$ indicating that if $i' \geq i$, then i' refers to a "later" state. We also require that $\mathcal{I}$ is directed with the idea that the different states converge towards "one limit" without reaching it[6].

The objects in the different states $\mathcal{M}_{i'}$ and $\mathcal{M}_i$ for $i' \geq i$ are connected by a relation $a_i \overset{s}{\mapsto} a_{i'}$ if $a_i \in \mathcal{M}_i$ and $a_{i'} \in \mathcal{M}_{i'}$. The intention is that $a_{i'}$ is a *successor* of a_i. The simplest case for FOL is that $\overset{s}{\mapsto}$ is an embedding $emb_i^{i'} : \mathcal{M}_i \to \mathcal{M}_{i'}$. In that situation we identify a_i with $emb_i^{i'}(a_i)$, so $\mathcal{M}_i \subseteq \mathcal{M}_{i'}$. For instance, $\mathbb{N}_i \subseteq \mathbb{N}_{i'}$ and $\mathbb{N}_i \ni n \overset{s}{\mapsto} n \in \mathbb{N}_{i'}$ for $n < i \leq i'$, which is a special case of a *direct system*. In the sequel we simply speak of a *system* if we refer to $(\mathcal{M}_\mathcal{I}, \overset{s}{\mapsto})$.

---

[6]This "not reaching the limit" applies for infinite index sets. It is obvious that any fixed finite set can be subsumed under this concept by considering a singleton index set $\mathcal{I} = \{*\}$, or another finite index set with greatest element, which then represents the limit state.

3.2 Logic

The set $\mathcal{L}$ of expressions, such as terms and formulas, is typically a free structure over a vocabulary. What is important here is that if $\mathcal{L}$ is infinite, it is consequently seen as indefinitely extensible. In all usual languages $\mathcal{L}$ one can exhaust its expressions by some measure of complexity, whereby the set $\mathbb{N}$ as index set suffices.[7] So $\mathcal{L} = (\mathcal{L}_k)_{k \in \mathbb{N}}$ with finite sets $\mathcal{L}_k$ and $\mathcal{L}_k \subseteq \mathcal{L}_{k'}$ for $k \leq k'$. Each model has an underlying carrier set $\mathcal{M}$, probably more. $\mathcal{M}$ is formalized as a system $\mathcal{M}_\mathcal{I} := (\mathcal{M}_i)_{i \in \mathcal{I}}$, for instance $\mathbb{N}$ is the system[8] $(\mathbb{N}_i)_{i \in \mathbb{N}}$.

3.2.1 A reinterpretation of the universal quantifier?

The reinterpretation of the universal quantifier, or better, its finitistic reading, is crucial here. Van Bendegem (1999) argues that a reinterpretation of the universal quantifier is indispensable and unassailable for a finitistic point of view:

> In the first place, a classical mathematician or logician will surely remark that, however clever this procedure might be, it still implies a reinterpretation of the universal quantifier. 'For all x, ...' does not have its classical meaning, for, in all cases, we are supposed to read 'For all $x \leq K, \ldots$'. [...] if one asks for a standard interpretation of the universal quantifier, then one presupposes the possibility of an infinite domain, hence one can never have such an interpretation in a finite domain. (p. 123)

A new interpretation makes it impossible to formulate the concept of a potential infinite in the object language. So whether an unbounded set is seen as actual or potential infinite cannot be stated in a formal system, e.g., as some axiom—for this reason Niebergall (2014) found no convincing formulation that a theory assumes merely the potential infinite, and not an actual infinite. Only in a formalization of the background theory it is possible to distinguish potential and actual infinity as different interpretations of the universal quantifier.

---

[7]Consider for instance a first-order language with free variables $x_0, x_1, \ldots$, a finite set of relation symbols, the connectives $\neg, \wedge, \vee, \rightarrow$ and the quantifiers $\forall, \exists$. Then $\mathcal{L}_k$ may consist of the finitely many formulas $\Phi(x_0, \ldots, x_{n-1})$ (i.e., free variables within $x_0, \ldots, x_{n-1}$) with $n \leq k$ and less than k connectives and quantifiers.

[8]We sometimes take the index set $\mathbb{N}^+ := \{1, 2, \ldots\}$ instead of $\mathbb{N}$ to avoid empty sets of objects. This is relevant only in HOL, but is an insignificant change.

3.2.2 Classical first-order logic

In a FOL the only subject that is considered to be infinite is the domain of discourse and the functions and relations on it, not the objects themselves. Our starting point for classical FOL is the usual Tarskian model theory. The variable assignment $a = (a_0, \ldots, a_{n-1})$, instantiating the free variables $x_0, \ldots, x_{n-1}$ in a formula Φ, are taken from $\mathcal{M}_C := \mathcal{M}_{i_0} \times \cdots \times \mathcal{M}_{i_{n-1}}$ (see Section 2.5). Relation R, the interpretation of an n-ary relation symbol R in the signature Σ, is a family $(R_C)_{C \in \mathcal{I}^n}$ with $R_C \subseteq \mathcal{M}_C$. It satisfies $R_C \subseteq R_{C'}$ for $C \leq C'$ with a pointwise order.[9]

The notion of validity becomes $\mathcal{M}_\mathcal{I} \models_\ll \Phi[a : C]$ with $\ll$ as an additional parameter. The interpretation of the logical connectives is as usual. The universal quantifier is interpreted by

$$\mathcal{M}_\mathcal{I} \models_\ll \forall x_n \Phi[a : C] \quad :\Longleftrightarrow \quad \mathcal{M}_\mathcal{I} \models_\ll \Phi[ab : Ci] \text{ for all } b \in \mathcal{M}_i \quad (1)$$

with an *indefinitely large* set $\mathcal{M}_i$, i.e., $i \gg C$. Therein, ab is the extension of the variable assignment a by b, and Ci the extension of the context C by index i. The existential quantifier $\exists x$ can be reduced to $\neg \forall x \neg$.

The interpretation thus uses a semantic reflection principle: An assertion about all infinitely many elements in $\mathcal{M}_\mathcal{I}$ is true if and only if it is true about a part $\mathcal{M}_i$ that reflects the whole infinite set in the current context C. In order to show that this interpretation is sound (relative to a usual inference system for classical FOL) and that the validity is independent of the chosen index $i \gg C$, one has to restrict the set of relations $\ll$ to those, which contain all witnesses of valid existential assertions. The construction is similar to the proof of the Löwenheim-Skolem theorem and Mycielski's original work (see Mycielski, 1986). The interpretation $\models_\ll$ for these relations $\ll$ is also *complete*. Moreover it is *locally finite* in the sense that a finite set of expressions requires a finite model only.

3.2.3 Intuitionistic first-order logic

This interpretation with reflection principle is also applicable to intuitionistic logic with Kripke models, or alternatively Beth models. Kripke models are based on a frame $(\mathcal{K}, \leq)$, being a preorder of *epistemic states* or *nodes*. Deviating from the index set $\mathcal{I}$ these frames are typically not directed—different branches generate alternatives, which may not "meet later".

---

[9] It is possible to add function symbols and functions, too.

Infinity is Not a Size

There is however a more important difference between the index sets $\mathcal{K}$ and $\mathcal{I}$: The set $\mathcal{K}$ represents *epistemic states* or *states of information*, whereas the index set $\mathcal{I}$ is related to the *ontological* side of the semantics. So we may ask whether an element a exists in $\mathcal{M}_i$, but if it exists there, it has already all of its properties (nothing new can be added to a in $\mathcal{M}_{i'}$ for $i' \geq i$). In comparison, if a exists at k, it has the properties known at k, but we may discover at $k' > k$ additional properties, even if the same objects exist at k and k'. Formally, a relation R in a Kripke structure is a family of sets R_C^k with k a node and C a context. The different requirements become $R_C^k(a) \iff R_{C'}^k(a)$ for $a \in \mathcal{M}_C \cap \mathcal{M}_{C'}$ on the one hand and the weaker requirement $R_C^k(a) \Rightarrow R_C^{k'}(a)$ for $k \leq k'$ and $a \in \mathcal{M}_C$ on the other hand.

Similar as for classical logic, one can define a forcing relation with a reflection principle and show soundness and completeness as well as locally finiteness.

3.2.4 Higher-order logic

HOL extends FOL in that it handles infinite objects. A finitistic point of view can deal with infinite objects, such as ω (the object representing the set of natural numbers in set theory), in two ways:

1. The object, say ω, may be regarded as a single abstract object. FOL can deal with these kind of objects only.

2. Or infinite objects are indefinitely extensible objects, analogously as sets are indefinitely extensible. Then there is no single object ω, only approximations of it, e.g., finite sets ω_i. HOL formalizes this concept as well.

In set theory, as in any other first-order theory, the universe is a system $(\mathcal{M}_i)_{i \in \mathcal{I}}$ with finite sets $\mathcal{M}_i$. The abstract set-objects $a \in \mathcal{M}_i$ become concrete in the background model due to the indefinitely increasing membership relation ϵ, which is a family with states $\epsilon_i \subseteq \mathcal{M}_i \times \mathcal{M}_i$. An object such as ω *represents* an increasing family $\omega_i := \{a \in \mathcal{M}_i \mid a \, \epsilon_i \, \omega\}$ for all $i \in \mathcal{I}$ with $\omega \in \mathcal{M}_i$. Thereby ω_i is a "real" set, i.e., a (finite) set in the background model, that increases if we enlarge $i \in \mathcal{I}$. There is no explicit concept in FOL that sees all of these ω_i as approximations of ω. The relation $\overset{s}{\mapsto}$ in a first-order model is always a function, that is, an object is not differentiated if the model increases (e.g., in HOL an approximation 1.41 of a real number differentiates to $1.410, 1.411, \ldots 1.419$ if the precision increases from 2 to 3 digits). To put it another way, if a (first-order) object occurs at some stage, it is already "complete" and is not an approximation that needs further differentiation.

HOL explicitly introduces approximations which are placed in relation by $\stackrel{s}{\mapsto}$. Infinite objects are only handled by their approximations, e.g., ω is the family of finite von Neumann ordinals n and $n \stackrel{s}{\mapsto} n'$ if $n \in n'$. Then ω exists only as this family, not as an abstract limit object—but in HOL there might be both, abstract objects and families of approximations. Moreover, each sufficiently large n plays the role of ω, whereby its size depends on the current context.

Similarly, a real number, for instance $\sqrt{2}$, can be given as an indefinitely extensible Cauchy sequence of rational numbers (more precisely an equivalence class thereof). Then a sufficiently good approximation by a rational number is a substitute for the real number. Additionally the real number may be given as abstract object of some base type, which is an element of a complete ordered field (defined axiomatically). But $\sqrt{2}$ is *not* the infinite Cauchy sequence, being an element of this field of real numbers.

An elegant way to formulate classical HOL is simple type theory (STT) (see, e.g., Farmer, 2008). A prerequisite for a model is an indefinitely extensible system with a limit construction in the style of a direct and inverse limit. Adding a rule for the universal quantifier (not only a constant) allows an interpretation of the universal quantifier as defined in (1).

In STT all infinite objects are formulated as (higher-order) functions. In order to see how STT handles them by approximations, consider functions on natural numbers. Instead of first building an actual infinite set $\mathbb{N}$ and then defining the function space $[\mathbb{N} \to \mathbb{N}]$ as consisting of "all" functions, $[\mathbb{N} \to \mathbb{N}]$ is approximated by finite function spaces $\mathcal{N}_{i \to j} := [\mathbb{N}_i \to \mathbb{N}_j]$, with $i, j \in \mathbb{N}^+$—we suggestively write $i \to j$ for such a pair of indices. These function spaces form a system $(\mathcal{N}_{i \to j})_{i,j > 0}$ with $f \stackrel{s}{\mapsto} f'$, whereby a function $f : \mathbb{N}_i \to \mathbb{N}_j$ is related to $f' : \mathbb{N}_{i'} \to \mathbb{N}_{j'}$ for $i \leq i'$ and $j \leq j'$ if f is the restriction of f' to $\mathbb{N}_i$. A function is itself an indefinitely extensible system $(f_{i \to j})_{(i,j) \in \mathcal{H}}$ with $f_{i \to j} : \mathbb{N}_i \to \mathbb{N}_j$ and a suitable index set[10] $\mathcal{H} \subseteq \mathbb{N}^+ \times \mathbb{N}^+$.

All other constructions to handle infinite objects in a finitistic way use, to our knowledge, completed infinities. A wide spread approach to handle infinite objects very generally is *domain theory*. A domain $\mathcal{D}$ is *directed*

---

[10] For simple functions from a first-order domain to another, e.g., $f : \mathbb{N} \to \mathbb{N}$, it is easy to find these index sets (e.g., $\mathcal{H} = \{i \to j \mid j > \max_{n < i}(f(n))\}$). For functionals of type 2 or higher, the definition of what is a suitable index set is indeed a challenge.

complete, which is a property that requires actual infinite sets in order to be non-trivial, since each finite directed set has automatically a greatest element. There is another, less known approach, the hyperfinite type structure (cf. Normann, Palmgren, & Stoltenberg-Hansen, 1999). It is based on a Fréchet product. In a Fréchet product two elements $(a_i)_{i \in \mathbb{N}}$ and $(b_i)_{i \in \mathbb{N}}$ are identified if they differ only w.r.t. finitely many indices i. This is obviously trivial if $\mathbb{N}$ is regarded as potential infinite, since all elements are identified then.[11]

3.3 Some meta-mathematical considerations

The understanding of the language as increasing is important for meta-mathematical properties. With this understanding some notions collapse, for instance *consistent* is the same as *finitely consistent*. As a consequence, all models are saturated. Interestingly, compactness did not become trivial, but uniformity: If we find a model at each stage of $(\mathcal{T}_k)_{k \in \mathbb{N}}$ (an increasing set of formulas with $\mathcal{T}_k \subseteq \mathcal{L}_k$), then there is a common model for all stages. This follows from the construction of the Henkin-model in a first-order completeness proof.

We will shortly discuss the presence of non-standard elements in FOL. What is new in our approach is that relation $\stackrel{s}{\mapsto}$ allows a distinction between standard and non-standard models. Systems with partial surjections $\stackrel{p}{\mapsto}$ (here $\stackrel{p}{\mapsto}$ is the inverse of $\stackrel{s}{\mapsto}$, indicating a *predecessor* relation) are called *standard*, all others *non-standard*. That is, in a standard model objects are not identified at later states whereas non-standard models allow an *identification* of objects when the elements increase, i.e., $a_0 \stackrel{s}{\mapsto} a$ and $a_1 \stackrel{s}{\mapsto} a$ for $a_0 \neq a_1$ both in $\mathcal{M}_i$. HOL additionally allows a *differentiation* in contrast to FOL, i.e., $a \stackrel{s}{\mapsto} a_0$ and $a \stackrel{s}{\mapsto} a_1$ for $a_0 \neq a_1$ both in $\mathcal{M}_i$.

Let us consider Peano arithmetic (PA). The non-standard elements arising in Henkin's completeness construction (see, e.g., Enderton, 2001) are not infinitely large numbers. They are natural numbers that cannot be seen as a number, or for which we do not know, at the current stage of the model construction, which number it is. These non-standard elements are usually considered to be *beyond* all natural numbers $n \in \mathbb{N}$. A reading of infinite as potential infinite only shows that the number is larger than all finitely

---

[11]Note that notions such as directed completeness or a Fréchet product are meaningful inside a theory if interpreted in indefinitely extensible models. They get a new reading in which the limit is indefinitely large, but not actual infinite. However, we cannot use this understanding to establish the model theory itself, if we do not want to already presuppose this new interpretation.

many numbers in the current context, which is easily satisfied by a (standard) natural number due to the indefinite extensibility.

Moreover, PA is a theory with equality, so $=$ is interpreted by the identity. In order to satisfy this condition in the common Henkin-construction, a switch to equivalence classes is done *after* all elements have been introduced to the model. The two processes—adding elements to the model on the one side and identifying them if an equality can be proven on the other—are seen as independently increasing. One is able to complete the first task (adding elements) before starting the second one (identifying them).

With a dynamic reading these processes must be done simultaneously in a direct limit construction with non-injective embeddings. So at each stage of the construction there are typically elements (closed terms) that are known to be provably equal to some n (i.e., the closed term S...S0 with n successor symbols S). At each stage there will always be some of these unknown elements due to open terms and also due to Gödel's first incompleteness theorem. In contrast, the usual iterative construction $(\mathbb{N}_i)_{i\in\mathbb{N}}$ of a standard model does not introduce non-standard numbers nor identify elements in further steps. Remember that completeness holds with respect to non-standard models (incl. the standard model), whereas categoricity holds w.r.t. the standard model only.

3.4 Application to the Background Model

The here presented model theory is applicable to its own background model if we apply the idea of indefinitely extensible sets to its meta theory and model. That is, the implicit background model of model theory that we applied in Section 3.2 makes use of extensible totalities, too.

In particular, the index set $\mathcal{I}$ is seen as indefinitely extensible (note that $\mathcal{I}$ does not contain objects of the investigated model $\mathcal{M}_\mathcal{I}$, but indices are part of the background model). This does not lead to an infinite regress. One might think that since we replaced $\mathbb{N}$ by $\mathbb{N}_\mathbb{N}$, then the index set must be replaced in the same way leading to $\mathbb{N}_{(\mathbb{N}_\mathbb{N})}$, and so on. But we do not have to perpetually replace $\mathbb{N}$ by $\mathbb{N}_\mathbb{N}$ since there never was a completed set $\mathbb{N}$ that had to be replaced. From the very beginning there were only indefinitely extensible sets with states $\mathbb{N}_j$, and this model theory made this explicit.

In a first reading, the reader may regard an index set $\mathbb{N}$ as actual infinite. If she accepts the new interpretation, the reader can go through the paper, but now with the new understanding of set $\mathbb{N}$. Then, if we mention set $\mathbb{N}_\mathbb{N} := (\mathbb{N}_i)_{i\in\mathbb{N}}$, the index set $\mathbb{N}$ refers to some stage $\mathbb{N}_j$ of the background model, i.e., $\mathbb{N}_\mathbb{N}$ is $(\mathbb{N}_i)_{i<j}$. The requirement w.r.t. the index j is that it must be sufficiently large to describe all investigated models $\mathcal{M}_\mathcal{I}$. Therein the locution "all" has to be read as "all indefinitely many" as defined formally in Formula (1).

4 Conclusion

We presented a natural way how to use a potential infinite in mathematics instead of completed infinities, without any restrictions on the inference system. Infinite sets are seen as indefinitely extensible, realized by an interpretation of the universal quantifier that uses a finitistic reflection principle.

References

Bremer, M. (2007). Varieties of finitism. *Metaphysica*, *8*(2), 131–148.

Dummett, M. (1994). What is mathematics about? *Mathematics and Mind*, 11–26.

Enderton, H. B. (2001). *A Mathematical Introduction to Logic*. Academic press.

Farmer, W. M. (2008). The seven virtues of simple type theory. *Journal of Applied Logic*, *6*(3), 267–286.

Fletcher, P. (1989). *Truth, Proof and Infinity*. Kluver Academic Publishers, Netherlands.

Hallett, M. (1984). *Cantorian Set Theory and Limitation of Size*. Clarendon Press Oxford.

Lavine, S. (2009). *Understanding the Infinite*. Harvard University Press.

Linnebo, Ø., & Shapiro, S. (2019). Actual and potential infinity. *Noûs*, *53*(1), 160–191.

Mostowski, M. (2003). On representing semantics in finite models. In *Philosophical Dimensions of Logic and Science* (pp. 15–28). Springer.

Mostowski, M., & Zdanowski, K. (2005). FM-representability and beyond. In *New Computational Paradigms* (pp. 358–367). Springer.

Mycielski, J. (1986). Locally finite theories. *Journal of Symbolic Logic*, *51*(1), 59–62.

Niebergall, K.-G. (2014). Assumptions of infinity. *Formalism and Beyond: On the Nature of Mathematical Discourse*, 229–274.

Normann, D., Palmgren, E., & Stoltenberg-Hansen, V. (1999). Hyperfinite type structures. *The Journal of Symbolic Logic*, *64*(3), 1216–1242.

Parikh, R. (1971). Existence and feasibility in arithmetic. *The Journal of Symbolic Logic*, *36*(03), 494–508.
Priest, G. (2013). Indefinite extensibility—dialetheic style. *Studia Logica*, *101*(6), 1263–1275.
Sereno, L. A. (1998). *Infinity and Experience* (Unpublished doctoral dissertation). Massachusetts Institute of Technology.
Shapiro, S., & Wright, C. (2006). All things indefinitely extensible. *Absolute Generality*, 255–304.
Tait, W. W. (1981). Finitism. *The Journal of Philosophy*, 524–546.
Tait, W. W. (2002). Remarks on finitism. *Reflections on the Foundations of Mathematics. Essays in Honor of Solomon Feferman, LNL, 15.*
Van Bendegem, J. P. (1994). Strict finitism as a viable alternative in the foundations of mathematics. *Logique et Analyse*, *37*(145), 23–40.
Van Bendegem, J. P. (1999). Why the largest number imaginable is still a finite number. *Logique et Analyse*, *42*(165–166), 107–126.

Matthias Eberl
Independent Researcher
Germany
E-mail: `matthias.eberl@mail.de`

Identity in Relevant Logics: A Relevant Predication Approach

NICHOLAS FERENZ[1]

Abstract: I construct a formal semantics for quantified **R** with identity, a logic which is obtained using the set of axioms for identity in relevant logics given by Kremer (1999). The semantics given here builds on the semantics of Mares and Goldblatt (2006).

Keywords: relevant logic, identity, relevant predication

1 Introduction

Kremer (1999) proposed an axiom system for identity in relevant logic motivated by J. Michael Dunn's project of relevant predication (see Dunn, 1987, 1990). Kremer did not prove his axiom system complete with respect to any formal semantics. Nonetheless, the axiom choice is motivated by Dunn's relevant predication, and "harmonises with our other relevance intuitions" (Kremer, 1999, p. 218). In 1992, Mares (1992) developed a semantics for identity in relevant logics based on Fine's semantics for quantified relevant logics. Mares and Goldblatt (2006) constructed an alternative approach to the semantics of quantified relevant logic that is more intuitive and simpler than the semantics of Fine (1988), which for many years were the only semantics available for quantified relevant logics. Building on the semantics of Mares and Goldblatt, I offer a new semantics for identity in relevant logic for the philosophically inspired axiomatization of Kremer. Soundness and completeness for **R** extended by Kremer's axiom system are given for the defined semantics.

2 Kremer and relevant indiscernibility

As identified by both Kremer (1999) and Mares (1992), traditional approaches to identity in relevant logic lead to the validity of some irrelevant implica-

---

[1] I would like to thank the organizers of this volume. For helpful comments, I'd like to thank the anonymous referees and the audiences of earlier presentations of this paper.

tions. On the *traditional interpretation of identity*, as labeled by Kremer, the identity relation holds between two terms if and only if they refer to the same individual. To avoid irrelevancies, this approach would require that a single variable refer to different individuals (1) at different worlds, and (2) at one and the same world. Without the former, $\mathcal{A} \to (x = y \to y = x)$ is valid. Without the later, $\mathcal{A} \to x = x$ is valid. On the *indiscernibility interpretation of identity*, the identity relation holds between two terms if and only if they are "interchangeable, *salva veritate*, in all contexts" (Kremer, 1999, p. 202). That is, $t_1 = t_2$ is true if and only if $\mathcal{A}[t_1/x] \leftrightarrow \mathcal{A}[t_2/x]$ is the case for *every* $\mathcal{A}$. This includes when $\mathcal{A}$ does not contain x, and thus we validate $x = y \to (\mathcal{A} \leftrightarrow \mathcal{A})$. Kremer's *relevant indiscernibility interpretation of identity* solves this problem, and motivates a set of axioms for identity.

The *relevant indiscernibility interpretation* (RI-interpretation) of identity, relies on Dunn's *relevant predication*. Dunn's relevant predication is proposed as a "way to sort out those properties that have an intimate life with an object from those that do not" (Dunn, 1987, p. 347). The thrust of Dunn's relevant predication can be shown by the example of Socrates and Reagan. Suppose that Socrates is wise. It follows that "if anyone is Socrates, then he is wise". Classically, and irrelevantly, it also follows that "if anyone is Reagan, then Socrates is wise". Here, there are two properties or predications; *being wise* and *being such that Socrates is wise*. The first appears a relevant property of Socrates, while the second is not relevant to Reagan. In fact, Dunn claims that the former sentence relies on *thinning* (axiom form $\mathcal{A} \to (\mathcal{B} \to \mathcal{A})$), which is a concept rejected in relevant logic (Dunn, 1987, p. 350).

To further clarify the concept of a relevant predication, it is worth contrasting it with a similar notion. We might, for example, say that a predicate holds of an object relevantly, where the relevance is generated contextually, and that a predicate holds relevantly of an object in a particular context, but not another. For example, *tasting acidic* might be relevant to a certain cup of coffee in the context of selecting a coffee to drink, but not in the context of measuring fluids. While this may be a useful notion, it is not what is meant here by relevant predication. Rather, relevant predicates (as they are used here) are predicates that hold relevantly of everything they hold of, regardless of context.[2] In other words, for every object, if a relevant predicate holds of that object, then the predicate represents a property that 'has an intimate

---

[2] There is a fairly straightforward way of introducing a kind of context-sensitivity by changing the set of relevant predicates. The logics defined below are always relative to a set of relevant predicates, and in application our choice of this set may be influenced by context.

Identity in Relevant Logics: A Relevant Predication Approach

life with' that object. For example, when an object has the property of *being wise*, this property is had relevantly (or perhaps essentially). Let us assume for the remainder of the paper that we are able to identify or choose a set of relevant predicates.

Kremer's RI-interpretation states that "$s = t$ can be interpreted, intuitively, as an infinite conjunction of biconditionals $(\mathcal{B}[s/x] \leftrightarrow \mathcal{B}[t/x]) \wedge (\mathcal{C}[s/x] \leftrightarrow \mathcal{C}[t/x] \wedge \ldots$, where $\mathcal{B}x, \mathcal{C}x$, etc., run through *only those formulas that express relevant properties*" (Kremer, 1999, p. 204). That is, two terms are equal when the same set of relevant predications are true of each of them. To this end, the axiom of indiscernibility chosen is

$$(\text{RI}): x = y \rightarrow (Gx \rightarrow Gy),$$

where G is a relevant predicate constant.

In addition to (RI), Kremer's interpretation motivates (REF), (SYM), and (TRANS).

(REF)	$x = x$
(SYM)	$x = y \rightarrow y = x$
(TRANS)	$x = y \rightarrow (y = z \rightarrow x = z)$

For the exact philosophical motivation behind this choice, and for an explication of the informal semantic interpretation motivating the RI axiom, the reader is directed to Kremer (1999). The interpretation Kremer offers, however, is not formalized by the semantics given here, and so explicating such interpretation is not relevant (with the page limit for the paper).

3 The logics $\mathbf{QR}^{\circ t}$ and $\mathbf{RQ}^{\circ t}$

We assume here a fairly standard definition for first order logics in a language with a single quantifier $\forall$, the constant t, and set of the logical connectives $\{\rightarrow, \neg, \wedge, \vee, \circ\}$. We take the existential quantifier to be defined by $\exists x =_{df} \neg \forall x \neg$. A variable assignment, $f \in U^\omega$, assigns to each variable an element of the domain of individuals U. We write $\mathcal{A}[\tau/x]$ for the well-formed formula obtained by replacing all free occurrences of x in $\mathcal{A}$ with τ, and $f[i/n]$ for the variable assignment just like f with the modification that it assigns object i to the nth variable. The logic $\mathbf{RQ}^{\circ t}$ is defined by axiom schemes (A1)–(A19) and rules MP, ADJ, and RIC. The logic $\mathbf{QR}^{\circ t}$ is obtained from $\mathbf{RQ}^{\circ t}$ by the deletion of the axiom scheme (A19), which is known as the axiom of extensional confinement.

(A1) $\mathcal{A} \to \mathcal{A}$
(A2) $(\mathcal{A} \to \mathcal{B}) \to ((\mathcal{C} \to \mathcal{A}) \to (\mathcal{C} \to \mathcal{B}))$
(A3) $(\mathcal{A} \to (\mathcal{B} \to \mathcal{C})) \to (\mathcal{B} \to (\mathcal{A} \to \mathcal{C}))$
(A4) $(\mathcal{A} \to (\mathcal{A} \to \mathcal{B})) \to (\mathcal{A} \to \mathcal{B})$
(A5) $(\mathcal{A} \wedge \mathcal{B}) \to \mathcal{A}$
(A6) $(\mathcal{A} \wedge \mathcal{B}) \to \mathcal{B}$
(A7) $((\mathcal{A} \to \mathcal{B}) \wedge (\mathcal{A} \to \mathcal{C})) \to (\mathcal{A} \to (\mathcal{B} \wedge \mathcal{C}))$
(A8) $((\mathcal{A} \to \mathcal{C}) \wedge (\mathcal{B} \to \mathcal{C})) \to ((\mathcal{A} \vee \mathcal{B}) \to \mathcal{C})$
(A9) $\mathcal{A} \to (\mathcal{A} \vee \mathcal{B})$
(A10) $\mathcal{B} \to (\mathcal{B} \vee \mathcal{A})$
(A11) $\mathcal{A} \wedge (\mathcal{B} \vee \mathcal{C}) \to ((\mathcal{A} \wedge \mathcal{B}) \vee (\mathcal{A} \wedge \mathcal{C}))$
(A12) $(\mathcal{A} \to \neg \mathcal{A}) \to \neg \mathcal{A}$
(A13) $\neg\neg \mathcal{A} \leftrightarrow \mathcal{A}$
(A14) $(\mathcal{A} \to \neg \mathcal{B}) \to (\mathcal{B} \to \neg \mathcal{A})$
(A15) t
(A16) $\mathcal{A} \leftrightarrow (t \to \mathcal{A})$
(A17) $(\mathcal{A} \to (\mathcal{B} \to \mathcal{C})) \leftrightarrow ((\mathcal{A} \circ \mathcal{B}) \to \mathcal{C})$
(A18) $\forall x \mathcal{A} \to \mathcal{A}[\tau/x]$, where x is free for τ in $\mathcal{A}$
(A19) $\forall x(\mathcal{A} \vee \mathcal{B}) \to (\mathcal{A} \vee \forall x \mathcal{B})$, where x is not free in $\mathcal{A}$

$$\text{(MP)} \frac{\vdash \mathcal{A} \to \mathcal{B} \quad \vdash \mathcal{A}}{\vdash \mathcal{B}} \qquad \text{(RIC)} \frac{\vdash \mathcal{A} \to \mathcal{B}}{\vdash \mathcal{A} \to \forall x \mathcal{B}}$$

$$\text{(ADJ)} \frac{\vdash \mathcal{A} \quad \vdash \mathcal{B}}{\vdash \mathcal{A} \wedge \mathcal{B}}$$

The RIC rule comes with the restriction that x does not occur free in $\mathcal{A}$.

Semantics for $QR^{\circ t}$ and $RQ^{\circ t}$

Here I briefly explicate the general frame semantics, and some of the results, of Mares and Goldblatt (2006) (with minor cosmetic changes). The set of admissible propositions, *Prop*, in the general frame semantics here will be a subset of the hereditary subsets (up-sets) of possible worlds. We will also call these *admissible propositions*. Operators on propositions are defined as follows. Induced by the ternary relation R introduced in full below, the operator $\Rightarrow$ on $\wp(K)$ is defined as, for every $X, Y \in \wp(K)$,

$$X \Rightarrow Y = \{w : \forall x \forall y (Rwxy \text{ and } x \in X \text{ implies } y \in Y)\}.$$

Identity in Relevant Logics: A Relevant Predication Approach

We can define a similar operation for the fusion operator:

$$X \cdot Y = \{w : \exists x \exists y (Rxyw \text{ and } x \in X \text{ implies } y \in Y)\}.$$

A unary operator $*$ on $\wp(K)$ is defined by lifting $*$ to sets of worlds. For every $X \in \wp(K)$, $X^* = \{w : w^* \notin X\}$. We also include the operators $\cup$ and $\cap$ on the powerset $\wp(K)$.

Mares and Goldblatt introduce an operation in order to model the universal quantifier. When *Prop* is a set of hereditary subsets of possible worlds, an operation $\sqcap$ of type $\sqcap : \wp\wp K \to \wp K$ such that, for every $S \subseteq \wp(K)$

$$\sqcap S = \cup \{X \in Prop : X \subseteq \cap S\}.$$

This operation is "motivated by the intuition that the sentence $\forall x \phi$ expresses the conjunction of all the sentences $\phi[a/x]$" (Goldblatt, 2011, p. 17). The problem with using the arbitrary conjunction of the sentences is that this conjunction is not guaranteed to be admissible, for there can be denumerable sentences $\phi[a/x]$ but we only guarantee that the binary conjunction of two members of *Prop* is also in *Prop*. The more complex $\sqcap$ operator solves this problem by guaranteeing the result of the operation is an admissible proposition.

Goldblatt (2011) explains that using the notion of entailment between propositions (of the sort considered),[3] $\sqcap S$ is the *weakest* member of *Prop* that entails every member of the S. The key point here is that $\sqcap_{i \in I} X_i$ is always a subset of $\cap_{i \in I} X_i$, and sometimes properly so. The latter is not guaranteed to be a member of *Prop*, unlike the former.

We also use a set *PropFun* of admissible propositional functions. That is, functions from value assignments for variables to admissible propositions. That is, propositional functions are of type $\phi \colon U^\omega \longrightarrow Prop$, where U is the domain of individuals in a model.

Definition 1 *For any two elements ϕ and ψ of PropFun, the functions $\phi \cap \psi, \phi \Rightarrow \psi, \phi \cdot \psi, \phi \cup \psi$, and ϕ^* of the same type are defined by, for every value assignment to the variables $f \in U^\omega$:*

$$(\phi \cap \psi)f = \phi f \cap \psi f \qquad (\phi \cup \psi)f = \phi f \cup \psi f$$
$$(\phi \Rightarrow \psi)f = \phi f \Rightarrow \psi f \qquad (\phi^*)f = (\phi f)^*$$
$$(\phi \cdot \psi)f = \phi f \cdot \psi f$$

---

[3] Where a proposition X entails a proposition Y if $X \subseteq Y$. We say that Y is *weaker* than X, and X *stronger* than Y.

When we have a propositional function ϕ, we will define the functions $\forall_n \phi$ of the same type for each $n \in \omega$ by the following:

$$(\forall_n \phi)f = \bigcap_{j \in I} \phi(f[j/n])$$

Definition 2 An **R**-frame is a tuple $\langle K, 0, R, * \rangle$, where K is a set, $0 \subseteq K$, $R \subseteq K^3$ and $*$ is a unary function on K, a binary relation $\leq$ on K is defined by $a \leq b$ iff $\exists x \in 0(Rxab)$ and the following conditions are satisfied:

- $\leq$ is a preorder
- 0 is closed upwardly under $\leq$
- $Rabc$ implies $Rbac$
- $Raaa$
- $Rabc$ implies Rac^*b^*
- $a^{**} = a$
- If $Rabc$ and $d \leq a$ then $Rdbc$
- $\exists x(Rabx \land Rxcd)$ implies $\exists x(Racx \land Rxbd)$

Definition 3 A **QR**$^{\circ t}$-frame is a tuple $\mathfrak{F} = \langle K, 0, R, *, U, Prop, PropFun \rangle$, where $\langle K, 0, R, * \rangle$ is as **R**-frame, U is a non-empty set, $Prop$ is a set of hereditary subsets of K based on $\leq$, $Prop$ contains 0, and $PropFun$ is a subset of the functions from U^ω to $Prop$, such that the C1–C4 hold:

C1 If X and Y are in $Prop$, then $X \cap Y, X \cup Y, X \Rightarrow Y, X \cdot Y, X^* \in Prop$

C2 The constant function ϕ_0—for every f, $\phi_0 f = 0$—is in $PropFun$

C3 $PropFun$ is closed under $\Rightarrow, *, \cap, \cup, \cdot$

C4 If $\phi \in PropFun$, then $\forall_n \phi \in PropFun$ for every $n \in \omega$

A **QR**$^{\circ t}$-frame is called *full* if the set $Prop$ contains *every* hereditary subset of K, and $PropFun$ contains *every* function from U^ω to $Prop$.

Definition 4 A pre-model $\mathcal{M}$ on a **QR**$^{\circ t}$-frame $\mathfrak{F}$ is given by an assignment function $|-|^\mathcal{M}$ that assigns

1. an element $|c|^\mathcal{M} \in U$ to each constant $c \in \mathcal{L}$;

2. a function $|P|^\mathcal{M} : U^n \longrightarrow \wp K$ to each n-ary predicate symbol $P \in \mathcal{L}$;

3. a propositional function $|\mathcal{A}|^\mathcal{M} : U^\omega \longrightarrow \wp K$ to each $\mathcal{L}$-formula $\mathcal{A}$.

Identity in Relevant Logics: A Relevant Predication Approach

The propositional function assigned to the atomic formula $P\tau_1, \ldots, \tau_n$, *is given by*

$$|P\tau_1, \ldots, \tau_n|^{\mathcal{M}} f = |P|^{\mathcal{M}}(|\tau_1|^{\mathcal{M}} f, \ldots, |\tau_n|^{\mathcal{M}} f)$$

for each $f \in U^{\omega}$. *When* $\mathcal{A}$ *is not atomic, the propositional function assigned to* $\mathcal{A}$ *is given by the following:*

$$|t|^{\mathcal{M}} = \phi_0 \qquad |\neg \mathcal{A}|^{\mathcal{M}} = (|\mathcal{A}|^{\mathcal{M}})^*$$
$$|\mathcal{A} \wedge \mathcal{B}|^{\mathcal{M}} = |\mathcal{A}|^{\mathcal{M}} \cap |\mathcal{B}|^{\mathcal{M}} \qquad |\mathcal{A} \vee \mathcal{B}|^{\mathcal{M}} = |\mathcal{A}|^{\mathcal{M}} \cup |\mathcal{B}|^{\mathcal{M}}$$
$$|\mathcal{A} \to \mathcal{B}|^{\mathcal{M}} = |\mathcal{A}|^{\mathcal{M}} \Rightarrow |\mathcal{B}|^{\mathcal{M}} \qquad |\mathcal{A} \circ \mathcal{B}|^{\mathcal{M}} = |\mathcal{A}|^{\mathcal{M}} \cdot |\mathcal{B}|^{\mathcal{M}}$$
$$|\forall x \mathcal{A}|^{\mathcal{M}} = \forall_x |\mathcal{A}|^{\mathcal{M}}$$

Corresponding to these propositional function assignments are "truth set" assignments, which assign a set of worlds to each proposition-function pair. Where $|\mathcal{A}|^{\mathfrak{M}}$ is a function, $|\mathcal{A}|^{\mathfrak{M}} f$ is a member of *Prop*.

We can then give the relation $\vDash_{\mathfrak{M}}$—or simply $\vDash$. The case for $\forall x \mathcal{A}$ is given, the rest are as expected for relevant logic, and can be found in (Mares & Goldblatt, 2006).

- $a, f \vDash \forall x \mathcal{A}$ iff there is an $X \in Prop$ such that $X \subseteq \bigcap_{g \in xf} |\mathcal{A}|^{\mathfrak{M}} g$ and $a \in X$

Here xf is the set x-variant of f—the variable assignments differing from f by at most the assignment to the variable x.

Definition 5 *A model based on* $\mathbf{QR}^{\circ t}$-*frame* $\mathfrak{F}$ *is a pre-model based on* $\mathfrak{F}$ *such that* $|p|^{\mathcal{M}} \in PropFun$ *for every atomic formula p.*

Given the closure conditions required of the models, $|\mathcal{A}|^{\mathcal{M}} \in PropFun$ and $|\mathcal{A}|^{\mathcal{M}} f \in Prop$, for every formula $\mathcal{A}$, and every variable assignment f (Mares & Goldblatt, 2006, Corollary 4.1).

Definition 6 *A formula* $\mathcal{A}$ *is valid in the model* $\mathcal{M}$ *if* $0 \subseteq |\mathcal{A}|^{\mathcal{M}} f$ *for every* $f \in U^{\omega}$. *Further,* $\mathcal{A}$ *is valid on a* $\mathbf{QR}^{\circ t}$-*frame* $\mathfrak{F}$ *if* $\mathcal{A}$ *is valid on every model based on* $\mathfrak{F}$. *Finally,* $\mathcal{A}$ *is valid on a class of frames if it is valid for every frame in the class.*

The models for $\mathbf{RQ}^{\circ t}$ require a further restriction.

Definition 7 *The set of* $\mathbf{RQ}^{\circ t}$*-models is the subset of* $\mathbf{QR}^{\circ t}$*-models that satisfy the following condition:*

$$X - Y \subseteq \bigcap_{a \in U} \phi(f[a/x]) \text{ implies } X - Y \subseteq (\forall_x \phi)f$$

for all $\phi \in PropFun$, *all variables* x, *all* $X, Y \in Prop$, *and all* $f \in U^\omega$.

Theorem 1 (Soundness and Completeness for $\mathbf{QR}^{\circ t}$ ($\mathbf{RQ}^{\circ t}$))
(1) The set of theorems of $\mathbf{QR}^{\circ t}$ ($\mathbf{RQ}^{\circ t}$) *are valid on the class of all* $\mathbf{QR}^{\circ t}$-*frames* ($\mathbf{RQ}^{\circ t}$-*frames*). *(2) If* $\mathcal{A}$ *is valid in the class of all* $\mathbf{QR}^{\circ t}$-*frames* ($\mathbf{RQ}^{\circ t}$-*frames*), *then* $\mathcal{A}$ *is a theorem of* $\mathbf{QR}^{\circ t}$ ($\mathbf{RQ}^{\circ t}$).

4 $\mathbf{QR}^{\circ t}$ and $\mathbf{RQ}^{\circ t}$ with Identity

The logic $\mathbf{QR}^{\circ t}_=$ ($\mathbf{RQ}^{\circ t}_=$) is obtained by adding the axioms (REF), (SYM), (TRANS), and (RI) to $\mathbf{QR}^{\circ t}$ ($\mathbf{RQ}^{\circ t}$), and expanding the language to contain the 2-place predicate for identity, written using infix notation. Further, the predicates are taken to be sorted into relevant and non-relevant predicates (using G and F, respectively, with subscripts), and with the simplifications that the relevant predicates are one-place predicates. We use RP to denote the set of relevant predicates.

Roughly two major additions to the Mares-Goldblatt semantics are needed for $\mathbf{QR}^{\circ t}_=$ ($\mathbf{RQ}^{\circ t}_=$). First, inspired by the semantic equivalence relation from Mares (1992), we define a function that determines a left-to-right identity relation at each world for each object in the domain. The function is of type $\Rightarrow : K \times U \longrightarrow \wp(U)$. We write $y \in \Rightarrow(a, x)$ to mean that at the world a, x is left-to-right identical to y. By Currying this function, we obtain a function $|\approx| : U^2 \longrightarrow \wp(K)$ such that

$$y \in \Rightarrow(a, x) \text{ iff } a \in |\approx|(x, y).$$

This function is similar to an atomic predicate, as both are functions from n-tuples of elements of U to a set of worlds.

The other substantial addition is required for the (RI) axiom. We need to single out both the set of admissible propositional functions to which the relevant predications are mapped, and the corresponding admissible propositions. Thus, let us write $PropFun^G$ to denote a subset of the admissible propositional functions to which it is acceptable to map relevant predicates. Similarly, we write $Prop^G$ to denote the admissible propositions onto which

Identity in Relevant Logics: A Relevant Predication Approach

$PropFun^G$ maps the relevant predications. Thus, a relevant predication Gx will only be mapped onto a member of $PropFun^G$. The way I propose to do this lets us identify what will be the free variable in a formula which will be mapped into $PropFun^G$. The first step is to name the elements of $PropFun^G$. The second step is to use the names of propositional function when assigning formulas to them in the model.

Step 1: For each predicate $G \in RP$, and each term τ, let $:G\tau:$ be a name. Then consider a surjective function assigning names to each member of $PropFun^G$.

Step 2: We additionally require that, for every $G \in RP$ and every term τ, the valuation function assigns $G\tau$ the element of $PropFun^G$ named by $:G\tau:$. This lets us identify propositional functions that *are* assigned to Gx in a model, which enables us to use the fact that x is free.

It will be helpful to have notation for propositional functions that differ from each other only by the propositions assigned to certain x-variants.

Definition 8 *Where Φ is a propositional function, let $\Phi[y/x]$, the propositional function that differs at most from Φ by what propositions are assigned to x-variants, be defined by*

$$\Phi[y/x]f = \Phi f[|y|f/x].$$

That is, $\Phi[y/x]f$ behaves exactly as Φf, except that the function f assigns $|y|f$ to the variable x. In other words, $\Phi[y/x]$ is as if all x's were substituted for y's. In fact, that is the point. The formulas assigned to $\Phi[y/x]$ will differ from that assigned to Φ by just this substitution. We prove this below.

Lemma 1 *$PropFun^G$ is closed under term substitution. That is, if Φ is in $PropFun^G$, then so is $\Phi[y/x]$, for all variables x and y.*

The proof is obvious from the definition.

Lemma 2 *The relevant predication Gx is mapped to $:Gx:$ iff Gy is mapped to $:Gx:[|y|f/x]$.*

Proof. We know that Gx is mapped to $:Gx:$ and that Gy is mapped to $:Gy:$. Thus, it suffices to show that $:Gx:[|y|f/x] =: Gy:$. Both Gx and Gy have one free variable. For any assignment g, $:Gx:[|y|g/x]$ and $:Gy:$ can only differ if g assigns a different value to x than it does y. It is easy to see that this is in fact not the case. With $:Gx:[|y|g/x]g$ iff $:Gx:g[|y|g/x]$ by definition, we have that g must assign $|y|g$ to x. But by definition, this is also the object that g assigns to y. □

Lemma 3 *For all $a \in K$ and $f \in U^\omega$ and $G \in RP$, $a \in: Gx : [|y|f/x]$ iff $a, f \vDash Gy$.*

Proof. The proof is straightforward from lemma 2 and the fact that $a, f \vDash Gx$ iff $a \in: Gx : f$. □

4.1 Models for $\mathbf{QR}^{ot}_{\underline{=}}$ and $\mathbf{RQ}^{ot}_{\underline{=}}$

Definition 9 *A frame for $\mathbf{QR}^{ot}_{\underline{=}}$ ($\mathbf{RQ}^{ot}_{\underline{=}}$) is a tuple*

$$\mathfrak{F} = \langle K, 0, R, *, U, Prop, PropFun, PropFun^G, \Rightarrow \rangle,$$

where $\langle K, 0, R, *, U, Prop, PropFun \rangle$ is a frame for $\mathbf{QR}^{ot}$ ($\mathbf{RQ}^{ot}$), and the following conditions are satisfied:

1. *$PropFun^G \subseteq PropFun$*
2. *For every term τ and $G \in RP$, $: G\tau :$ names an element of $PropFun^G$*
3. *$a \in 0 \Rightarrow x \in \Rightarrow(a, x)$*
4. *$a \leq b \ \& \ y \in \Rightarrow(a, x) \Rightarrow x \in \Rightarrow(b, y)$*
5. *$Rabc \ \& \ y \in \Rightarrow(a, x) \ \& \ z \in \Rightarrow(b, y) \Rightarrow z \in \Rightarrow(c, x)$*
6. *$y \in \Rightarrow(a, x) \ \& \ Rabc \ \& \ b \in: Gx : f \Rightarrow c \in: Gx : [|y|f/x]f$, for every $G \in RP$ and $f \in U^\omega$.*

Definition 10 *A model for $\mathbf{QR}^{ot}_{\underline{=}}$ ($\mathbf{RQ}^{ot}_{\underline{=}}$) is frame $\mathfrak{F}$ for $\mathbf{QR}^{ot}_{\underline{=}}$ ($\mathbf{RQ}^{ot}_{\underline{=}}$) plus a valuation $|-|^{\mathfrak{M}}$, where $|-|^{\mathfrak{M}}$ is an assignment as in the models for $\mathbf{QR}^{ot}$ ($\mathbf{RQ}^{ot}$), such that it assigns an element of PropFun to every atomic formula, and with the exceptions/additions:*

- *where $\mathcal{A}$ is the atomic $\tau_1 = \tau_2$, the propositional function assigned to $\mathcal{A}$ is given by*

$$|\tau_1 = \tau_2|^{\mathfrak{M}} f = |\approx|^{\mathfrak{M}}(|\tau_1|^{\mathfrak{M}} f, |\tau_2|^{\mathfrak{M}} f),$$

where $|\approx|^{\mathfrak{M}}$ is the function $|\approx|$ obtained from $\Rightarrow$.

- *when $\mathcal{A}$ is $G\tau$ and $G \in RP$, $|G\tau|^{\mathfrak{M}} =: G\tau :$*

The last condition states that the relevant predicates are mapped onto their corresponding elements of $PropFun^G$.

The $\vDash$ relation is much like it was, but for the case of identity, which is

$$a, f \vDash \tau_1 = \tau_2 \text{ iff } a \in |\tau_1 = \tau_2|^{\mathfrak{M}} f$$

Identity in Relevant Logics: A Relevant Predication Approach

Satisfaction and validity are defined as usual. Hereditary and Semantic Entailment lemmas are proven adding a new case to the induction for identity. We record the lemmas here.

Lemma 4 (Heredity)   *If $a \leq b$ and $a, f \vDash \mathcal{A}$, then $b, f \vDash \mathcal{A}$.*

Lemma 5 (Semantic Entailment)   *$\mathcal{A} \to \mathcal{B}$ is satisfied by a variable assignment f in a model $\mathfrak{M}$ iff, for every $a \in K$, if $a, f \vDash \mathcal{A}$, then $a, f \vDash \mathcal{B}$.*

Theorem 2 (Soundness for $\mathbf{QR}_{=}^{\circ t}$ (and $\mathbf{RQ}_{=}^{\circ t}$))   *All of the theorems of $\mathbf{QR}_{=}^{\circ t}$ ($\mathbf{RQ}_{=}^{\circ t}$) are valid in every $\mathbf{QR}_{=}^{\circ t}$-model ($\mathbf{RQ}_{=}^{\circ t}$-model).*

Proof. To show this involves going through the lemmas proven in (Mares & Goldblatt, 2006), verifying inductions with the new base case for identity. These new base cases are straightforward, and will be left for the reader. Here we show that the (REF), (SYM), (TRANS), and (RI) axioms are valid in every $\mathbf{QR}_{=}^{\circ t}$ (and $\mathbf{RQ}_{=}^{\circ t}$) model.

(**REF**): We are required to show that $a, f \vDash \tau_1 = \tau_1$, for every $a \in 0$, and every $f \in U^\omega$. Take an arbitrary f and $a \in 0$. By the definition of models we have that $\tau_1 \in \Rightarrow(a, \tau_1)$, and thus that $a \in |\approx|^{\mathfrak{M}}(|\tau_1|^{\mathfrak{M}} f, |\tau_1|^{\mathfrak{M}} f)$. This immediately gives $a \in |\tau_1 = \tau_1|^{\mathfrak{M}} f$, which gives us $a, f \vDash \tau_1 = \tau_1$.

Given the link between $\Rightarrow$ and $|\approx|^{\mathfrak{M}}$, the remaining cases will use only the latter.

(**SYM**): Suppose that $a, f \vDash \tau_1 = \tau_2$ for some a and f. Thus, $a \in |\tau_1 = \tau_2|^{\mathfrak{M}} f$. That is, $a \in |\approx|^{\mathfrak{M}}(\tau_1, \tau_2)$. Given that $a \leq a$, by the definition of our models we get that $a \in |\approx|^{\mathfrak{M}}(\tau_2, \tau_1)$, and so $a \in |\tau_2 = \tau_1|^{\mathfrak{M}} f$. The result follows by Semantic Entailment.

(**TRANS**): Suppose that $a, f \vDash \tau_1 = \tau_2$. Then $a \in |\approx|^{\mathfrak{M}}(\tau_1, \tau_2)$. Suppose further that $Rabc$ and $b, f \vDash \tau_2 = \tau_3$. Thus $b \in |\approx|^{\mathfrak{M}}(\tau_2, \tau_3)$. By a condition imposed on the models, we have that $c \in |\approx|^{\mathfrak{M}}(\tau_1, \tau_3)$, in other words $c, f \vDash \tau_1 = \tau_3$. The result follows by Semantic Entailment.

(**RI**): Suppose that $a, f \vDash x = y$. For reductio, let $a, f \nvDash Gx \to Gy$. Then $Rabc$ for some b, c such that $b, f \vDash Gx$ but $c, f \nvDash Gy$. From $b, f \vDash Gx$ we get that $b \in |Gx| f$. As Gx is relevant predication, it is mapped onto the element of $PropFun^G$ named : Gx :, and so we have that $b \in : Gx : f$. We can now apply the frame condition for RI and infer that $c \in : Gx : [y/x] f$. By lemma 2, we get that $c \in : Gy : f$, and so $c, f \nvDash Gy$, giving us our contradiction. Thus $a, f \vDash Gx \to Gy$, and the result follows from Semantic Entailment. $\square$

Completeness

The proof of completeness in this section will again follow Mares and Goldblatt (2006). For notational convenience, let $\Gamma \gg_{\mathbb{L}} \Delta$ mean that there for some $\mathcal{A}_1, \ldots, \mathcal{A}_n \in \Gamma$ and $\mathcal{B}_1, \ldots, \mathcal{B}_m \in \Delta$ such that $\vdash_{\mathbb{L}} (\mathcal{A}_1 \wedge \cdots \wedge \mathcal{A}_n) \rightarrow (\mathcal{B}_1 \vee \cdots \vee \mathcal{B}_m)$, where Γ and Δ are sets of formulas and $\mathbb{L}$ is a logic.

Definition 11 *A pair* (Γ, Δ) *is* $\mathbb{L}$-*independent if and only if* $\Gamma \not\gg \Delta$. *An* $\mathbb{L}$-*theory is a set of formulas* Γ *such that if* $\Gamma \gg \mathcal{A}$, *then* $\mathcal{A} \in \Gamma$, *for every* $\mathcal{A}$. *A theory* Γ *is* prime *if and only if, if* $\mathcal{A} \vee \mathcal{B} \in \Gamma$, *then either* $\mathcal{A} \in \Gamma$ *or* $\mathcal{B} \in \Gamma$. *A theory* Γ *is* regular *if and only if it contains every theorem of* $\mathbb{L}$.

Lemma 6 *If* $\mathcal{A} \in \Gamma$ *and* $\vdash_{\mathbb{L}} \mathcal{A} \rightarrow \mathcal{B}$, *then* $\mathcal{B} \in \Gamma$, *for* $\mathbb{L} = \mathbf{QR}_{=}^{\circ t}$ *or* $\mathbb{L} = \mathbf{RQ}_{=}^{\circ t}$.

The proof of this lemma is trivial. Finally, the extension lemma, is needed to show that, if a formula is not a theorem, then there is a regular prime theory that does not contain the formula.

Lemma 7 (Extension) *If* (Γ, Δ) *is* $\mathbb{L}$-*independent, there there is some prime theory* Γ' *such that* $\Gamma \subseteq \Gamma'$ *and* (Γ', Δ), *for* $\mathbb{L} = \mathbf{QR}_{=}^{\circ t}$ *or* $\mathbb{L} = \mathbf{RQ}_{=}^{\circ t}$

Definition 12 (Canonical Frames) *A canonical frame for* $\mathbf{QR}_{=}^{\circ t}$ ($\mathbf{RQ}_{=}^{\circ t}$) *is a tuple*

$$\langle K_c, 0_c, R_c, *_c, U_c, Prop_c, PropFun_c, PropFun_c^G, \Rightarrow_c \rangle, \text{ where}$$

- K_c *is the set of all prime theories.*
- 0_c *is the set of all regular prime theories.*
- R_c *is defined by* $R_c abc$ *iff* $\{\mathcal{A} \circ \mathcal{B} : \mathcal{A} \in a \,\&\, \mathcal{B} \in b\} \subseteq c$.
- $*_c$ *is defined by* $a^* = \{\mathcal{A} : \neg \mathcal{A} \notin a\}$.
- U_c *is the infinite set of constants* Con.
- $\Rightarrow_c$ *is defined by* $\Rightarrow_c (a, \tau_1) = \{\tau_i : \tau_1 = \tau_i \in a\}$.
- *For every closed formula* $\mathcal{A}$, $||\mathcal{A}||_c$ *is defined to be the set* $\{a \in K : \mathcal{A} \in a\}$.
- $Prop_c$ *is defined as the set* $\{||\mathcal{A}||_c : \mathcal{A} \text{ is a closed formula}\}$.

Identity in Relevant Logics: A Relevant Predication Approach

- Given a variable assignment f, the value fn is a constant. Substituting each variable in a formula $\mathcal{A}$ with the constant assigned to it by a variable assignment f results in a closed formula which will be denoted $\mathcal{A}^f$. Therefore $\mathcal{A}^f = \mathcal{A}[f0/x_0, \ldots, fn/x_n, \ldots]$.

- To each formula $\mathcal{A}$, there is a corresponding function $\phi_{\mathcal{A}}$ of type $U^\omega \longrightarrow K$ given by $\phi_{\mathcal{A}} f = ||\mathcal{A}^f||_c$. $PropFun_c$ is the set of all function $\phi_{\mathcal{A}}$, where $\mathcal{A}$ is a formula.

- Moreover, we define $PropFun_c^G$ to be the set of all functions $\Phi_{\mathcal{A}}$ where $\mathcal{A}$ is a relevant predication—that is, is of the form $G\tau$ for some term τ and some $G \in RP$. Further, $\Phi_{\mathcal{A}} \in PropFun_c^G$ will also have the name $: \mathcal{A} :$. Thus, by the definition of $PropFun_c$, we have that for all relevant predications, $: G\tau : f = ||G\tau^f||_c$.[4]

Definition 13 *A canonical model for* $\mathbf{QR}_=^{\circ t}$ *(*$\mathbf{RQ}_=^{\circ t}$*) is composed of the canonical frame* $\mathfrak{F}_c$, *plus a valuation* $|-|_c^{\mathfrak{M}}$, *such that*

- $|c|_c = c$, for every constant symbol c.
- $|P(c_0, \ldots, c_n)|_c = ||P(c_0, \ldots, c_n)||_c$.
- *The valuation is extended to all well-formed formulas as before.*

Many of the lemmas proven on the way to completeness in (Mares & Goldblatt, 2006) involve inductions which now have a new base case for identity. Again, these cases are straightforward, and will be left for the reader. Here we demonstrate that the canonical frame satisfies the conditions for identity introduced above.

Lemma 8 *The canonical frame is a frame.*

Proof. By the arguments of Mares and Goldblatt (2006), we can demonstrate everything except the conditions for $\Rightarrow$ and $PropFun_c^g$. Conditions 1. and 2. from Definition 9 are obvious from the definition of the canonical frame. For the remaining conditions:

3. Suppose that $a \in 0$, and thus $a \in ||t||_c$. As t is a closed formula, $t \in a$. The formula $t \to x = x$ is a theorem, so $x = x \in a$. Thus, $x \in \Rightarrow(a, x)$, as required.

---

[4]Using this, we get that $b \in : Gx : f$ iff $Gx^f \in b$.

4. Suppose that $a \leq b$ and $y \in \Rightarrow(a,x)$. It follows that $a \subseteq b$ and $x = y \in a$. By the axiom $x = y \rightarrow y = x$ we get that $y = x \in a$, and thus $y = x \in b$. By definition, $x \in \Rightarrow(b,y)$, as required.

5. Suppose that $Rabc$, $y \in \Rightarrow(a,x)$ and $z \in \Rightarrow(b,y)$. By definition, $x = y \in a$ and $y = z \in b$, and so $x = y \circ y = z \in c$. It is also a theorem that $(x = y \circ y = z) \rightarrow x = z$, so $x = z \in c$, which by definition means $z \in \Rightarrow(c,x)$, as required.

6. Suppose that $x = y^f \in a$, $R_c abc$, and $b \in : Gx : f$ for $G \in RP$. It is a theorem that $(x = y \circ Gx) \rightarrow Gy$. By the definition of R_c and the fact that we can further infer that $Gx^f \in b$, we have that $Gy^f \in c$. From this we can infer that $b \in : Gy : f$, which by lemma 3 is equivalent to $b \in : Gx : [|y|f/x]$, as required. □

Corollary 1 *The canonical model is a model.*

Lemma 9 (Truth Lemma) *For any formula, $\mathcal{A}$, $\mathcal{A} = \phi_\mathcal{A}$. In other words, for every f, $|\mathcal{A}|f = ||\mathcal{A}||_c$. That is, $a, f \models \mathcal{A}$ iff $\mathcal{A}^f \in a$.*

Theorem 3 (Completeness for $\mathbf{QR}^{\circ t}_{\underline{=}}$ ($\mathbf{RQ}^{\circ t}_{\underline{=}}$)) *If $\mathcal{A}$ is valid in every $\mathbf{QR}^{\circ t}_{\underline{=}}$-model ($\mathbf{RQ}^{\circ t}_{\underline{=}}$-model), then $\mathcal{A}$ is a theorem of $\mathbf{QR}^{\circ t}_{\underline{=}}$ ($\mathbf{RQ}^{\circ t}_{\underline{=}}$).*

The proofs of the previous three stated results are as in (Mares & Goldblatt, 2006), with minor changes such as adding a new base case to inductions for the identity atomic formulas.

5 Concluding remarks

Kremer's axiom choices are in the background of the logic **R**, and have not been shown to keep their philosophical motivation in weaker relevant logics. There is work to be done in selecting axioms for other relevant logics. In particular (1) Kremer's reasons for rejecting certain axioms may not hold in weaker logics, and (2) the Logic of Entailment has additional requirements for its axiom system. In (Shramko, 1994, p. 109), it is claimed that **E** is a better logic for relevant predication. Shramko's suggestion is brief and, for example, does not give any details as to which of RI or its permuted form—a relevant substitution axiom—should be adopted.

To briefly describe the problem identity poses in **E**, note that (RI) and (TRANS) are of the form of an atomic formula entailing an entailment. This

Identity in Relevant Logics: A Relevant Predication Approach

is a form that commits fallacies of modality, which are anathema to the philosophical motivation of **E**. Kremer's relevant indiscernibility interpretation suggests that these axioms may be safely added to **E** when all terms are rigid designators. In these cases, identity statements are *necessitives* (see Anderson & Belnap, 1975), so the axioms do not commit fallacies of modality. Moreover, the approach to identity in this paper must be altered if it is to work in **E**. The current approach uses fusion, for which we get the fallacy of modality '$\mathcal{A} \to (\mathcal{B} \to (\mathcal{A} \circ \mathcal{B}))$'.[5] It remains an open question how to add Kremer's relevant indiscernibility interpretation's identity relation to **E**. This may be done by showing that no theorems of quantified **E** (minus fusion) plus Kremer's axioms commit fallacies of modality, and may be done by altering the proofs given in (Anderson & Belnap, 1975) for propositional **E**.

References

Anderson, A. R., & Belnap, N. D. (1975). *Entailment: The Logic of Relevance and Necessity*. Princeton: Princeton University Press.

Anderson, A. R., Belnap, N. D., & Dunn, J. M. (1992). *Entailment: The Logic of Relevance and Necessity (Vol. 2)*. Princeton: Princeton University Press.

Dunn, J. M. (1987). Relevant predication 1: The formal theory. *Journal of Philosophical Logic, 16*, 347–381.

Dunn, J. M. (1990). Relevant predication 2: Intrinsic properties and internal relations. *Philosophical Studies, 60*, 177–206.

Fine, K. (1988). Semantics for quantified relevance logics. *The Journal of Philosophical Logic, 17*, 22–59.

Goldblatt, R. (2011). *Quantifiers, Propositions and Identity*. Cambridge: Cambridge University Press.

Kremer, P. (1997). Dunn's relevant predication, real properties and identity. *Erkenntnis, 47*, 37–65.

Kremer, P. (1999). Relevant identity. *Journal of Philosophical Logic, 28*, 199–222.

Mares, E. (1992). Semantics for relevant logics with identity. *Studia Logica, 51*, 1–20.

Mares, E., & Goldblatt, R. (2006). An alternative semantics for quantified relevant logic. *The Journal of Symbolic Logic, 71*, 163–187.

---

[5]The author thanks an anonymous reviewer for pointing this out.

Shramko, Y. (1994). Relevant properties. *Logic and Logical Philosophy*, 2, 103–116.

Nicholas Ferenz
University of Alberta, Department of Philosophy
Canada
E-mail: `ferenz@ualberta.ca`

Modelling Reflective Equilibrium with Belief Revision Theory

ANDREAS FREIVOGEL[1]

Abstract: This article brings together two different topics: reflective equilibrium (RE) and belief revision theory (BRT). RE is a popular method of justification in many areas of philosophy, it involves a process of mutual adjustments striving for a state of coherence, but it lacks formally rigorous elaborations and faces severe criticism. To elucidate core elements of RE and provide a solid basis to address objections, a formal model of RE within BRT is presented. A fruitful starting point to the formalization of RE is Olsson's coherentist interpretation of semi-revision, but it does not come with a comparative notion of stability. This paper develops an account of comparative stability in an RE setting for belief changing operations. The operations are useful to characterize RE states as well as processes, and they are shown to satisfy postulates of rational belief change.

Keywords: reflective equilibrium, belief revision theory, coherence, semi-revision, consolidation

1 Introduction

Reflective equilibrium (RE) is a deliberative method of justification that plays a prominent role in methodological discussions in many areas of philosophy. The term was coined by John Rawls (1971), but the idea can be traced to Nelson Goodman (1955). RE is commonly understood as a *state of coherence* among *commitments* ('judgments') about a subject matter and a *theory* containing systematic elements ('principles'). Every so often, RE also refers to the *process*, in which commitments and systematic elements are mutually adjusted striving to establish coherence among them. Commitments are systematized by theories that account for them, and in

---

[1] The research for this paper is part of the project "How far does reflective equilibrium take us? Investigating the power of a philosophical method" (Swiss National Science Foundation grant no. 182854 and German Research Foundation grant no. 412679086). I would like to thank Claus Beisbart and Georg Brun as well as two anonymous reviewers for helpful discussions and valuable comments.

turn, the commitments are revised in view of theories. The idea of RE was discussed in a broad spectrum of philosophical research, including ethics, political philosophy, epistemology, rationality theory, and logic. Despite its widespread popularity, RE is subject to criticism (for a recent overview of objections and rejoinders see Tersman, 2018) and it has not been extensively studied in a formally rigorous manner (a recent exception is Beisbart, Betz, & Brun, in press). The lack of formal models of RE impairs the prospect of addressing worries that RE is too weak as a method of justification.

If we understand commitments and theories as sets of beliefs represented by sentences and approximate mutual adjustments with belief revising operations, belief revision theory (BRT), the formal study of rational belief change, may prove to be a useful framework for formalization. BRT provides a wide range of belief changing operations and ties them to rationality postulates via representation theorems.[2] Although BRT predominantly aims at operations that establish or maintain consistency, ideas about coherence and equilibrium have occasionally been taken up. For example, Gärdenfors (1990) characterized coherence with consistent and deductively closed belief sets. His proposal was criticized for being ill-suited to capture the coherence of epistemic states (see, e.g., Hansson & Olsson, 1999). Nonetheless, closed and consistent belief sets have been associated with RE states, but the few allusions to RE refer at best to a highly generalized account of RE (see, e.g., Hansson, 2000; Rott, 2001). These accounts are too imprecise to apply BRT results to RE in a way that would fertilize the philosophical debate because they do not represent commonly expected core elements of RE like a basic distinction of commitments and theory.

This paper aims to alleviate these problems by imposing an RE inspired structure on existing operations of belief change. The upshot of the present work is a new philosophical application of BRT providing a precise formal model of RE that allows for a systematic discussion of merits and pitfalls of a hitherto more or less vaguely conceived method. The paper is organized as follows. Olsson's account of coherentist semi-revision and consolidation based on stability sets, as well as his findings are introduced in Section 2 and shown to provide some insights about RE states. In Section 3, Olsson's vaguely characterized notion of stability is accommodated to RE desiderata yielding a stability relation allowing to compare positions, i.e., pairs of commitments and theories. Section 4 reconstructs consolidation and semi-

---

[2]For the sake of terminological continuity I will speak of 'belief', although 'acceptance' of a proposition may be more appropriate (see Rott, 2001, ch. 1.1). So understood, beliefs serve as a rough approximation of typical RE ingredients such as judgments or principles.

revision operations from stability sets relative to a position. I show that the operations satisfy Olsson's postulates for consolidation and semi-revision and use them to characterize RE states. Section 5 is a short outlook to RE processes involving these operations.

2 Olsson's coherentist interpretation of semi-revision

Let $\mathcal{L}$ be a propositional language with usual connectives ($\neg, \wedge, \vee, \rightarrow$). Lower-case Greek letters (except γ) represent formulas over elements of $\mathcal{L}$ and upper-case Roman letters denote subsets of $\mathcal{L}$. Cn is a classical consequence operator[3], and the falsum $\bot$ is used as a constant symbol for a contradiction in $\mathcal{L}$. A set A is called *consistent* if and only if $\bot \notin Cn(A)$, otherwise it is *inconsistent*. Olsson (1997) generalizes consistency by introducing a *stability set* $\mathcal{S} \subseteq \mathcal{P}(\mathcal{L})$, i.e., a set of sets of sentences from $\mathcal{L}$. A set A is called *stable* if and only if $A \in \mathcal{S}$, otherwise A is *unstable*.

Definition 1 (Olsson, 1997) $A \subseteq \mathcal{L}$ is coherent *with respect to a stability set $\mathcal{S}$ if and only if (i) $A \in \mathcal{S}$, and (ii) A is consistent.*

If a sentence $\alpha \in K$ is up for removal from a set $K \subseteq \mathcal{L}$, the usual strategy is to consider the maximal subsets of K that do not imply α. If we restrict this set of remainders to a stability set $\mathcal{S}$, we arrive at the set of all maximal stable subsets of K that do not imply α, denoted by $K \bot_\mathcal{S} \alpha$:

Definition 2 (Olsson, 1997) $K' \in K \bot_\mathcal{S} \alpha$ *if and only if*

(i) $K' \subseteq K$,

(ii) $K' \in \mathcal{S}$ and $\alpha \notin Cn(K')$,

(iii) if $K' \subset K'' \subseteq K$, then $K'' \notin \mathcal{S}$ or $\alpha \in Cn(K'')$.

If we take the falsum $\bot$ for α in the definition of a stable remainder set, $K \bot_\mathcal{S} \bot$ denotes the set of all maximal coherent subsets of K. Olsson ensures that there is always a maximal coherent subset of a given set K by imposing very general constraints on a stability set to accommodate his proposal to many kinds of epistemic stability:

(Coherent Void) $\emptyset \in \mathcal{S}$

---

[3]In particular, Cn satisfies inclusion, monotonicity, iteration, deduction and compactness (for a formal presentation, see Hansson, 1999, p. 26).

(Existence assumption) For every set $A \subseteq K$ that is coherent, there exists a set $X \supseteq A$ such that $X \in K \perp_S \perp$.

The existence assumption is only required for infinite cases, but for the rest of this paper, I will resort to finite belief bases as an expressive and adequate representation of belief states of a rationally bounded agent.

Since $K \perp_S \alpha$ may contain more than one element, a (two-place) *selection function* γ_K models an agent's choice. For a given K and α, $\gamma_K(K \perp_S \alpha)$ yields a non-empty subset of $K \perp_S \alpha$, if $K \perp_S \alpha$ is non-empty, otherwise, $\gamma_K(K \perp_S \alpha) = \{K\}$ (Hansson, 1999, p. 105). Explicit reference to K in the subscript of γ may be omitted if the context is unambiguous.

Definition 3 (Hansson, 1997; Olsson, 1997) *Let γ be a two-place selection function and S a stability set. The operation $\div_S \perp$ such that $K \div_S \perp = \bigcap \gamma_K(K \perp_S \perp)$ for all K is the* partial meet consolidation *based on γ and S. The* partial meet semi-revision *(based on γ and S) of K by a sentence α is given by $(K \cup \{\alpha\}) \div_S \perp$.*

Theorem 1 (Olsson, 1997) *An operation ? is an operation of partial meet semi-revision if and only if ? satisfies for all K:*

(Coherence for Identity) *If $K?\alpha = K \cup \{\alpha\}$, then $K?\alpha$ is coherent*

(Inclusion) $K?\alpha \subseteq K \cup \{\alpha\}$

(Strong Coherent Relevance) *If $\beta \in K$, but $\beta \notin K?\alpha$, then there is some K' with $K?\alpha \subseteq K' \subseteq K \cup \{\alpha\}$ such that*

   (i) *K' is coherent, and*

   (ii) *K'' is not coherent for all K'' such that $K' \cup \{\beta\} \subseteq K'' \subseteq K \cup \{\alpha\}$.*

(Pre-expansion) $(K \cup \{\alpha\})?\alpha = K?\alpha$

(Internal Exchange) *If $\alpha, \beta \in K$, then $K?\alpha = K?\beta$.*

Theorem 2 (Olsson, 1998) *An operation $>$ is an operation of partial meet consolidation if and only if $>$ satisfies for all K:*

(Coherence for Identity) *If $K> = K$, then $K>$ is coherent*

(Inclusion) $K> \subseteq K$

(Strong Coherent Relevance) *If $\beta \in K$, but $\beta \notin K>$, then there is some K' with $K> \subseteq K' \subseteq K$ such that*

(i) K' is coherent

(ii) K'' is not coherent for all K'' such that $K' \cup \{\beta\} \subseteq K'' \subseteq K$.

Partial meet semi-revision deviates from the 'standard' operation of partial meet revision as it does not satisfy the success postulate, i.e., $\alpha \in K?\alpha$ may not hold. Moreover, ? does not guarantee coherent outcomes because the intersection of coherent sets may not be coherent in general.[4] The twofold approach to belief changing operations in BRT is often reflected by the use of two different symbols, e.g., $>$ for the axiomatic characterization of consolidation and $\div_S$ for its explicit construction. However, there is traditionally only one symbol in case of semi-revisions. For the sake of notational continuity, I will also resort to use ? for the construction of semi-revision, i.e., $K?\alpha = (K \cup \{\alpha\}) \div_S{^\perp}$.

Observation 1 (collected from Olsson, 1997)   *Let ? and $\div_S{^\perp}$ be operations of partial meet semi-revision and consolidation for a stability set S. Then the following hold:*

(i) K is coherent if and only if $K \div_S{^\perp} = K$.

(ii) If K is coherent and $\alpha \in K$, then $\alpha \in K?\alpha$. (Coherentist Stability)

(iii) K is coherent if and only if for each $\alpha \in K$, $\alpha \in (K \setminus \{\alpha\})?\alpha$. (Mutual Support)

Applying Olsson's Account to Reflective Equilibrium   Observation 1 enables an insightful characterization of RE states that goes beyond their metaphorical description found in the classical literature on RE, e.g., "agreement" (Goodman, 1955, p. 64) or "everything fitting together" (Rawls, 1971, p. 21). (i) and (ii) characterize the state of equilibrium in RE as *invariance* of a base under consolidation and semi-revision with its elements. The operations are reflective in the sense of being internal and not requiring input beliefs that are external to the base. (i) and (ii) are also in line with Gärdenfors' description of a rational belief state that "is in equilibrium under all

---

[4] In addition to the *subtractive* consolidation and revision operations considered in this article, one might extend the RE model to include an *additive* strategy that is based on minimal supersets instead of maximal subsets (Olsson, 1998).

forces of internal criticism." (Gärdenfors, 1988, p. 10). Furthermore, the characterization is faithful to the widespread idea that a state of RE is not imperturbably stable and may break down if new external input is considered (e.g., Rawls, 1971, p. 20). Finally, (iii) captures the idea of mutually supporting elements in RE in a weak sense. Every element is supported by every other element in that a semi-revision is successful because the addition of an element to the rest does not impair the stability of the base.

At this point, one might be tempted to identify RE states with coherent belief bases, but this move would be premature. Firstly, if RE is understood on purely coherentist terms, it falls prey to the standard objections against coherentism, namely that fictions may be perfectly coherent or that justification may be created *ex nihilo*. One might worry that coherence could easily be achieved by streamlining one's prejudices by eliminating contradictions and conclude that RE is too weak as a method of justification. Secondly, Olsson does not specify the stability set because he wants to allow for various accounts of epistemic stability. As long as coherence is not elaborated, it is of little help to render RE less vague. Finally, Olsson treats coherence categorically, which is inadequate to model more or less coherent, intermediate stages of an RE process. In the next section, I refine Olsson's account so that we do not lose the representation theorems and observations capturing core elements of RE, but are more accurate to additional RE features.

3 Stability relations from RE desiderata

This section presents a new start and introduces the basics to construct consolidation and semi-revision for an RE setting. As a first step towards an accurate characterization of RE within BRT, we need to recognize that the process of RE commonly involves the adjustment of two sets of sentences: a set of *commitments* and a set of principles, called *theory*.[5] Commitments and elements of a theory may be distinguished by their role in an RE process. Commitments provide the presystematic starting point to the process of RE, delineate the subject matter, and serve as a revisable touchstone for theories. In turn, theories systematize commitments and account for them through inference, explanation, unification, etc.

We associate a belief state of an agent with an ordered pair $\Phi = \langle C, T \rangle$, called *position*, where C and T are finite sets of sentences of $\mathcal{L}$. Note that

---

[5] A popular approach to RE includes *background theories* as the third ingredient to *wide* RE (Daniels, 1979), but its consideration would go beyond the scope of this paper.

my use of 'theory' departs from the classical identification with consistent and deductively closed belief sets in BRT, exploiting the representational and computational advantages of belief base approaches (for an overview of belief base operations, see, e.g., Fermé & Hansson, 2018, ch. 6). The belief base corresponding to a position Φ is given by $\bigcup \Phi = C \cup T$.

The next step in the refinement of Olsson's account for RE overcomes the shortcoming of treating coherence categorically, which is not adequate if one is interested in studying processes in which intermediate stages may be more or less stable. To surmount this problem and to allow for a comparative notion of stability, I will impose more structure on the stability that reflects further requirements and desiderata of RE.

For this purpose let us introduce a *stability relation* on the set of positions.[6] Formally, we equip the set of pairs of finite subsets of $\mathcal{P}(\mathcal{L})$ with a binary relation $\lesssim$, where $\Phi \lesssim \Psi$ reads as "Ψ is at least as stable as Φ". The strict part of the stability relation, $\Phi < \Psi$, holds, if $\Phi \lesssim \Psi$ but not $\Psi \lesssim \Phi$. Finally, the equivalence relation $\sim$ means that "... is equally stable as ..." and it pertains to Φ and Ψ if and only if both $\Phi \lesssim \Psi$ and $\Psi \lesssim \Phi$ hold. Which properties may be expected of $\lesssim$ to impose some order on positions?

(i) $\lesssim$ is reflexive and transitive.

(ii) $\Phi < \langle \emptyset, \emptyset \rangle$ if and only if $\bigcup \Phi$ is inconsistent.

Similar to Olsson's requirement of coherent void, (ii) provides a minimal point of reference that separates positions with inconsistent bases from those that are more stable because their bases are consistent. One might object that in the presence of few or "mild" contradictions, an inconsistent belief base may be preferable to believing nothing. One could meet this objection by specifying degrees of inconsistencies and replacing Cn with a non-classical consequence operator, but both tasks exceed the scope of this article. (i) secures that $\lesssim$ is a *preorder*, but this does not provide enough structure. In particular, positions may not be comparable, i.e., neither $\Phi \lesssim \Psi$ nor $\Psi \lesssim \Phi$ may hold. For this case we may strengthen the stability relation to be a *total* preorder. That is, in addition to reflexivity and transitivity, the stability relation is also required to satisfy totality. Let Φ and Ψ be positions:

(Totality) $\Phi \lesssim \Psi$ or $\Psi \lesssim \Phi$

---

[6]One might expect that the stability of a positions is defined in terms of its behaviour with respect to its potential revisions. Here, I consider stability (and what this could amount to in an RE setting) to be prior and use it for the construction of potential revisions.

Postulating totality for $\lesssim$ entails that the relation is *trichotomous*, i.e., exactly one of $<, >$, and $\sim$ holds for any two positions.

Totality may not be a plausible requirement of a stability relation in an RE setting, but let us make this assumption for simplicity's sake. Moreover, totality may result as a byproduct of how other RE aspects are modelled as follows. Recent and elaborate accounts of RE (e.g., Baumberger & Brun, 2020; Beisbart et al., in press) include additional desiderata, which go beyond mere consistency or unspecified coherence considerations. On the one hand, the theory of a position should do justice to epistemic goals including well-known virtues of theories in science (e.g., simplicity, explanatory power, scope, etc.; Kuhn, 1977). Their configuration depends on the pragmatic-epistemic objectives for a specific application of RE. Beisbart et al. (in press) put it more concretely and confine themselves to three RE desiderata for a formal model of RE in the framework of the theory of dialectical structures. The first desideratum demands a *systematic* theory rather than a jumbled collection of ad-hoc principles. Next, the commitments should respect the starting point of the process, i.e., the initial commitments about a subject matter. This desideratum of *faithfulness* ensures that the subject matter is not changed during an RE process and that a position is tethered to initially credible commitments. Finally, the coherentist aspect of "agreement" or "fit" is captured by a desideratum called *account*.

This list of RE desiderata helps to specify stability and coherence for RE. Instability is a tendency to leave a position due to the pressure of unsatisfied or imbalanced RE desiderata. As long as there is no optimal balance of RE desiderata, there is room for improvement, and hence, there is some incentive for the agent to revise a position. These desiderata are operationalizable with simple means provided by the BRT framework. A simple way to define measures for account, systematicity, and faithfulness in a finite setting is to compare sizes. Systematicity of some theory T may depend on syntactical simplicity, crudely the size of T or the number of conjuncts in a canonical conjunctive normal form representing T, and its scope, i.e., the number of literals entailed by T. Faithfulness may be based on the extent to which the current commitments deviate from the initial commitments, expressed by the symmetric difference or some other form of distance between them. For the desideratum of account, the so called "fit" or "agreement" of theory and commitments in RE can be spelled out by requiring that the commitments are inferable from the theory. For this paper, the classical consequence operator Cn is a simple and suitable candidate, and we aim for $C \subseteq Cn(T)$ in an optimal state. Hence, account may be operationalized by measuring the overlap of commitments and the logical consequences of the theory given by the size of their intersection. The measures stand in need of normalization, but such details need not be at the center of this project.

Note that these desiderata may clash with each other and force an agent to make trade-offs. For example, an agent may have to give up on systematicity or impair faithfulness to increase the accounting of a theory for the commitments. Formally, a linear combination of weighted measures for the desiderata may model these trade-offs, yielding a single, numerical function of overall stability, which induces a total stability relation $\lesssim$ on positions (relative to some initial commitments C_0). The details of measuring RE desiderata and their trade-offs require further motivation and allow for many refinements (for a concrete implementation, see Beisbart et al., in press). However, the fact that we can measure RE desiderata in a simple way and transform them into a total stability ordering is sufficient to achieve the goal of having an RE inspired structure, on which belief changing operations can be defined.

4 Stability sets, operations and RE states

A last hurdle needs to be cleared to perform operations of belief change based on a stability relation. The selection function γ defined above operates on remainder sets, i.e., sets of subsets of $\mathcal{P}(\mathcal{L})$ in contrast to the stability relation, which is defined over pairs from $\mathcal{P}(\mathcal{L}) \times \mathcal{P}(\mathcal{L})$. We bridge this divide by defining a stability set relative to a position. Formally, let

$$\mathcal{S}_\Phi = \{\bigcup \Psi \mid \Phi \lesssim \Psi\}.$$

Intuitively, $\mathcal{S}_\Phi$ contains the bases of all positions that are at least as stable as Φ. Note that $\mathcal{S}_\Phi$ is no longer equipped with a stability relation and the distinction between principles and commitments is lost.

To construct operations in the same vein as Olsson, we need to decide which stability sets are adequate for a semi-revision or consolidation of a base B corresponding to a position Φ. Unfortunately, $\mathcal{S}_\Phi$ is not adequate, since the consistency of B is sufficient to guarantee $B \div_{\mathcal{S}_\Phi} \bot = B$, since $B \in \mathcal{S}_\Phi$ due to the reflexivity of the stability relation, which would reduce invariance under consolidation to a mere consistency check. The same would hold for semi-revisions of B with its elements. These operations become more useful if specific elements are removed from a position to construct a stability set. To this purpose, we define $\Phi - \alpha = \langle C \setminus \{\alpha\}, T \setminus \{\alpha\} \rangle$ for a position $\Phi = \langle C, T \rangle$ and a sentence α. $\mathcal{S}_{\Phi-\alpha}$ consists of bases of positions which are at least as stable as $\Phi - \alpha$. $\alpha \in B = \bigcup \Phi \in \mathcal{S}_{\Phi-\alpha}$ implies that α is an element of a base corresponding to a position, which is at least as stable as $\Phi - \alpha$. Based on the notion of a comparative stability relation we are able to reintroduce a categorical notion of coherence. $B = \bigcup \Phi$ is coherent *with respect to a stability set* $\mathcal{S}_{\Phi-\alpha}$ if and only if B is consistent and $B \in \mathcal{S}_{\Phi-\alpha}$. B is called coherent *tout court* if and only if B is coherent with respect to $\mathcal{S}_{\Phi-\alpha}$ for all $\alpha \in B$.

Observation 2 *The properties of a stability relation $\lesssim$ entail the following:*

(i) $B \perp_{\mathcal{S}_{\Phi-\alpha}}$ is non-empty for $B = \bigcup \Phi$ and $B = \bigcup \Phi \cup \{\alpha\}$ for any sentence α.

(ii) If $\Phi \lesssim \Psi$, then $\mathcal{S}_\Psi \subseteq \mathcal{S}_\Phi$ and analogously for the strict part.

(iii) $\bigcup \Phi$ is consistent if and only if $\mathcal{S}_\Phi \subseteq \mathcal{S}_{\langle \emptyset, \emptyset \rangle}$.

Proof. (i) Let $B = \bigcup \Phi$. If $\Phi - \alpha \lesssim \Phi$, then $B \in \mathcal{S}_{\Phi-\alpha}$. If B is consistent, we have $B \in B \perp_{\mathcal{S}_{\Phi-\alpha}}$. Otherwise, B is inconsistent and $\Phi < \langle \emptyset, \emptyset \rangle$ implies $\emptyset \in \mathcal{S}_{\Phi-\alpha}$ by transitivity. Thus, $B \perp_{\mathcal{S}_{\Phi-\alpha}}$ is non-empty. If $\Phi < \Phi - \alpha$ it is not guaranteed that we have $B \in \mathcal{S}_{\Phi-\alpha}$, so let us consider $B' = \bigcup (\Phi - \alpha)$, which is a subset of B and an element of $\mathcal{S}_{\Phi-\alpha}$ due to reflexivity. If B' is consistent, it is an element of $B \perp_{\mathcal{S}_{\Phi-\alpha}}$ if $B \notin \mathcal{S}_{\Phi-\alpha}$. If B' is inconsistent, an argument as above applies. In any case $B \perp_{\mathcal{S}_{\Phi-\alpha}}$ is non-empty. For $B = \bigcup \Phi \cup \{\alpha\}$ a similar proof can be given.

(ii) Let $\Phi \lesssim \Psi$, and $B \in \mathcal{S}_\Psi$. Consequently $B = \bigcup \mathcal{X}$ for some $\mathcal{X}$ with $\Psi \lesssim \mathcal{X}$. By transitivity of $\lesssim$, we have that $\Phi \lesssim \mathcal{X}$. Thus, $B = \bigcup \mathcal{X} \in \mathcal{S}_\Phi$.

(iii) One direction of the biconditional follows from (ii) and the requirement that $\langle \emptyset, \emptyset \rangle \lesssim \Phi$ if Φ is consistent. For the other direction, assume for a contradiction that $\mathcal{S}_\Phi \subseteq \mathcal{S}_{\langle \emptyset, \emptyset \rangle}$, but that $\bigcup \Phi$ is inconsistent. Consequently, we have $\Phi < \langle \emptyset, \emptyset \rangle$ and by the strict part of (ii), $\mathcal{S}_{\langle \emptyset, \emptyset \rangle} \subset \mathcal{S}_\Phi$ holds. This implies $\mathcal{S}_\Phi \subset \mathcal{S}_\Phi$, which is a contradiction. $\square$

As a consequence of Observation 2 (i), we can define selection functions with respect to the stability set relative to a position. Belief changing operations on a base $B = \bigcup \Phi$ corresponding to a position Φ are constructed as follows

$$B \div_{\mathcal{S}_{\Phi-\alpha}} = \bigcap \gamma(B \perp_{\mathcal{S}_{\Phi-\alpha}})$$
$$B ? \alpha = \bigcap \gamma((B \cup \{\alpha\}) \perp_{\mathcal{S}_{\Phi-\alpha}})$$

Note that the reference to a stability set $\mathcal{S}_{\Phi-\alpha}$ is omitted in B?α if there is no danger of ambiguity.

Observation 3 *Let $B = \bigcup \Phi$ be a base corresponding to a position Φ and let α be a sentence.*

(i) $B \div_{\mathcal{S}_{\Phi-\alpha}} \bot$ satisfies Olsson's consolidation postulates of Theorem 1.

(ii) $B?\alpha$ satisfies Olsson's postulates of semi-revision of Theorem 2.

A proof of this observation closely follows Olsson's proofs from construction to postulates, replacing unspecified stability sets $\mathcal{S}$ by $\mathcal{S}_{\Phi-\alpha}$. Internal exchange reads as $B?_{\mathcal{S}_{\Phi-\alpha}}\beta = B?_{\mathcal{S}_{\Phi-\alpha}}\delta$ for $\beta, \delta \in B$, since $B?\alpha = B?\beta$ may not hold for different stability sets. Observation 3 shows that operations constructed with stability sets are indeed consolidations and semi-revisions with one caveat: the base B, on which is operated, is fixed by the position Φ and may not be replaced by an arbitrary base since our definition of stability set does not rely on Olsson's coherent void and existence assumptions. To the purpose of providing tailor-made operations for an RE setting, this result is all we need since it is desirable for a piece-meal RE process that belief changing operations are based on the current position instead of a single stability set for all bases.

Observation 4 *Let $B = \bigcup \Phi$ be the base for a position Φ.*

(i) B is coherent tout court if and only if $B \div_{\mathcal{S}_{\Phi-\alpha}} \bot = B$ for all $\alpha \in B$.

(ii) If B is coherent tout court and $\beta \in B$, then $\beta \in B?\alpha$ for all $\alpha \in B$.

(iii) If B is coherent tout court, then $\alpha \in B \setminus \{\alpha\}?\alpha$ for all $\alpha \in B$.

Proof. (i) If B is coherent tout court, B is coherent with respect to $\mathcal{S}_{\Phi-\alpha}$ for all $\alpha \in B$, i.e., B is consistent and $B \in \mathcal{S}_{\Phi-\alpha}$. Consequently, $B \bot_{\mathcal{S}_{\Phi-\alpha}} \bot = \{B\}$, and hence $B \div_{\mathcal{S}_{\Phi-\alpha}} \bot = \bigcap \gamma(B \bot_{\mathcal{S}_{\Phi-\alpha}} \bot) = B$. For the other direction, assume that $B \div_{\mathcal{S}_{\Phi-\alpha}} \bot = B$ for all $\alpha \in B$. The postulate coherence for identity implies that B is coherent with respect to $\mathcal{S}_{\Phi-\alpha}$ for all $\alpha \in B$, and thus B is coherent tout court.

(ii) Let B be coherent tout court. In case of $\alpha = \beta \in B$, we have that $\beta \in B = B \div_{\mathcal{S}_{\Phi-\beta}} \bot = (B \cup \{\alpha\}) \div_{\mathcal{S}_{\Phi-\alpha}} \bot = B?\alpha$ due to (i). For $\alpha, \beta \in B$, $\alpha \neq \beta$, assume for contradiction that $\beta \notin B?\alpha = \bigcap \gamma(B \cup \{\beta\} \bot_{\mathcal{S}_{\Phi-\alpha}} \bot)$. The postulate of strong coherent relevance implies that there is B', $B?\alpha \subseteq B' \subseteq B \cup \{\alpha\}$ such that B' is coherent with respect to $\mathcal{S}_{\Phi-\alpha}$, and B'' is not coherent with respect to $S_{\Phi-\alpha}$ for all B'' such that $B' \cup \{\beta\} \subseteq B'' \subseteq B \cup \{\alpha\}$. If B'' is incoherent because it is inconsistent, B cannot be coherent tout court, which is a contradiction. Otherwise, consider $B'' = B \cup \{\alpha\} = B$ incoherent because $B'' \notin S_{\Phi-\alpha}$, which also entails that B is not coherent tout court.

(iii) follows from (i), since revisions with own elements reduce to consolidations. $\square$

Observation 4 establishes the properties of RE states from Section 1 for bases which are coherent tout court: equilibrium as invariance under consolidation, coherentist stability, and mutual support. In contrast to Olsson's approach, the stability sets are tailor-made for an RE setting capturing additional core elements of RE (e.g., the distinction of commitments and theories) and reflecting various RE desiderata. Informally, coherence tout court means that a base B is stable for small perturbations of its position Φ in many directions, expressed by $\Phi - \alpha$ for $\alpha \in B$.

So far, core elements of RE states become clear with coherence tout court, but is it sufficient for RE? The problem is, that coherence tout court is easily attainable for small positions including the limiting case of $\langle \emptyset, \emptyset \rangle$. However, complete suspension of belief is an implausible RE state defying the idea of RE to provide justification for views about a subject matter.

There is a formal solution to this problem. Observation 2 (ii) entails that the stability sets S_Φ are totally ordered by set-inclusion reflecting the original stability relation on positions in reverse, meaning that $\Phi \lesssim \Psi$ is reflected by $S_\Psi \subseteq S_\Phi$. Reflexivity, transitivity and totality with respect to set inclusion follow directly. While anti-symmetry does not hold for the stability relation over positions, it does so for stability sets of equally stable positions. Φ is as stable as Ψ if and only if, $\Phi \lesssim \Psi$ and $\Psi \lesssim \Phi$, and hence $S_\Psi \subseteq S_\Phi$ and $S_\Phi \subseteq S_\Psi$, which implies $S_\Psi = S_\Phi$.

The total ordering of stability sets allows us to consider minimal elements, i.e., a stability set S_Φ such that there is no other stability set S' with $S' \subset S_\Phi$. If such a stability set exists, it is unique due to totality. A sufficient condition to produce this situation is a position Φ, for which S_Φ contains only $\bigcup \Phi$. Then $S_\Phi = \{\bigcup \Phi\}$ is minimal with respect to set inclusion since there is no empty stability set. What kind of properties do such a position and its minimal stability set exhibit?

Observation 5 *Let Φ be such that $S_\Phi = \{\bigcup \Phi\}$. Then $B = \bigcup \Phi$ is coherent tout court.*

Proof. B is consistent, so we need to show that $B \in \mathcal{S}_{\Phi-\alpha}$ for all $\alpha \in B$. If $\Phi - \alpha \lesssim \Phi$, then $\mathcal{S}_\Phi \subseteq \mathcal{S}_{\Phi-\alpha}$ and hence, $B \in \mathcal{S}_{\Phi-\alpha}$. Otherwise $\Phi < \Phi - \alpha$, which implies $\mathcal{S}_{\Phi-\alpha} \subset \mathcal{S}_\Phi$ contradicting the minimality of $\mathcal{S}_\Phi$. □

We could identify RE states with positions (differing only in their division of commitments and theory) which realize a minimal stability set. The existence of a minimal stability set and its uniqueness is a theoretical option, but probably an unrealistic assumption. In practice, we may require coherence tout court for RE, but also demand that the RE desiderata are satisfied to a sufficient degree. For example, we could require full account, i.e., that all commitments are entailed by the theory so that we have $C \subseteq Cn(T)$ (for more requirements of RE states, see Beisbart et al., in press).

5 An outlook to RE processes

This final section is a sketch of how to model an RE process of mutual adjustments of commitments and theories with operations of belief change. The starting point of an RE process is given by $\Phi_0 = \langle C_0, T_0 \rangle$. C_0 are the initial commitments about a subject matter and the initial theory T_0 is assumed to be empty. A process of RE is a sequence of positions: $\Phi_0, \Phi_1, \ldots$ In each step, an agent may try to systematize her current commitments with an adjustment of her theory, or adjust her commitments to her current theory. In both cases, the agent may revise her belief base $B = \bigcup \Phi_k$ with any belief $\alpha \in \mathcal{L}$ or consolidate her base in an attempt to increase stability. This allows us to model the belief change of the base during an RE step with a single BRT operation. The process may terminate in an RE endpoint if a position is coherent tout court and satisfies additional RE desiderata to a sufficient degree.

Figure 1 depicts an agent's transition from her current position $\Phi_k = \langle C_k, T_k \rangle$ to an updated one in step k. First, the base $B_k = \bigcup \Phi_k = C_k \cup T_k$ and the corresponding stability set $\mathcal{S}_k = \mathcal{S}_{\Phi_k}$ are built up from Φ_k. Then the operation of belief change acts on B_k yielding a new base B_{k+1}. If the operations occur on the level of belief bases, how does an agent get to a new position? The new base could be divided into a new commitment-theory-pair $\langle C_{k+1}, T_{k+1} \rangle = \Phi_{k+1}$ by intersecting the new belief base B_{k+1} with the original counterparts plus possibly a belief α since semi-revision and consolidation satisfy inclusion postulates. In case of a theory adjustment by an operation of semi-revision with α, the new position is given by $T_{k+1} = B_{k+1} \cap (T_k \cup \{\alpha\})$ and $C_{k+1} = B_{k+1} \cap C_k$.

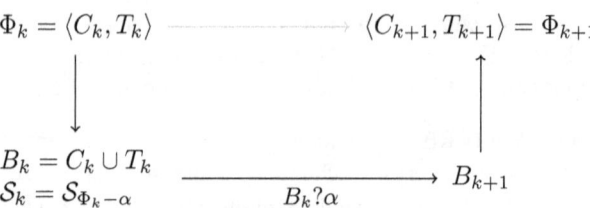

Figure 1: The structure of an adjustment step for a semi-revision by α.

How do subsequent RE steps relate? In general, Φ_{k+1} is not guaranteed to be at least as stable as Φ_k because the construction of new commitments and theory may yield less stable positions. We might overcome this shortcoming by selecting a new position directly from $\mathcal{S}_{\Phi_k}$, but this would move us away from BRT operations on belief bases towards revision procedures for positions. However, there is a weaker result concerning the bases of positions. Assume that $B_{k+1} = \bigcup \Phi_{k+1} \neq \bigcup \Phi_k = B_k$. If the transition from Φ_k to Φ_{k+1} involves a consolidation with respect to $\mathcal{S}_{\Phi_k - \alpha}$ for some $\alpha \in B_k$, either $B_k \notin \mathcal{S}_{\Phi_k - \alpha}$ or $\bot \in Cn(B_k)$ holds. In the latter case we have $\Phi_k < \langle \emptyset, \emptyset \rangle \lesssim \Phi_{k+1}$ since consolidation yields a consistent B_{k+1}. In the former case, $B_k \notin \mathcal{S}_{\Phi_k - \alpha}$ implies $\Phi_k < \Phi_k - \alpha$. Thus, B_{k+1} will be a subset of bases corresponding to positions that are more stable than Φ. Similarly, if B_k and B_{k+1} differ for a step with a semi-revision by an external sentence α, the resulting base is a subset of bases corresponding to positions that are at least as stable as Φ_k, or in other words, these positions cannot be less stable than Φ_k.

Consequently, this sketch of RE processes, involving consolidation and semi-revisions, indicates that on the level of bases, worsening in terms of stability is prevented. This is an interesting insight and a starting point to inquiring the monotonicity of RE processes, and it brings us closer to a position to address specific objections against RE from BRT itself, for example that RE processes cannot reach fixed points due to a suspected non-monotonicity (Bonevac, 2004). The prospect of addressing objections to RE more precisely and in a philosophically insightful way vindicates the fruitfulness of my approach that applies BRT to RE.

6 Conclusion

I hope to have provided a fruitful basis for further research of many open questions at the nexus of BRT and RE. The highly needed provision and exploration of examples would force to develop an account of how an agent chooses her next operation and inputs in an RE process. The question then is whether RE processes reliably attain RE states. This leads to the consideration of further worries that RE processes may be path-dependent and may result in diverging outcomes, yet another unanswered objection to RE. At least, we now have a firm, formal basis for the upcoming tasks. I have refined the coherentist interpretation of consolidation and semi-revision, adapting it to an RE setting. Stability sets relative to positions are extractable from a stability relation, which reflects various RE desiderata that go far beyond mere consistency. This allows for a formal characterization of RE states as invariance under consolidation operations and spell out mutual support with semi-revisions. Furthermore, the mutual adjustments of commitments and theory in an RE process can be modelled to involve consolidation and semi-revision. The refined operations satisfy postulates of rational belief change and this fact strengthens RE as a method of justification.

References

Baumberger, C., & Brun, G. (2020). Reflective equilibrium and understanding. *Synthese*.

Beisbart, C., Betz, G., & Brun, G. (in press). Making reflective equilibrium precise. A formal model. *Ergo*.

Bonevac, D. (2004). Reflection without equilibrium. *Journal of Philosophy*, *101*(7), 363–388.

Daniels, N. (1979). Wide reflective equilibrium and theory acceptance in ethics. *The Journal of Philosophy*, *76*(5), 256–282.

Fermé, E., & Hansson, S. O. (2018). *Belief Change: Introduction and Overview*. Cham: Springer International Publishing.

Gärdenfors, P. (1988). *Knowledge in flux: Modeling the dynamics of epistemic states*. The MIT press.

Gärdenfors, P. (1990). The dynamics of belief systems: Foundations vs. coherence theories. *Revue Internationale de Philosophie*, *44*(172 (1)), 24–46.

Goodman, N. (1955). *Fact, Fiction, and Forecast*. Cambridge, Massachusetts: Harvard University Press.

Hansson, S. O. (1997). Semi-revision. *Journal of Applied Non-Classical Logics*, *7*(1–2), 151–175.

Hansson, S. O. (1999). *A Textbook of Belief Dynamics. Theory Change and Database Updating*. Dordrecht: Kluwer Academic Publishers.

Hansson, S. O. (2000). Coherentist contraction. *Journal of Philosophical Logic*, *29*(3), 315–330.

Hansson, S. O., & Olsson, E. J. (1999). Providing foundations for coherentism. *Erkenntnis*, *51*(2–3), 243–265.

Kuhn, T. S. (1977). Objectivity, value judgment, and theory choice. In *The Essential Tension: Selected Studies in Scientific Tradition and Change* (pp. 320–39). Chicago: University of Chicago Press.

Olsson, E. J. (1997). A coherence interpretation of semi-revision. *Theoria*, *63*(1–2), 105–134.

Olsson, E. J. (1998). Making beliefs coherent: The subtraction and addition strategies. *Journal of Logic, Language and Information*, *7*(2), 143–163.

Rawls, J. (1971). *A Theory of Justice*. Cambridge, Massachusetts: Harvard University Press.

Rott, H. (2001). *Change, Choice and Inference: A Study of Belief Revision and Nonmonotonic Reasoning*. Oxford University Press.

Tersman, F. (2018). Recent work on reflective equilibrium and method in ethics. *Philosophy Compass*, *13*(6), e12493.

Andreas Freivogel
University of Bern, Institute of Philosophy
Switzerland
E-mail: `andreas.freivogel@philo.unibe.ch`

A Useful Four-Valued Logic with Indefinite and Privative Negations: Ammonius and Belnap on Term Negations

JOSÉ DAVID GARCÍA CRUZ[1]

Abstract: In this work we present a logic with three operations that represent simple, indefinite and privative negation. This logic is intended to be a formal reconstruction of Ammonius' logic of privation.

Keywords: Ammonius Hermiae, FDE, indefinite term, privative term

1 Introduction

An indefinite term is a term with attached negation—for example "not-man"—such that it does not produce a proposition and is not a simple term like "man". Aristotle calls them indefinite in the absence of a better name (*De In. 16a 30*). Privative terms are terms with a prefix negative particle representing a kind of absence with respect to predication, for example "unjust". These terms are mentioned by Aristotle in *Prior Analytics I, 47*, and also in *De In. 19b 22–24* privative and indefinite terms are presented in the context of opposition theory.[2] These two kind of terms are not negations in the same sense as the propositional negation is, just consider the three propositions:

---

[1] I would like to thank to anonymous referees for his excellent comments and suggestions. Special thanks to Manuel Correia and Sofia Lombardi for their unconditional support and willingness to discuss these issues. This work is supported by Agencia Nacional de Investigación y Desarrollo (ANID-Chile) and John Templeton Foundation. This work is part of the project: *The logic of prophetical conditionals: Prophetical language, divine communication, and human freedom 61559-3*.

[2] Another passages in which privation is commented by Aristotle are: *Met. V, 12. 1019 b, Met. V, 22. 1022 b 22, Met. XII, 4. 1070 a 30*. In specific in *1022 b 22* is presented a kind of privation relative to privative α, corresponding to the mentioned prefix negative particle.

José David García Cruz

a) "Socrates is not just"
b) "Socrates is not-just"
c) "Socrates is unjust"

Only a) could be considered as a real negation because b) and c) are positive propositions, and our logical language is not able to represent indefinite and privative particles. Ammonius—an Aristotle's commentator (Blank, 1996; Busse, 1985)—suggests that the three propositions are different ways of making negations of the same proposition, and also presents in his commentary to Aristotle's *De Interpretatione* a theory to represent the logical relations between propositions with simple, indefinite and privative terms.

The main aim of this paper is to present a formal reconstruction of Ammonius' ideas about this topic. We present an expansion of the logic FDE (see, e.g., Belnap, 1977a, 1977b) with the introduction of two more negations, which can be considered as weakened negations of the usual one of FDE. These connectives have been designed to represent *indefinite* and *privative* negations. We propose a reconstruction of Ammonius' exposition of the logic of indefinite and privative terms, the main consequence of our analysis is a semantical explanation of logical relations between negative, indefinite and privative formulas.

This reconstruction aligns with the data stipulated by Ammonius when presenting his analysis of Aristotle's *De Interpretatione*, in that sense our reconstruction is not only an imposition of an empty formalism on philosophical theses, rather, we intend to make a philosophical interpretation of Ammonius' ideas, assisted by a logical system capable of representing the relations that Aristotle's commentator presents. In that sense our analysis focuses on a) the meaning of indefinite and privative negations in terms of FDE semantics; b) the meaning of double negation, in specific double mixed negations (e.g., "Socrates is not unjust").

The plan of the paper is as follows. In the following section, we present the basic historical and philosophical elements of the problem of the logic of indefinite and privative negations based on (Correia, 2006b) and (Correia, 2006a). We briefly analyze the main ideas of Ammonius Hermiae. In the third part, we present our FDE expansion with indefinite and privative negations, we explain the semantics of these two negations in the light of Ammonius' comments and considering the semantical properties stipulated by him, in that sense we justify the inclusion of FDE in this problem. Finally, in the fourth part, we will focus on the opposition and implication relations between indefinite and privative formulas.

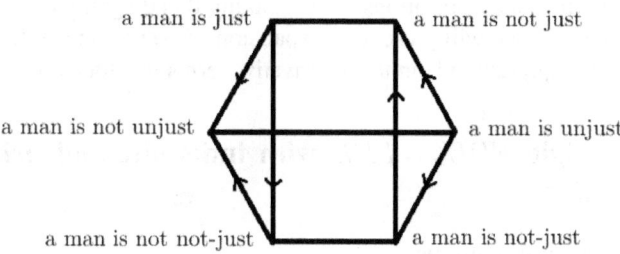

Figure 1: Ammonius' Hexagon

2 Ammonius on indefinite and privative negations

The problem of the logical relation between propositions with simple, indefinite and privative terms dates back to Aristotle's *De Interpretatione*. In *19 b 19*, Aristotle states that these kind of propositions are related by consequence and by contradiction. Many commentators interpret this passage in different ways (see, e.g., Correia, 2006a, 2006b, 2017) but at this respect the most original interpretation was that Ammonius Hermiae presents in his Commentary to Aristotle *De Interpretatione* (Blank, 1996), (Busse, 1985, pp. 162–164). The main thesis of his analysis is that privation is more specific than negation.

In Figure 1 is depicted Ammonious' hexagon of simple, indefinite and privative propositions. His interpretation is synthesized in this diagram in which black arrows represent implication and black lines represent contradiction.

There are two main consequences of Ammonious' analysis: 1) affirmative privation implies both simple negative and affirmative indefinite propositions; 2) on the other side, negative privative propositions are implied by both simple afirmative and negative indefinite propositions. As we can see in the diagram, the proposition "a man is unjust" implies both "a man is not just" and "a man is not-just", that is due to the fact that privation is applied to less entities that negation (Busse, 1985, pp. 162–164). This diagram is a replication of Manuel Correia's reconstruction of this theory (see, e.g., Correia, 2006a, 2006b, 2017), and is related to other very known opposition structures that have been studied in (Béziau, 2012), (Blanche, 1953), (Demey & Smessaert, 2016b), (Demey & Smessaert, 2016a), (Smessaert, 2012) and (Smessaert, 2009). This diagram synthesizes Ammonious' conception of the way in which these propositions relate. Taking this diagram as a guide, in the next section, we will present an expansion of FDE and we define two negations to represent indefinite and privative terms in a four valued logic.

3 The logic FDE_{ip}: FDE with indefinite and privative negations

3.1 Syntax and semantics

The language $\mathcal{L}$ is defined as usual from a denumerable collection of variables $Var = \{A, B, C, \ldots\}$ and a collection of logical connectives $C = \{\wedge, \vee, \neg, \sim, -\}$, to represent conjunction, disjunction, simple negation, indefinite negation, and privative negation, respectively. Four-valued semantics is composed of models of the form $\mathcal{M} = \langle \mathcal{L}, V, D^+, v \rangle$, with a collection of truth values $V = \{\mathbf{f}, \mathbf{n}, \mathbf{b}, \mathbf{t}\}$ (representing truth-values *only false*, *neither true nor false*, *both true and false*, and *only true*), a sub-collection of designated values $D^+ = \{\mathbf{b}, \mathbf{t}\}$, and a mapping from variables to truth values $v : Var \longrightarrow V$, called valuation. The collection V is partially ordered, in that sense we have the following: $\mathbf{f} < \mathbf{b} < \mathbf{t}$, and $\mathbf{f} < \mathbf{n} < \mathbf{t}$. If we consider also an alternative order with $\mathbf{n} <_k \mathbf{f} <_k \mathbf{b}$, and $\mathbf{n} <_k \mathbf{t} <_k \mathbf{b}$, we are facing with Belnap-Dunn approximation and truth lattice (Belnap, 1977b).[3] The conditions for logical connectives are as follows:

1. $v(\varphi \wedge \psi) = inf_<(v(\varphi), v(\psi))$

2. $v(\varphi \vee \psi) = sup_<(v(\varphi), v(\psi))$

3. $v(\neg \varphi) = \begin{cases} \mathbf{t}, & \text{iff } v(\varphi) = \mathbf{f} \\ v(\varphi), & \text{otherwise} \end{cases}$

4. $v(\sim \varphi) = \begin{cases} \mathbf{t}, & \text{if } v(\varphi) = \mathbf{f} \\ \mathbf{f}, & \text{if } v(\varphi) \in D^+ \\ v(\varphi), & \text{otherwise} \end{cases}$

---

[3]The symbol k in $\mathbf{t} <_k \mathbf{b}$ denotes the knowledge (or information/approximation) order for truth values. This paper focuses only on the veritative interpretation of FDE, leaving the informational one to future work.

A	$\neg A$	$\sim A$	$-A$
t	f	f	f
b	b	f	f
n	n	n	f
f	t	t	t

Table 1: Simple, indefinite and privative negation

5. $v(-\varphi) = \begin{cases} \mathbf{t}, \text{ if } v(\varphi) = \mathbf{f} \\ \mathbf{f}, \text{ otherwise} \end{cases}$

The truth tables for conjunction and disjunction are as usual, and for negations Table 1 shows them.[4] Logical consequence and logical truth are defined in a standard way as preservation of designated values. Let $\mathfrak{M}$ be a collection of models, we say that a formula φ is a logical truth (and we will write $\Vdash \varphi$) if $v(\varphi) \in D^+$, $\forall v \in \mathcal{M}, \forall \mathcal{M} \in \mathfrak{M}$. Given a collection Σ of formulas, we say that a formula φ is a logical consequence of Σ, and we write $\Sigma \Vdash \varphi$, when $\forall \beta \in \Sigma$, $v(\beta) \leq v(\varphi)$, $\forall v \in \mathcal{M}, \forall \mathcal{M} \in \mathfrak{M}$.

The next section presents some comments about the meaning of our indefinite and privative negation in the light of *FDE* relational semantics and Ammonius' ideas about indefinite and privative terms.

3.2 The meaning of indefinite and privative negations

We will begin by presenting an analysis of the three negations with the help of the relational interpretation of this logic. For this analysis we follow in detail Priest's (2008) exposition. The four-valued interpretation of the previous section can be reinterpreted as a relational semantics as follows. Let *Var* be a non-empty collection of variables and a partially ordered collection of truth values $\{0, 1\}$ (false and true). Valuations are determined simply by $r \subseteq Var \times \{0, 1\}$, and represent four possible assignments for each formula. Formulas could be only true (**t**), only false (**f**), true and false (**b**), and neither true nor false (**n**). These four truth values may be represented in relational semantics as follows:

---

[4]There could be other similar systems such as expansions of *FDE* obtained by adding Baaz' Delta operator, or exclusion negation (Omori & Wansing, 2017, pp. 1030–1032). On the other hand, The resulting systems could have some similarities with the system $B_4^{\rightarrow}$ studied by Sergei Odintsov in (Odintsov, 2005). I thank an anonymous reviewer for the observation.

	1	2	3	4
a)	$Ar1$	$Ar0, Ar1$	$\text{not}(Ar0), \text{not}(Ar1)$	$Ar0$
	$\neg Ar0$	$\neg Ar1, \neg Ar0$	$\text{not}(\neg Ar1), \text{not}(\neg Ar0)$	$\neg Ar1$
b)	$Ar1$	$Ar0, Ar1$	$\text{not}(Ar0), \text{not}(Ar1)$	$Ar0$
	$\sim Ar0$	$\sim Ar0$	$\text{not}(\sim Ar1), \text{not}(\sim Ar0)$	$\sim Ar1$
c)	$Ar1$	$Ar0, Ar1$	$\text{not}(Ar0), \text{not}(Ar1)$	$Ar0$
	$-Ar0$	$-Ar0$	$-Ar0$	$-Ar1$

Table 2: Relational semantics for negations

$v(A) = \mathbf{t}$ iff $Ar1$ (A is related only with true)
$v(A) = \mathbf{f}$ iff $Ar0$ (A is related only with false)
$v(A) = \mathbf{b}$ iff $Ar1$ and $Ar0$ (A is related with true and false)
$v(A) = \mathbf{n}$ iff not $(Ar1)$ and not $(Ar0)$ (A is related with nothing)

This reading allows us to analyze the meaning of the negations that we have defined to represent the indefinite and privative negation in two-valued terms. Table 2 shows, in relational terms, the three negations displayed in the form of rules in which the upper part of each one represents a condition for the formula A, and the lower part its corresponding output for the negation in question. This table can be read as a horizontal version of the truth tables in Table 1.

The row labeled with "a)" represents the truth table definition of the usual FDE negation, row "b)" the indefinite negation and row "c)" the privative negation. The numbers from 1 to 4 represent possible valuations for each negation. For example, quadrant "1a)" represents the case where $v(A) = \mathbf{t}$ and $v(\neg A) = \mathbf{f}$, quadrant "2a)" the case where $v(A) = \mathbf{b}$ and $v(\neg A) = \mathbf{b}$, and so on. If we carefully analyze this table, we can see that columns 1 and 4 represent the "normal" cases of inversion of truth values that we know in classical logic, so those cases we will not mention much. The interesting cases are those of columns 2 and 3.

We will start by analyzing the way the usual FDE negation operates and then move on to the corresponding proposed negations. Suppose that the proposition in question is "Socrates is just", which we have formalized with the variable A. Therefore, in the cases "1a)" and "4a)" what we have is that if "Socrates is just" relates to truth then the negation "Socrates is not just" relates to falsehood, and vice versa. The case "2a)" therefore, implies that the proposition "Socrates is just" is related to truth and falsehood simultaneously[5], when A relates to truth the negation relates to falsehood, and vice versa, therefore the lower part of the quadrant "2a)" contains the relations that $\neg A$ maintains with respect to A relations. That is, as A relates to both, $\neg A$ also.

Continuing with case "3a)", semantics allows us to keep A unrelated to any of the values, that is, if there is no information about the proposition "Socrates is just", there is no need to assign it any value, and for that same reason its negation is also unrelated to either truth or falsehood.[6] These cases are expressing, according to our interpretation, the behavior of propositional negation, which we represent as $\neg A$ and corresponds to the negation of the copula in Ammonius' terms, i.e., "Socrates is not just" or "No: Socrates is just". We will now consider the cases of the indefinite negation.

As in the previous case, quadrants "1b)" and "4b)" represent the desired inversion in classical negation, therefore, in these cases we omit the analysis assuming that the reader has no qualms about accepting these cases. The interesting cases are "2b)" and "3b)", again. Quadrant "2b)" is analogous to the case of the previous negation, and a similar justification could be expected, but there are at least two reasons to present a different condition. Table 2 in quadrant "2b)" shows us that when $Ar1$ and $Ar0$ the negation only relates A with 0. There are two main reasons for that meaning. Firstly, because this negation must imply the previous one, i.e., indefinite negation implies propositional negation. This is an algebraic justification, the truth value must be lower with respect to truth values order, in other words, the

---

[5]This may seem incongruous, but consider that the relations described here represent states of information, and suppose that there are two sources of information, one of them offers us the data that "Socrates is just" is true, because for that source, Socrates is the most just person in Athens. On the other hand, the second source of information offers us the fact that "Socrates is just" is false, because this source of information has seen Socrates perform an unjust act. Considering this situation, there is no reason to think that we should stop reasoning in the face of these two contradictory data, the classical way out would be to trivialize the logical system and imply any conclusion. In this case, semantics allows us to accept these inconsistent cases, and to offer a coherent interpretation for the negation. It is not the purpose of this paper to present a justification of FDE, so we will not argue further for this four-value semantics. We consider that with the above example it is clear that at least it is plausible to think of situations like the one described there and therefore our argument is relevant philosophically speaking.

[6]In a similar way to the previous case, and as we mentioned in the previous note, consider an analogous situation but now both sources of information have no data about Socrates, nor about his qualities, therefore, is it coherent to stop reasoning in that situation? We consider that no, because this semantics allows us to assume that if there is no data on Socrates, the proposition "Socrates is just" lacks value in this state, possibly our ignorance can be overcome and the value of the proposition updated. Classical semantics would keep us obliged to assign some value in this case, but we have no evidence, or information to know which one.

formula must be related to fewer values. That is, in this case the negation $v(\neg A) = \mathbf{b}$ because $v(A) = \mathbf{b}$, and since $\sim A$ implies $\neg A$, the latter must have a smaller or equal truth value, if it has the same one, the indefinite negation collapses into propositional negation, and it would not make sense to define them as two different negations. Therefore, $v(\sim A) = \mathbf{f}$, since $\mathbf{f} < \mathbf{b}$ according to the order in V. Secondly, suppose that we are in the previous situation with two inconsistent sources of information about the proposition "Socrates is just". In this case we are dealing with the proposition "Socrates is non-just". Ammonious' analysis states that "non-just" is more specific than "not just", that means that "Socrates is just" and "Socrates is non-just" are not true at the same time but could be false together. Therefore, from the fact that $Ar0$ we do not imply that $\sim A$ relates to 1, and that restriction give us this condition.

The case of quadrant "3b)" is also interesting, in states without information about the proposition, indefinite negation also lacks relation with some truth value. But in this specific quadrant there is something else to add in this justification with respect to indefinite negation. It is possible to identify indefinite negation with a kind of complementation, in the sense that the term 'unjust' means everything but just, and precisely because of this, it has sometimes been called *infinite predicate* or *infinite term* (see, e.g, Cavini, 1985; Meiser, 1887; Soreth, 1972). Reference made by indefinite terms is infinite or embraces infinitely many objects, i.e., the quality 'blue' may be what one refers to as 'unjust' or it may be 'hardness', or anything else (excluding 'justice').

In that sense, when we relate A to nothing, what happens with its indefinite negation is that it also takes the value $\mathbf{n}$, representing the fact that it is also not related to any value. This is so, because the complement of 'just' may include more qualities that may not be attributed to Socrates, an also because despite the fact that it is a way of affirming with truth, this way of affirming is equivocal.[7] Therefore, it is not conclusive that the proposition can be related to any truth value, and there is no information for supporting the truth or falsehood of its negation.[8] Something different happens with privation, and now we analyze this case.

---

[7]For example, if we say that "Socrates is unjust" is true, we may say that Socrates is everything but just, and that includes the quality "white", so, we are saying equivocally that "Socrates is white" is truth, but not in a univocal way, because unjust is not only the quality white.

[8]Later on this is analyzed in detail with something we will call *logical emptiness* in the discussion of accidental and necessary predication.

Indefinite and Privative Negations

First, let us start with case "2c)". This case is similar to "2b)" but here we can establish some specifications. If there is information about the truth and falsehood of "Socrates is just", then the privative proposition "Socrates is unjust" cannot be true; because the privative and the affirmative cannot be true at the same time, although they can be false at the same time. The main reason is that justice is an accidental quality and the proposition formed is contingent, these two facts cause that between justice and injustice there may be a third quality attributable to Socrates, which cancels the presence of both. In other words, Socrates can be neither just nor unjust, therefore, both can be false simultaneously, and hence the privative negation relates to 0 in this case. While agreeing on this point with the indefinite negation, in the following case it must differ for two reasons again. Firstly, because the three negations are related by consequence[9], and secondly, for a distinction between necessary and accidental predication. We will focus on the latter, but before offering this justification we introduce some nuances necessary to continue.

We will stop for a moment to present the distinction between necessary and accidental predication in our language, and to consider the anomalous cases of privation. In our formalism we have not introduced anything to represent this distinction, but from now on we will use two new constants to represent two types of propositions: $\mathbb{T}$ and $\mathbb{N}$, the propositional constant of necessary predication and the empty propositional constant, respectively.

We will use the first one to represent all those propositions that are true by definition, i.e., formulas that are only true. They are all equivalent, and for that reason we decided to represent them with a constant. In a different way, any variable of the language is a proposition with accidental predication, these represent propositions like those of our example ("Socrates is just"), because they can be related to truth and falsehood (or both, or none).

This specification makes coherent the existence of the negation of a proposition with necessary predication. Under this semantics, the three negations are identified if they negate the constant $\mathbb{T}$, and this reflects the idea proposed by Ammonius, of maintaining that the indefinite and privative terms are co-extensive only when there is necessary predication (Correia, 2017, p. 320). For the same reason, in cases of accidental predication negations are different and do not collapse one into the other. This makes it clear how our reconstruction formalizes this theory, but what about those

---

[9] As we have said in Section 2, the main thesis of Aristotle and Ammonius about simple, indefinite, and privative negations is that they are related by consequence. In our interpretation we have the following implications: $-\varphi \Vdash \sim\varphi \Vdash \neg\varphi$

T	N
t	n

Table 3: Truth tables for $\mathbb{T}$ and $\mathbb{N}$

spurious cases of predication where the quality that is intended to be deprived does not belong to the subject? This question leads us to the second constant.

If we name the constant $\mathbb{T}$ *logical truth*, its simple, indefinite, and privative negation will be a *logical falsity*. Following this reasoning, any formula of the language non equivalent to any of these two constants, will be a *logical contingency*, but also there is another kind of formulas, that we will call *logical emptiness* denoted as $\mathbb{N}$.[10] Truth tables for $\mathbb{T}$ and $\mathbb{N}$ are depicted in Table 3.

The constant of *logical emptiness* will be used to denote such propositions that are positive but establish meaningless predication, since the quality in question is outside the range of applicable predicates, e.g., "the stone is just", "the melon sees", etc. These propositions would usually be considered as false, since the quality in question is not a possible predicate of a subject, but, we consider that these propositions should maintain the "**n**" value, which we use to establish that the proposition in question is related neither to truth nor to falsehood.

That is to say, in a situation where the proposition "the melon sees" is used, if it is accepted that it is false, we must also accept that its negation—any of them—is true by definition. This does not seem to be the case, and it conflicts with our current reasoning. We consider it more coherent to accept that these types of propositions are rather 'empty', or if you like without meaning, without truth value, etc.

Table 4 shows the relations of the constants $\mathbb{T}$ and $\mathbb{N}$ with their negations, and in the case of simple and indefinite negation the relevant justifications have been given, but in the case of privative negation the table throws the value "**f**", this brings us back to our question, the relational analysis of quadrant "3c)" of the Table 2. Why is the privation of an empty proposition false?

---

[10]The addition of these constants causes the expressivity of the system to be extended, as well mentioned in the paper. Moreover, it should be noted that these constants are not definable in the expansion suggested in (Omori & Wansing, 2017, p. 1029), perhaps they are definable in a different expansion, which is not explored here. What is relevant about these constants is their incidence in the justification of the privative negation, which, in a borderline case such as the one pointed out with the empty constant, does not take the value "**n**" as an infectious value. These difficulties and problems are left open to be dealt with in future work.

Indefinite and Privative Negations

T	¬T	∼T	−T	N	¬N	∼N	−N
t	f	f	f	n	n	n	f

Table 4: Negations of necessary predication and logical emptiness

To return to one of the examples, consider the proposition "the melon sees" denoted by N, simple negation of this proposition is "the melon does not see". The affirmative proposition intends to predicate a quality that is outside those applicable to melons, it is rather a quality assignable to animals, and in a strict sense to agents that fulfill certain specific characteristics. That is to say, this proposition is a *logical emptiness*.

Simple negation, intends to dissociate this quality, and for the same reason that the affirmative proposition, should be considered meaningless, without value, etc. Something similar happens with the indefinite negation, since "the melon is non-seer"—which would seem to be a true proposition—is also a meaningless proposition, since it seeks to link the subject with everything except a quality that by default does not correspond to the subject. That is to say, besides the fact that the quality in question does not correspond to the subject, by linking it with the complement of this quality, it does not link the subject 'only' with the qualities that are assignable to melons, since this negation is 'infinite' or complimentary, within this infinity or complementation there will be found another collection of qualities that are not proper to melons, therefore, predication continues to be ambiguous.

Finally, why is "melon is blind" false? The main characteristic of the privation is the following. A subject must be capable of possessing that which he is intended to be deprived, in the opposite case, the privation is false. For example, Socrates may be deprived of vision, because he is able to see, but the stone is not able to see, therefore it is false that it is deprived of vision. This characteristic condition of privation justifies this case, when an affirmative proposition *A* has no relation to some value (i.e., is a logical emptiness), the privative negation is related to falsehood. With this, the analysis of the three negations is complete. The following section presents the opposition and implication relations of the formulas with simple, indefinite and privative negation.

José David García Cruz

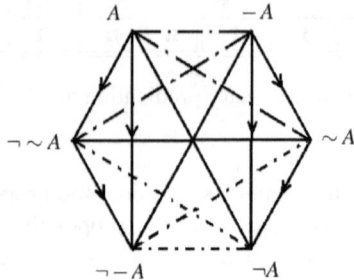

Figure 2: Ammonius' hexagon in FDE_{ip}

4 Ammonius' hexagon of opposition and implication in FDE_{ip}

To sum up, in this section we will present Ammonious' hexagon in our *FDE* expansion. In Figure 1 we have seen Ammonius' hexagon, with the respective relations between simple, indefinite and privative propositions, which, according to Aristotle's commentator, are maintained. In this part we will present a formal version of the same hexagon and explore some relations that are implied according to Ammonius' ideas about privation and negation. In Figure 2 we present our version of the hexagon, in which the opposition relations are explicitly represented by the standard code (see, e.g., Demey & Smessaert, 2016a, 2016b; Smessaert, 2009, 2012).[11]

The previous hexagon (Figure 1) depicted the relations of contradiction in a horizontal-like form, and this caused a loss of balance between implication relations since these did not maintain a uniform order. For example, on the right side, the relation between "a man is not just" and "a man is not-just" goes in the opposite direction to the relations on the left side. On the right side the implication between "a man is not-just" and "a man is not unjust" goes in the opposite direction to the two arrows on the same side. We consider that although the relations are correct, to give coherence to the diagram, they should be ordered concerning implication, preserving the order from the universal to the particular. That means that the two "more general" ones go in the superior nodes, the intermediate ones in the middle, and the weaker ones

---

[11]Contrariety:_._._._._._._ , Subcontrariety:_._._._._._._._ .

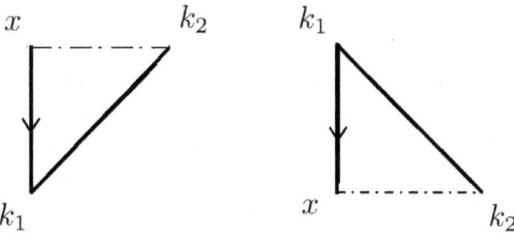

Figure 3: Contrariety and Subcontrariety

in the nodes from below. This gives us the configuration of Figure 2, and it corresponds to Ammonius' theory of the order of negations concerning the specificity of the substances to which they apply (Busse, 1985, pp. 162–164). This is the first modification we suggest for Ammonius' hexagon.

The second variation corresponds to the relations between the propositions that are neither contradictory nor implicative among themselves. If we look at Figure 1 there are some relations lost.[12] According to the order in which they have been configured and according to our semantics, the remaining relations are contrariety and subcontrariety. If we analyze it from the traditional Aristotelian perspective of oppositions (see, e.g., Demey & Smessaert, 2016b, p. 239), the requirements to have contrariety and subcontrariety are two: implication and contradiction. If we have two contradictory formulas k_1 and k_2, and one of them, let us say k_1, is implied by a third x, we have that x and k_2 are contrary; on the other hand, we have two contradictory k_1 and k_2, and one of them, let's say k_1, implies a third x, we have that x and k_2 are subcontrary. Figure 3 illustrates this reasoning. Therefore, the remaining relations are contrariety and subcontrariety, that explain how simple, indefinite and privative propositions relate.

This reasoning is consistent with the semantics we have outlined, and therefore the hexagon would be as proposed in Figure 2. A last consideration is possible, the simple combinatorics allows us to stipulate a group of formulas with which to form an expansion of the hexagon into more complex structures. There is nothing to prevent such an expansion, but it is interesting to consider what meaning to attach to the resulting formulas.

---

[12]For example, the relation between "a man is just" and "a man is unjust", or between "a man is not unjust" and "a man is not just", etc.

5 Conclusion

We have presented an FDE expansion to formalize the logic of privation of Ammonius Hermiae. In this logic, we have defined three different negations intended to represent privative, indefinite, and negative propositions following Ammonius' theory. As a result, we have shown that these three negations form a hexagon of opposition and implication. Finally, we have suggested a different configuration of Ammonius' hexagon that have some representational advantages over the first one presented.

References

Belnap, N. (1977a). How a computer should think. In G. Ryle (Ed.), *Contemporary Aspects of Philosophy*. Oriel Press.

Belnap, N. (1977b). A useful four-valued logic. In J. M. Dunn & G. Epstein (Eds.), (pp. 55–76). D. Reidel.

Béziau, J.-Y. (2012). The power of the hexagon. *Logica Universalis*, 6(1–2), 1–43.

Blanche, R. (1953). Sur l'opposition des concepts. *Theoria*, 19(3), 89–130.

Blank, D. (1996). *Ammonius On Aristotle On Interpretation 1 - 8. Ancient Comentators of Aristotle, R. Sorabji (Ed.)*. Duckworth.

Busse, A. (1985). *Ammonii In Aristotelis De Interpretatione Comentarius*. Reimer.

Cavini, W. (1985). La negazione di frase nella logica greca. In S. Stern, A. Hourani, & V. Brown (Eds.), *Studi su Papiri Greci di Logica e Medicina*. Accademia Toscana di Scienze e Lettere "La Colombaria".

Correia, M. (2006a). The proto-exposition of aristotelian categorical logic. In J.-Y. Béziau & G. Basti (Eds.), *The Square of Opposition: A Cornerstone of Thought. Studies in Universal Logic* (pp. 21–34). Birkhauser.

Correia, M. (2006b). ¿Es lo mismo ser no-justo que ser injusto? Aristóteles y sus comentaristas. *Méthexis*, 24, 41–51.

Correia, M. (2017). Aristotle's squares of opposition. *South American Journal of Logic*, 3, 313–326.

Demey, L., & Smessaert, H. (2016a). The interaction between logic and geometry in aristotelian diagrams. *Diagrammatic Representation and Inference, Diagrams*, 9781, 67–82.

Demey, L., & Smessaert, H. (2016b). Metalogical decorations of logical diagrams. *Logica Universalis*, 10(2–3), 233–292.

Meiser, K. (1887). *Anicii Manlii Severini Boeti Commentarii in Librum Aristotelis PERI ERMHNEIAS.* Teubner.
Odintsov, S. (2005). The class of extensions of nelson's paraconsistent logic. *Studia Logica*(80), 291–320.
Omori, H., & Wansing, H. (2017). 40 years of FDE: An introductory overview. *Studia Logica, 105*(6), 1021–1049.
Priest, G. (2008). *An introduction to Non-Classical Logics: From If to Is.* Oxford University Press.
Smessaert, H. (2009). On the 3D visualisation of logical relations. *Logica Universalis, 3*(2), 303–332.
Smessaert, H. (2012). The classical aristotelian hexagon versus the modern duality hexagon. *Logica Universalis, 6*(1-2), 171–199.
Soreth, M. (1972). Zum infiniten Prädikat im zehnten Kapitel der Aristotelischen Hermeneutik. In S. Stern, A. Hourani, & V. Brown (Eds.), *Islamic Philosophy and the Classical Tradition* (pp. 389–424).

José David García Cruz
Pontifical Catholic University of Chile, Institute of Philosophy
Chile
E-mail: `jdgarcia2@uc.cl`

Exploring a Result by Ghilardi: Projective Formulas vs. the Extension Property

IRIS VAN DER GIESSEN[1]

Abstract: Ghilardi (1999, 2000) presents in his papers on unification in IPC and several classical modal logics an important theorem that is used in the study of admissible rules. The theorem connects the extension property of Kripke models to projective formulas. In this paper, we investigate Ghilardi's bisimulation proof method used for classical modal logics and we present a small simplification of the solution. Our investigation of the key elements of Ghilardi's proof provides an explanation of the close relationship between bisimulation and the extension property via so-called extension structures.

Keywords: modal logic, extension property, projective formulas, admissible rules

1 Introduction

In this paper we examine an important result established by Ghilardi that provides a robust connection between *projective formulas* and the *extension property*. This is a characterization of a syntactic property of formulas in terms of a semantic property of Kripke models. Ghilardi first develops this characterization for IPC (Ghilardi, 1999, Theorem 5 of Section 2). Later he proves it for many well-known classical modal logics extending K4, among them S4 and GL (Ghilardi, 2000, Theorem 2.2). He used this result in the study of unification in logic. Ghilardi shows that unification in IPC and several modal logics is finitary, which means that the set of 'maximal' unifiers is finite.

The purpose of this paper is to provide a new explanation of Ghilardi's proof of the connection between projectivity and the extension property. We hope that our investigation will help researchers who are not familiar

---

[1] I would like to thank Rosalie Iemhoff, Amir Tabatabai and Raheleh Jalali for a lot of helpful discussions. Support by the Netherlands Organisation for Scientific Research under grant 639.073.807 is gratefully acknowledged.

with the result. In his first paper, Ghilardi provides a strategy that works for IPC. Here we are interested in the more general method that is based on bisimulation which is used for classical modal logic. Among researchers, the paper is considered to be both beautiful and difficult. The current paper is an attempt to clarify the ideas of Ghilardi. We indicate the key elements of Ghilardi's proof which provides an explanation of the close relationship between bisimulation and the extension property. We introduce so-called *extension structures* to explain this relationship.

Surprisingly, our analysis reveals an additional benefit in terms of a shortening of the solution. To prove projectivity from the extension property, Ghilardi constructs a unifier that is a concatenation of substitutions. We will argue that in the classical modal case, the concatenation of substitutions can be shortened. We would like to stress that this is a minor simplification and is still strongly based on Ghilardi's proof strategy. This paper does not reveal new big results, it rather provides a new explanation of Ghilardi's proof and it provides some examples, which will hopefully help the reader to understand the result.

An important reason to investigate Ghilardi's papers is that the equivalence between projectivity and the extension property is a very useful tool in the field of admissible rules. Admissible rules are those rules under which the set of theorems of a logic is closed. More precisely, a rule A/B is said to be admissible if every unifier of A is also a unifier for B. For example, the rule $\Box A/A$ is an admissible rule in many modal logics. Admissible rules are interesting to study, because they give insight in the structure of the logic in terms of consequence relations (see Iemhoff, 2016, for an introduction).

Projective formulas play an important role in the study of admissible rules. Projective formulas are formulas A for which admissibility and derivability are the same in the sense that rule A/B is admissible if and only if it is derivable. It can be complicated to show that a certain formula is projective directly from its definition. It is often easier to prove the extension property of a class of Kripke models. Ghilardi's result provides the useful semantic characterization for projectivity in terms of the extension property.

Ghilardi's result is successfully applied in constructing bases for admissible rules. See for examples the papers of Iemhoff (2001), Jeřábek (2005) and Iemhoff and Metcalfe (2009) for logics including K4, GL, S4 and IPC. In addition, Ghilardi points out that an algorithm (provided in his papers) computing the finitely many maximal unifiers yields a new solution to Friedman's problem (Friedman, 1975): admissibility in IPC is decidable. This was first proved by Rybakov (1984).

Projective Formulas vs. the Extension Property

The paper is structured as follows. In Section 2 we introduce semantic terminology, including the definitions of the extension property and bisimulation. Section 3 treats the definition of projective formulas. Section 4 describes the key elements in Ghilardi's method, introduces the notion of extension structures and explains how the substitutions can be shortened. We end with a short conclusion.

2 Kripke semantics and the extension property

Following Ghilardi (2000), we consider classical modal logics that are sound and complete with regard to finite Kripke models. We consider the modal language with constant $\bot$, propositional variables $p, q, \ldots$, connectives $\wedge, \vee, \rightarrow$ and modal operator $\square$. We often use the term *atoms* to mean propositional variables. If A is a formula, $\neg A, \Diamond A$ and $\boxdot A$ are defined as $A \rightarrow \bot, \neg\square\neg A$ and $A \wedge \square A$, respectively. $F(p_1, \ldots, p_m)$ denotes the set of all formulas built from proposition letters $p_1, \ldots, p_m$. We consider *normal modal logics* L, which is a set of formulas containing all classical tautologies, K-axiom $\square(A \rightarrow B) \rightarrow \square A \rightarrow \square B$, and is closed under modus ponens (if A in L and $A \rightarrow B$ in L, then B in L), uniform substitution and necessitation (if A in L then also $\square A$ in L). Following Ghilardi, we write $A \vdash_L B$ to mean that $A \rightarrow B \in L$. We are interested in normal extensions of K4.

We deal with Kripke models that are defined on the basis of a finite transitive *frame*, which are structures (W, R) where W is a finite set of worlds equipped with a *transitive* relation R. We assume that those frames have a minimal element ρ, called a *root*, satisfying $\rho R w$ for each $w \neq \rho$. This minimal element does not have to be unique, but from now on we work with pointed frames (W, R, ρ) where the root is specifically specified. The *cluster* $cl(w)$ of a point w is the equivalence class of w under the equivalence relation $\sim_R$ defined as

$$w \sim_R v \text{ iff } wR^+v \text{ and } vR^+w,$$

where R^+ stands for the relation $R \cup id$. Note that for irreflexive frames, $\#cl(w) = 1$ for all $w \in W$. We also define the relation

$$wR^> v \text{ iff } wRv \text{ and not } vRw.$$

A *Kripke model* is a triple (W, R, V) where (W, R) is a frame and V is the *valuation*, which is a function $V : W \times Atoms \rightarrow \{0, 1\}$. We use letters

K, M to indicate Kripke models. We usually implicitly restrict the domain of the valuation to atoms which play a role in question and say that K is a model *over atoms* $\{p_1, \ldots, p_m\}$ if the domain is restricted to those atoms. We often write $w \in K$ to mean $w \in W$ when $K = (W, R, V)$. We write $K(w) := \{p \mid V(w, p) = 1\}$, the set of all atoms that hold in w. We extend the valuation to a forcing relation $\Vdash$ as usual:

$K, w \Vdash p$	iff $V(w, p) = 1$,
$K, w \Vdash \bot$	never,
$K, w \Vdash A \wedge B$	iff $K, w \Vdash A$ and $K, w \Vdash B$,
$K, w \Vdash A \vee B$	iff $K, w \Vdash A$ or $K, w \Vdash B$,
$K, w \Vdash A \to B$	iff $K, w \Vdash A$ implies $K, w \Vdash B$,
$K, w \Vdash \Box A$	iff for all v such that wRv; $K, v \Vdash A$.

We write $K \models A$ to mean $K, w \Vdash A$ for every $w \in K$ and say that K *satisfies* A. Since we consider rooted transitive models, $K, \rho \Vdash \Box A$ if and only if $K \models \Box A$. We denote K_v for the submodel of K generated by v. We let v be the root of K_v. We say that model K *almost satisfies* A if $K_w \models A$ for all w except for $w \in cl(\rho)$.

A frame is said to be an *L-frame* if for every model K based on that frame and every formula $A \in L$, we have that $K \models A$. We call K an *L-model* if K is based on an L-frame. We write Mod_L to be the set of all L-models and $Mod_L(A)$ to be the set of all L-models that satisfy A in the root.

Ghilardi makes two assumptions about the logic L. For the purpose of this paper we only have to require the first assumption, which is completeness with respect to the described finite models:

Assumption 1 *For all formulas A, B, we have $A \vdash_L B$ if and only if $Mod_L(A) \subseteq Mod_L(B)$.*

Ghilardi's second assumption is needed for the unification results (Ghilardi, 1999, 2000). This assumption is known as L being *extensible*, where the construction of attaching a new root to L-frames again yields an L-frame. This assumption is also crucial in the field of admissible rules (Iemhoff & Metcalfe, 2009; Jeřábek, 2005). Examples of logics satisfying these assumptions are K4, S4, GL and S4.Grz.

A *variant* of an L-model K is an L-model K', such that they have the same frame and their valuation agree on all worlds except for possibly worlds $w \in cl(\rho)$.

Projective Formulas vs. the Extension Property

Definition 1 *A class $\mathcal{K}$ of L-models over $\{p_1, \ldots, p_n\}$ is said to have the* extension property *if for every L-model K, if $K_w \in \mathcal{K}$ for each $w \notin cl(\rho)$, then there is a variant K' of K such that $K' \in \mathcal{K}$.*

We are interested in the extension property of classes $Mod_L(\Box A)$. The extension property states that we can turn models that almost satisfy $\Box A$ into a model of $\Box A$. In Section 4, we will see that there is a close relationship between the extension property and bisimulation. Here we only give the definitions.

Definition 2 *Let K, M be Kripke models. The notion for two models K, M together with points $k \in K$ and $m \in M$ being n-bisimilar is defined recursively and we denote it by $K_k \sim_n M_m$.*

$K_k \sim_0 M_m$ iff $K(k) = M(m)$ *(k and m satisfy the same atoms).*
$K_k \sim_{n+1} M_m$ iff $K(k) = M(m)$ *and for all k' such that kRk' there exists an m' such that mRm' and $K_{k'} \sim_n M_{m'}$, and vice versa.*

Note that $K_k \sim_l M_m$ implies $K_k \sim_n M_m$ for all $l \geq n$. For each n, $\sim_n$ is an equivalence relation. We denote the equivalence classes by $[K_k]_n$. For fixed n, the number of equivalence classes is bounded.

Proposition 1 *Consider models over $\{p_1, \ldots, p_m\}$. Define $N(0) := 2^m$ and $N(n+1) := 2^{N(n)+m}$. The number N of possible $\sim_n$ equivalence classes is smaller or equal to $N(n)$.*

Proof. See (Visser, 1996). □

The *modal degree* $d(A)$ of a formula A is defined inductively as follows: $d(\bot) = d(p) = 0$, for atoms p, $d(A_1 \wedge A_2) = d(A_1 \vee A_2) = d(A_1 \to A_2) = \max\{d(A_1), d(A_2)\}$ and $d(\Box A) = d(A) + 1$. The relation between bisimilar models and modal degree is explained in the following theorem.

Theorem 1 *Let K, M be models over $\{p_1, \ldots, p_m\}$. We have $K_k \sim_n M_m$ if and only if for each formula B with atoms in $\{p_1, \ldots, p_m\}$ with $d(B) \leq n$ we have $K, k \Vdash B \Leftrightarrow M, m \Vdash B$.*

Proof. See (Ghilardi, 2000). □

3 Projective formulas

In this section we introduce substitutions and the notion of projective formula. This can also be read in (Ghilardi, 2000), but we use other notation. A *substitution* is a function $\sigma : \{p_1, \ldots, p_m\} \to F(q_1, \ldots, q_l)$. This function can be extended to a function with domain $F(p_1, \ldots, p_m)$ by

$$\sigma(A(p_1, \ldots, p_m)) = A(\sigma(x_1)/x_1, \ldots, \sigma(x_m)/x_m).$$

From now on, we identify σ with this extension. The composition of $\sigma : F(p_1, \ldots, p_m) \to F(q_1, \ldots, q_l)$ and $\tau : F(q_1, \ldots, q_l) \to F(r_1, \ldots, r_k)$ is defined by $\tau\sigma(p) = \tau(\sigma(p))$.

A *unifier* for a formula A built from atoms $p_1, \ldots, p_m$ is a substitution $\sigma : F(p_1, \ldots, p_m) \to F(q_1, \ldots, q_l)$ such that

$$\vdash_L \sigma(A).$$

We are only interested in unifiers where domain and codomain are the same.

Definition 3 *A formula of the form $\Box A$ with proposition letters $p_1, \ldots, p_m$ is* projective *in L if there exists a unifier σ for it such that*

$$\Box A \vdash_L p \leftrightarrow \sigma(p) \tag{1}$$

for all proposition letters p_i. We call σ a projective unifier.

Using the substitution axiom, it is easy to prove that condition (1) is equivalent to

$$\Box A \vdash_L B \leftrightarrow \sigma(B) \tag{2}$$

for all formulas B in proposition letters $\{p_1, \ldots, p_m\}$.

Ghilardi builds suitable substitutions adopting this property in the following way. Let $\{p_1, \ldots, p_m\}$ be the atoms occurring in A. Let a be a subset of those atoms; the substitution $\sigma_a : F(p_1, \ldots, p_m) \to F(p_1, \ldots, p_m)$ is defined as:

$$\sigma_a^{\Box A}(p) = \begin{cases} \Box A \to p & \text{if } p \in a, \\ \Box A \wedge p & \text{if } p \notin a. \end{cases}$$

From now, we omit the superscript and just write σ_a when $\Box A$ is clear from the context. It is easy to see that $\Box A \vdash_L \sigma_a(p) \leftrightarrow p$. We sometimes call those substitutions *simple*. Ghilardi defines substitution $\theta := \sigma_{a_1} \cdots \sigma_{a_s}$ where $a_1, \ldots, a_s$ is any fixed ordering on the subsets of $\{p_1, \ldots, p_m\}$. Since the simple substitutions are closed under condition (2), we know that θ also satisfies condition (2).

4 Connecting projectivity to the extension property

In this section we investigate the important theorem that connects projectivity to the extension property. Recall that L is a logic extending K4 that satisfies the finite model property (Assumption 1).

Theorem 2 (Ghilardi, 2000)   *Formula $\Box A$ is projective in L if and only if $Mod_L(\Box A)$ has the extension property.*

We are interested in the difficult direction of this theorem, which is from right to left. For a proof for the other direction we refer to (Ghilardi, 2000). We give an analysis of Ghilardi's proof and we will identify key elements of his method.

We fix some notation that we use in the rest of the paper. Let $\Box A$ be a formula with atoms from $\{p_1, \ldots, p_m\}$. Assume that $Mod_L(\Box A)$ has the extension property. Suppose that $d(A) \leq n$. Let N be the number of different equivalence classes of n-bisimilar models and let N' be the number for $(n-1)$-bisimilar models. The goal is to prove that $\Box A$ is projective.

In short, Ghilardi proves that θ^{2N} is a projective unifier for $\Box A$. Number N belongs to n-bisimilar equivalence classes, but we will show that it suffices to use $(n-1)$-bisimilar classes. Number N' is smaller than N, so this results in the shorter concatenation $\theta^{2N'}$. If we carefully read the proof of Theorem 3, we actually conclude that $\theta^{2(N'+1)}$ is the projective unifier for $\Box A$.

The first ingredient in the proof makes a bridge between substitutions in syntax and semantic operations in models. Ghilardi gives the following definition of the semantic operator σ^* on models based on substitution σ:

$$\sigma^*(K), w \Vdash p \iff K, w \Vdash \sigma(p).$$

Note that σ^* only changes the valuation in the model. From now on we abuse terminology and call σ^* a substitution on models. This is a first step to connect the extension property to projectivity because the first is a property of semantics and the latter of syntax. We give some properties of σ^*.

Lemma 1   *Let A be a formula and let σ be a substitution. For every Kripke model K, we have*
  (i) $\sigma^*(K) \models A$ iff $K \models \sigma(A)$,
  (ii) *and for every substitution τ, $(\tau\sigma)^*(K) = \sigma^*(\tau^*(K))$.*

Point (ii) shows that the order of substitutions σ and τ reverses.

Ghilardi defines the useful substitutions σ_a, which we already defined in Section 3. We already saw that they are closed under condition (2), a key condition for $\Box A$ being projective. Now we only have to search for a suitable combination of those σ_a's and prove that this is a unifier for $\Box A$. However, finding the right concatenation is the hard part of the proof.

The extension property will guide us in the right direction for finding the correct concatenation of σ_a's. The method consists of several steps. We start with two relatively simple lemmas. For proofs see (Ghilardi, 2000, Lemma 2.1 and 2.3).

Lemma 2 *Let $\Box A$ be a formula with atoms in $\{p_1, \ldots, p_m\}$ and let K be an L-model. Suppose $a \subseteq \{p_1, \ldots, p_m\}$. We have*
  (i) $(\sigma_a^*(K))(w) = K(w)$ if $K_w \models \Box A$,
  (ii) $(\sigma_a^*(K))(w) = a$ if $K_w \not\models \Box A$, and
  (iii) $\sigma_a^* \sigma_a^* = \sigma_a^*$.

In words, the first two points of the lemma say that the atoms forced in a world w stay the same (in case $K_w \models \Box A$), or become exactly the atoms in a (in case $K_w \not\models \Box A$).

Lemma 3 *Let $\Box A$ be a formula with atoms in $\{p_1, \ldots, p_m\}$ and suppose that $\mathrm{Mod}_L(\Box A)$ has the extension property. Let K be a model that almost satisfies $\Box A$. Then there is a set $a \subseteq \{p_1, \ldots, p_m\}$ such that $\sigma_a^*(K) \models \Box A$.*

We combine the ingredients so far and sketch a proof idea to find a unifier for $\Box A$. We will see that this naive idea is not sufficient and that we need more. For simplicity, one can think of models without any clusters. We want to find a unifier θ that is a concatenation of σ_a's. In other words, we want to show that $\vdash_L \theta(\Box A)$. Using the completeness theorem and Lemma 1, we want to show that $\theta^*(K) \models \Box A$ for each L-model K. Let K be an L-model. We start at the leafs of the model and work our way down to the root. In each step we want to apply a σ_a that gives us a model in which more nodes validate $\Box A$. Consider a world w that almost satisfies $\Box A$, i.e., $K, w \not\Vdash \Box A$ and $K, v \Vdash \Box A$ for all $wR^> v$. By Lemma 3 there is a valuation a such that w satisfies the atoms from a and $\sigma_a^*(K), w \Vdash \Box A$. We pick σ_a and apply it to our model. This strategy sounds promising, because we can go through all the nodes and apply a substitution that works for that node. Define θ on the basis of all those substitutions to yield $\theta^*(K) \models \Box A$. The big problem is that the definition of our θ depends on K, so we cannot define a good sequence of σ's that works *for all* models K. Doing induction on the depth of the model will not solve the problem, because the depth is not bounded.

Projective Formulas vs. the Extension Property

The key idea is to connect the extension property to bisimulation of models. Ghilardi defines $\theta := \sigma_{a_1} \cdots \sigma_{a_s}$ where $a_1, \ldots, a_s$ is any fixed ordering on the subsets of $\{p_1, \ldots, p_m\}$. He shows that θ^{2N} is a projective unifier, where N is the number of the different n-bisimilar models. Ghilardi defines four important ingredients: frontier points, a rank, homogeneous models and the minimal rank (the last is our terminology).

- $f_K[\Box A] := \{w \in K \mid K_w \not\models \Box A \text{ and } \forall v(wR^>v \Rightarrow K_v \models \Box A)\}$ is the set of *frontier points*.
- The *rank* of a model K is defined as

$$r(K) := \#\{[K_w]_n \mid \rho R w \text{ and } K_w \models \Box A\}.$$

- Model K is *homogeneous* if $r(K_w) = r(K_v)$ for each w, v with $K, w \not\Vdash \Box A$ and $K, v \not\Vdash \Box A$.
- $\mu(K) := \min\{r(K_w) \mid K_w \not\models \Box A\}$, which we call the *minimal rank*.

Frontier points are the points w such that K_w almost satisfies $\Box A$. As observed above, for each frontier point we can use the extension property (Lemma 3) to find a σ_a such that $\Box A$ becomes true in that frontier point. For different frontier points there can be different σ_a's that work. However, after one application of θ, all frontier points are turned into points that satisfy $\Box A$. The next step is to find the new frontier points and apply θ again. Ghilardi shows that after two applications of θ, the minimal rank grows strictly. One θ covers irreflexive nodes and the other θ reflexive nodes. The minimal rank is bounded by N, therefore $K \models \theta^{2N}(\Box A)$ for all models K.

Figure 1 sketches the idea of the frontier points in a model. Each curved line represents the set of frontier points, which lowers after two applications of θ^*. There are at most N steps of $\theta^*\theta^*$ in the picture.

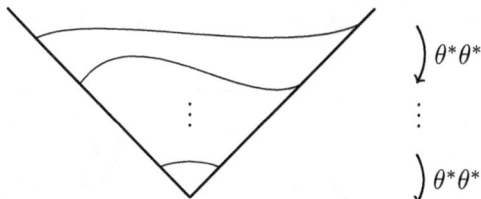

Figure 1: Lines of frontier points.

We keep the same idea in mind, but we propose to change the definition of the rank and give another approach for the homogeneous models. Those elements are highly based on Ghilardi's method. With our investigation, we want to address the important role of the frontier points and the link between bisimilar models and the extension property. The idea is to identify different so-called *extension structures* in the extension property of $Mod(\Box A)$. Those extension structures are identified using bisimulation. In turn, each extension structure will correspond to a simple substitution σ_a which are again the building blocks for θ. We will see that $2(N'+1)$ applications of θ is enough, where again N' is the number of different $(n-1)$-bisimulation equivalence classes.

Before we explore the new method, we give some examples to see that in many cases a short substitution suffices to act as a projective unifier for $\Box A$ and that this depends on the nature of the extension property of $Mod(\Box A)$.

Example 1 Let A be of the form $p \to B$ for some formula B and atom p. Formula $\Box A$ has the extension property, because for each model K that almost satisfies $\Box A$, we can find a variant K' in which no atom is forced in the root. This works independently of the shape of K. So $K' \models \Box A$. This means that $\sigma_\emptyset^*(K) \models \Box A$ for each K, so $\sigma_\emptyset$ is a projective unifier of $\Box A$.

Also for box-free formulas with the extension property, one σ suffices as projective unifier. In general, if the extension property does not depend on the models above the root, one σ suffices.

Example 2 Consider formula $A = (\Box p \to p) \land (p \to \Box p)$. For simplicity, we work with tree-like models. There are multiple cases of the extension property. If all nodes above the root satisfy p, extend it with a node where p also holds. This is illustrated below in the first two pictures, where in the first picture there are no submodels above the root. Note that A is true for each reflexive leaf in the tree. If there is at least one node in which p does not hold, extend the models with a node where p does not hold, illustrated in the last two models.

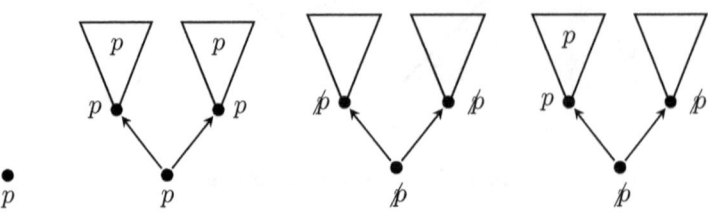

Projective Formulas vs. the Extension Property

We want to know which sequence of σ's turns each model into a model that satisfies $\Box A$. Let K be a model. We can first apply σ_p^* that belongs to the left two pictures. By Lemma 2, if $K_w \models \Box A$, then the atoms forced in w in model $\sigma_p^*(K)$ stay the same, and if $K_w \not\models \Box A$, then the only atom forced in w is p. Moreover, for each world w in $\sigma_p^*(K)$ such that $\sigma_p^*(K_w) \not\models \Box A$ we have that there is at least one node v above w such that $\sigma_p^*(K_v) \not\models p$. So all these nodes belong to the third or fourth picture. Now we can take $\sigma_\emptyset^*$ to conclude $\sigma_\emptyset^* \sigma_p^*(K) \models \Box A$. Hence, $\sigma_p \sigma_\emptyset$ is a projective unifier for $\Box A$.

Example 3 Formula $B = (\Box \neg p \to \neg p) \land (\neg p \to \Box \neg p)$ is the substitution instance of A from the previous example where $\neg p$ is substituted for p. The corresponding extensions are now as follows:

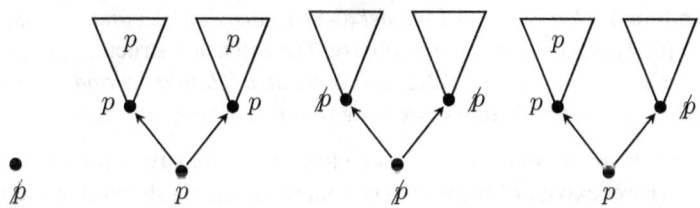

Now we see that $\sigma_p^* \sigma_\emptyset^*$ turns each model in a model that satisfies $\Box B$. Therefore $\sigma_\emptyset \sigma_p$ is a projective unifier for $\Box B$. Note that here the σ's depend on B, so now σ_p means $\sigma_p^{\Box B}$.

The examples illustrate that the set a of atoms forced in the extended root depends on the structure of the models above it. In addition, we distinguish between the extended root being reflexive or irreflexive. This results in different *extension structures* defined in Definition 4.

We will formalise our method. Recall that we work with formula $\Box A$ with atoms from $\{p_1, \ldots, p_m\}$ and $d(A) \leq n$. Let N' be the number of $(n-1)$-bisimilar equivalent models. Let us introduce our ingredients.

- We keep the same notion of *frontier points*.
- Define the *bisimulation set of K* as

$$B(K) = \{[K_w]_{n-1} \mid \rho R w \text{ and } K_w \models \Box A\}.$$

- The *rank* $r(K)$ is the cardinality of $B(K)$.
- We call a frontier point w *B-minimal* in K, if $r(K_w) \leq r(K_v)$, for all other frontier points v in K.

The bisimulation set of K is a subset of the set of all equivalence classes of $(n-1)$-bisimilar models that satisfy $\Box A$. Because we work with transitive models, we have the following important fact: $B(K) \subseteq B(\sigma_a^*(K))$ for every $a \subseteq \{p_1, \ldots, p_m\}$. And so $r(K) \leq r(\sigma_a^*(K))$. The rank is bounded by N'.

Example 4 Consider Example 2 and Example 3. The box-depth of formulas A and B is 1. So the different bisimulation sets depend on 0-bisimulation. Therefore we have to look at the atoms that are forced in the nodes above the root. There are four bisimulation sets which we write as $\emptyset, \{p\}, \{\not{p}\}$ and $\{p, \not{p}\}$. They correspond from left to right to the pictures in the examples.

The applied substitution in the examples is the same for reflexive and irreflexive roots, but this is not the case in general.

Definition 4 Let K be an L-model almost satisfying $\Box A$ and suppose that $\mathrm{Mod}_L(\Box A)$ has the extension property. The *extension structure* of K (with respect to $\Box A$) is the pair $(B(K), \cdot)$ of its bisimulation set and $\cdot = \mathrm{i}$ if the root of K is irreflexive and $\cdot = \mathrm{r}$ if the root is reflexive.

Each bisimulation set may define two extension structures, depending on the (ir)reflexivity of the root. The following lemma shows that the same substitutions work for models with the same extension structure. We will see that each extension structure gives rise to a *corresponding substitution*.

Lemma 4 Let K_1, K_2 be two models that almost satisfy $\Box A$. Assume that they have the same extension structure. Then for each a, $\sigma_a^*(K_1) \models \Box A$ if and only if $\sigma_a^*(K_2) \models \Box A$.

Proof. Let ρ_1 and ρ_2 be the roots of K_1 and K_2. The models have the same extension structure, so $B(K_1) = B(K_2)$ and ρ_1 and ρ_2 are both irreflexive or reflexive. By Lemma 2.3 of (Ghilardi, 2000), it is enough to consider models K_i with $cl(\rho_i)$ being a singleton. Suppose $\sigma_a^*(K_1) \models \Box A$. We will show that $\sigma_a^*(K_1) \sim_n \sigma_a^*(K_2)$. From this it follows from Theorem 1 that $\sigma_a^*(K_2) \models A$, since $\sigma_a^*(K_1) \models A$ and $d(A) \leq n$. Then also $\sigma_a^*(K_2) \models \Box A$, hence $\sigma_a^*(K_2) \models \Box A$.

Now we prove $\sigma_a^*(K_1) \sim_n \sigma_a^*(K_2)$. We have $\sigma_a^*(K_1) \sim_0 \sigma_a^*(K_2)$ by Lemma 2 (ii). First assume that ρ_1 and ρ_2 are irreflexive. Suppose $\rho_1 R_1 w$. Root ρ_1 is irreflexive so $w \neq \rho_1$ and therefore $K_{1,w} \models \Box A$. So

$$[K_{1,w}]_{n-1} \in B(K_1) = B(K_2).$$

Hence there is a v such that $\rho_2 R_2 v$, $K_{2,v} \models \Box A$ and $K_{2,v} \sim_{n-1} K_{1,w}$. By Lemma 2 (i) we have $\sigma_a^*(K_{2,v}) = K_{2,v}$ and $\sigma_a^*(K_{1,w}) = K_{1,w}$ and so

Projective Formulas vs. the Extension Property

$\sigma_a^*(K_{2,v}) \sim_{n-1} \sigma_a^*(K_{1,w})$. The other direction is analogous. Therefore $\sigma_a^*(K_1) \sim_n \sigma_a^*(K_2)$.

Now suppose that ρ_1 and ρ_2 are reflexive. We show by induction $\sigma_a^*(K_1) \sim_k \sigma_a^*(K_2)$, for $k = 0, \ldots, n$. We have $\sigma_a^*(K_1) \sim_0 \sigma_a^*(K_2)$ by Lemma 2 (ii). Take w such that $\rho_1 R_1 w$. If $w \neq \rho_1$ do the same as in the irreflexive case. If $w = \rho_1$, define $v = \rho_2$. By induction hypothesis we have $\sigma_a^*(K_1) \sim_{k-1} \sigma_a^*(K_2)$. Now pick v such that $\rho_2 R_2 v$. This case is symmetric of the previous one, so we can apply a similar argument. Hence $\sigma_a^*(K_1) \sim_k \sigma_a^*(K_2)$ for each $k = 0, \ldots, n$. □

Note that there can be multiple substitutions that can correspond to an extension structure, but there is at least one by Lemma 3. For each extension structure we fix such a substitution and call it the *corresponding substitution* to that extension structure. Note that different extension structures can be identified by the same substitution σ_a. We write σ_i and σ_r denoting the corresponding substitutions to the irreflexive and, respectively, reflexive extension structure of some bisimulation set.

Lemma 5 shows the connection between extensions of reflexive and irreflexive nodes under certain criteria. Informally, the substitution σ_r corresponding to a reflexive extension also works for the irreflexive extension with the same bisimulation set under these criteria.

Lemma 5 *Let K_1, K_2 be two models, with roots ρ_1, ρ_2, that almost satisfy $\Box A$. Let ρ_1 be reflexive and ρ_2 irreflexive. Suppose $B(K_1) = B(K_2)$. If $\sigma_a^*(K_1) \models \Box A$ and $B(K_1) = B(\sigma_a^*(K_1))$, then also $\sigma_a^*(K_2) \models \Box A$.*

Proof. Similarly to the proof of the previous lemma, we will show that $\sigma_a^*(K_1) \sim_n \sigma_a^*(K_2)$. From this it follows that $\sigma_a^*(K_2) \models \Box A$.

By Lemma 2.3 of (Ghilardi, 2000), it is enough to consider models K_i with $cl(\rho_i)$ being a singleton. We have $\sigma_a^*(K_1) \sim_0 \sigma_a^*(K_2)$ by Lemma 2 (ii). We must show that for all w such that $\rho_1 R_1 w$ there exists v such that $\rho_2 R_2 v$ and $\sigma_a^*(K_{1,w}) \sim_{n-1} \sigma_a^*(K_{2,v})$ and vice versa. First take w such that $\rho_1 R_1 w$. If $w \neq \rho_1$, we proceed in the same way as for the irreflexive case in the proof of Lemma 4. If $w = \rho_1$, we use the assumption $\sigma_a^*(K_1) \models \Box A$ to see that

$$[\sigma_a^*(K_1)]_{n-1} \in B(\sigma_a^*(K_1)) = B(K_1) = B(K_2).$$

There is a v such that $\rho_2 R_2 v$, $K_{2,v} \models \Box A$ and $K_{2,v} \sim_{n-1} \sigma_a^*(K_{1,\rho_1})$. By Lemma 2 (i), we have $K_{2,v} = \sigma_a^*(K_{2,v})$ and so $\sigma_a^*(K_{2,v}) \sim_{n-1} \sigma_a^*(K_{1,\rho_1})$.

Now pick v such that $\rho_2 R_2 v$. This case is easier than the previous one and is left to the reader. Therefore $\sigma_a^*(K_1) \sim_n \sigma_a^*(K_2)$. □

Now we present the key lemma. Recall that $\theta := \sigma_{a_1} \cdots \sigma_{a_s}$ where $a_1, \ldots, a_s$ is any fixed ordering on the subsets of $\{p_1, \ldots, p_m\}$. The lemma states that after two applications of θ^*, the B-minimal rank of the new frontier points increases. Intuitively, one θ covers the corresponding irreflexive substitutions σ_i's and the other the corresponding reflexive substitutions σ_r's. In the following we use the notation $\theta_j^* := \sigma_{a_j}^* \sigma_{a_{j-1}}^* \cdots \sigma_{a_1}^*$, where we define θ_0^* to be the empty substitution, i.e., $\theta_0^*(K) = K$ for each model K.

Lemma 6 *Let K be a model and let w be a B-minimal frontier point in K. Then for each frontier point v in $\theta^*\theta^*(K)$ below w we have that $B(K_w) \subset B(\theta^*\theta^*(K_v))$. Consequently, $r(K_w) < r(\theta^*\theta^*(K_v))$.*

Proof. Let K be a model with B-minimal frontier point w. Let v be a frontier point in $\theta^*\theta^*(K)$ below w. Note that $B(K_w) \subseteq B(\theta^*\theta^*(K_v))$. Suppose $B(K_w) = B(\theta^*\theta^*(K_v))$. We will prove that it implies $\theta^*\theta^*(K_v) \models \Box A$, and so v cannot be a frontier point in model $\theta^*\theta^*(K)$.

Observe that $B(K_w) \subseteq B(K_v) \subseteq B(\theta^*\theta^*(K_v))$, so $B(K_w) = B(K_v)$. Consider all v' above v such that $K_{v'} \not\models \Box A$ (this includes w itself). Since w is B-minimal and $B(K_{v'}) \subseteq B(K_v)$, these v''s satisfy $B(K_w) = B(K_{v'})$ as well. Also note that for each such v' and each index j we have

$$B(K_w) = B(K_{v'}) \subseteq B(\theta_j^*(K_{v'})) \subseteq B(\theta_j^*\theta^*(K_{v'})) \subseteq B(\theta^*\theta^*(K_v)).$$

Therefore, $B(K_w) = B(\theta_j^*(K_{v'})) = B(\theta_j^*\theta^*(K_{v'}))$ for each j. We have two cases: all v' are irreflexive or there is at least one that is reflexive.

Let start with the first case. Here w is irreflexive. By Lemma 3 we have, $\sigma_{a_j}^*(K_w) \models \Box A$ for some j. Note that also $\theta_j^*(K_w) \models \Box A$. This σ_{a_j} is the irreflexive substitution σ_i corresponding to $B(K_w)$. For each v' above v we will prove $\theta_j^*(K_{v'}) \models \Box A$. We proceed by induction on the maximal length of sequences $v' R x_1 R \ldots R x_k$ where x_k is a frontier point in K. If the length equals 1, then v' is a frontier point in K. If $\theta_{j-1}^*(K_{v'}) \models \Box A$, then also $\theta_j^*(K_{v'}) \models \Box A$ by Lemma 2 (i). If $\theta_{j-1}^*(K_{v'}) \not\models \Box A$, we know that v' is a frontier point in $\theta_{j-1}^*(K)$. Since $B(K_w) = B(\theta_{j-1}^*(K_{v'}))$, we can apply Lemma 4 to conclude $\theta_j^*(K_{v'}) \models \Box A$. Suppose now the length is $l > 1$. By induction hypothesis we know that v' is an irreflexive point for which all its successors satisfy $\Box A$ in $\theta_j^*(K)$. If $\theta_j^*(K_{v'}) \models \Box A$ we are done. If not, since $B(K_w) = B(\theta_j^*(K_{v'}))$ we know by Lemma 4 that $\sigma_{a_j}^* \theta_j^*(K_{v'}) \models \Box A$. Hence, by Lemma 2 (iii), $\theta_j^*(K_{v'}) \models \Box A$. Therefore by Lemma 2 (i), we have $\theta^*\theta^*(K_v) \models \Box A$.

Projective Formulas vs. the Extension Property

Now we turn to the second case. We consider model $\theta_j^*(K)$, where j is defined in such a way that σ_{a_j} is the irreflexive substitution σ_i corresponding to $B(K_w)$. In case there is no corresponding irreflexive substitution, we define $j = 0$. If $\theta_j^*(K_v) \models \Box A$ we are done. If not, we will see further in the proof that all frontier points in $\theta_j^*(K)$ above v are reflexive. Fix such a frontier point w'. Let σ_{a_h} be the corresponding reflexive substitution σ_r to $B(K_w)$. Note that $B(K_w) = B(\theta_j^*(K_{w'}))$, so $\sigma_{a_h}^* \theta_j^*(K_{w'}) \models \Box A$. We prove for all v' above v that $\theta_h^* \theta^*(K_{v'}) \models \Box A$. We do so by induction on the maximal length of $v' R x_1 \ldots x_{k-1} R x_k$ where x_i's do not belong to the same cluster and x_k is a frontier point in $\theta_j^*(K)$. If the length equals 1, then v' is a frontier point in $\theta_j^*(K)$ (v' may equal w'). Frontier point v' must be reflexive, because suppose v' was irreflexive. Recall that $B(K_w) = B(\theta_j^*(K_{v'}))$. By Lemma 4 and Lemma 2 (iii) it would follow that $\theta_j^*(K_{v'}) = \sigma_{a_j}^* \theta_j^*(K_{v'}) \models \Box A$. And so v' would not be a frontier point in $\theta_j^*(K)$. Thus v' is reflexive. If $\theta_{h-1}^* \theta^*(K_{v'}) \models \Box A$, we are done. If not, since $B(\theta_j^*(K_{w'})) = B(\theta_{h-1}^* \theta^*(K_{v'}))$, we can apply Lemma 4 to conclude $\theta_h^* \theta^*(K_{v'}) \models \Box A$. Now suppose the length is $l > 1$. By induction hypothesis, all the successors of v' not in the cluster of v' satisfy $\Box A$ in $\theta_h^* \theta^*(K)$. Again, if $\theta_h^* \theta^*(K_{v'}) \models \Box A$, we are done. If not, we have two cases. If v' is reflexive we can apply Lemma 4, because $B(\theta_j^*(K_{w'})) = B(\theta_h^* \theta^*(K_{v'}))$. If v' is irreflexive, we apply Lemma 5, because $B(\theta_j^*(K_{w'})) = B(\theta_h^* \theta^*(K_{v'}))$ and $B(\theta_j^*(K_{w'})) = B(\sigma_{a_h}^* \theta_j^*(K_{w'}))$. In both cases we obtain $\sigma_{a_h}^* \theta_h^* \theta^*(K_{v'}) \models \Box A$, hence $\theta_h^* \theta^*(K_{v'}) \models \Box A$. This concludes $\theta^* \theta^*(K_v) \models \Box A$. □

Consider again Figure 1 illustrating the frontier lines. Lemma 6 shows that the B-minimal rank of the frontier lines increases after each step of $\theta^* \theta^*$ in the picture. We show in the final theorem that there are at most $N' + 1$ of these steps. And so a concatenation of $2(N' + 1)$ θ's forms a projective unifier for $\Box A$. As mentioned before, Ghilardi uses $2N$ θ's. From a close look at the induction proof of Lemma 2.8 from (Ghilardi, 2000), we think that he would conclude $2(N + 1)$ instead of $2N$ θ's. The rank is indeed bounded by N, but it may start at 0, which contributes to an extra application of θ. However, this is not so important. We even think that a more clever proof can show that $2N'$ applications is sufficient in our case.

Note that if $d(A) = n = 0$, and thus A is box-free, $(n-1)$-bisimulation is undefined. In that case one θ will suffice. More precisely, only one σ_a will be enough, namely its classical propositional valuation making A true (compare to Example 1).

Theorem 3 $(\theta^*)^{2(N'+1)}(K) \models \Box A$ *for all models* K.

Proof. From Lemma 6 it follows with induction on $l \leq N'$, that the rank of the B-minimal frontier points in $(\theta^*)^{2l}$ is greater than or equal to l. Note that the rank can be 0, so the B-minimal rank can start at 0. Since the rank is bounded by N', we have that $(\theta^*)^{2(N'+1)}(K)$ does not contain any frontier points. Therefore $(\theta^*)^{2(N'+1)}(K) \models \Box A$. $\square$

5 Conclusion

This paper provides an extensive examination of Ghilardi's proof method of the connection between projective formulas and the extension property for several modal logics extending K4. The result plays an important role in the fields of unification and admissible rules. We provide an explanation of the close relationship between bisimulation and the extension property on the basis of extension structures.

It should be mentioned that the method only works for transitive models. For instance, it follows from (Jeřábek, 2015) that it is not possible to establish the same property for modal logic K. In terms of admissibility there are a lot of open questions for K.

We hope that this study will give more insight into the beautiful work of Ghilardi. It may clarify some aspects of Ghilardi's work which may be helpful for further research in the field of unification and admissibility. An interesting direction would be to establish a similar result for intuitionistic modal logics.

References

Friedman, H. (1975). One hundred and two problems in mathematical logic. *The Journal of Symbolic Logic, 40*(2), 113–129.

Ghilardi, S. (1999). Unification in intuitionistic logic. *The Journal of Symbolic Logic, 64*(2), 859–880.

Ghilardi, S. (2000). Best solving modal equations. *Annals of Pure and Applied Logic, 102*(3), 183–198.

Iemhoff, R. (2001). On the admissible rules of intuitionistic propositional logic. *The Journal of Symbolic Logic, 66*(1), 281–294.

Iemhoff, R. (2016). Consequence relations and admissible rules. *Journal of Philosophical Logic, 45*, 327–348.

Iemhoff, R., & Metcalfe, G. (2009). Proof theory for admissible rules. *Annals of Pure and Applied Logic*, *159*(1–2), 171–186.

Jeřábek, E. (2005). Admissible rules of modal logics. *Journal of Logic and Computation*, *15*(4), 411–431.

Jeřábek, E. (2015). Blending margins: The modal logic K has nullary unification type. *Journal of Logic and Computation*, *25*(5), 1231–1240.

Rybakov, V. (1984). A criterion for admissibility of rules in the model system S4 and the intuitionistic logic. *Algebra and Logic*, *23*, 369–384.

Visser, A. (1996). Uniform interpolation and layered bisimulation. In P. Hájek (Ed.), *Gödel '96: Logical foundations of mathematics, computer science and physics—Kurt Gödel's legacy, Brno, Czech Republic, August 1996, proceedings* (Vol. 6, pp. 139–164). Berlin: Springer-Verlag.

Iris van der Giessen
Utrecht University, Department of Philosophy and Religious Studies
The Netherlands
E-mail: i.vandergiessen@uu.nl

Serious Statements and Plans

JOHN T. KEARNS

Abstract: In this paper, I understand an actual, natural, language to be an activity in which someone engages by performing language acts, either by producing words and sentences to speak, write, or think, or by reading or listening with understanding to someone else. *Illocutionary acts* are the complete, concrete language acts that people perform when they address or write to one another, or when they are thinking things through for themselves. They are like the "units" of speech or language.

A straightforward illocutionary act can be constituted by performing another language act, a *locutionary act*, with a certain force. For example, an assertion can be performed by making a statement, a locutionary act, while accepting that statement for representing things as they are. Assertions, denials, and acts of supposing a statement to be or not to be the case, are *assertive* illocutionary acts. *Directive* (illocutionary) acts constitute another category. These acts are designed to get an addressee to do or not do something. *Commissives*, in which a person commits herself to do or not do something, constitute a third category.

Statements are the locutionary acts used to perform assertive illocutionary acts. The locutionary acts used to perform directive acts are (second-person) *plans*, like this one: 'Mark, please get up from your seat and close the door.' First-person plans are used to perform commissive illocutionary acts. A statement is *satisfied*, or true, if it represents things as they are, and false if it fails to do this. A plan is satisfied, or *implemented*, if it is intentionally carried out.

There are some puzzling or "paradoxical" locutionary acts that might be thought to pose a problem for logic and logical theories. These include statements made with sentences like the following:

> This statement is true.   This statement is not true.

where, in each case, the speaker uses the phrase 'This statement' to refer, or attend, to the statement she is making. There are also plans which are puzzling in much the same way. Examples are these plans intended for a woman named Grace:

> Grace, please implement this plan.
> Grace, please see that this plan is not implemented.

John T. Kearns

In each case, the speaker uses 'this plan' to refer to the plan he is formulating.

In this paper, I show that the puzzling statements and plans are defective, and cannot properly be used to perform assertive or directive illocutionary acts. It would be senseless, and irrational, to regard them as serious candidates for performing illocutionary acts.

Keywords: speech act, language act, illocutionary acts, assertive acts, directive acts, illocutionary logic, liar paradox

1 A language is a kind of activity

A *speech act* or *language act* is a meaningful act that someone performs by using one or more expressions. A person can produce the expressions she uses by speaking aloud, by writing or typing, or by thinking with words or sentences. These expressions are tokens, not types. A person can also use tokens produced by someone else, by listening to and understanding these expressions, or by reading. It is common to write and speak as if all language acts are performed by persons who produce the tokens they use. To simplify matters, I often do this as well.

An *illocutionary act* is a kind of language act, typically one in which someone uses a sentence or sentential clause to carry out a certain job, like making a promise or making a request, threatening someone or giving her advice, asserting that something is the case, or asking another person what time it is. Illocutionary acts are the "units" of speech or language, they are the complete, concrete language acts that people perform when they address or write to one another, or when they are thinking things through for themselves.

Illocutionary logic studies the kind of real-life arguments that are carried out by performing illocutionary acts. My papers (Kearns, 2006) and (Kearns, 2017) are examples of logical investigations that focus on illocutionary acts rather than on statements or propositions. In the present paper I show how a better understanding of illocutionary acts solves or dissolves an old logical puzzle.

An actual, or ordinary, language is a kind of activity, much like tennis or baseball is a kind of activity. People engage in this activity by performing language acts, especially illocutionary acts.

A relatively straightforward illocutionary act can be constituted by performing another language act, a *locutionary act*, with a certain *force*, or in a certain *manner*. For example, someone can produce and use a sentence to make a *statement*, which is a locutionary act, while *accepting* that statement for presenting or representing things as they are. Her statement is made/performed *with the force of assertion*. Acts of asserting, denying, or supposing a statement to be the case or not to be the case, are examples of *assertive illocutionary acts*. These acts don't require an audience. A person can perform an assertive act when she is alone, or she can address her act to someone else. (The 'locutionary,' 'illocutionary' terminology is explained in Austin, 1965.)

Assertive illocutionary acts are one of the five categories of illocutionary acts in John Searle's (1969) taxonomy. *Directive (illocutionary) acts* constitute another category. A person who *orders* someone else, or who *asks* the someone else, or who *advises* that someone to carry out, or to refrain from carrying out, some action, is performing a *directive* illocutionary act. Its directive character is due to its being designed and intended to get the addressee to do or not do something. Directive acts require an addressee. You can't ask someone to pass the salt if there is no someone there.

A third of Searle's categories is of *commissive acts*. In an act of this kind, a person commits herself to carry out, or to refrain from carrying out, some further act. Promises may be the most well-known kind of a commissive act, but I can commit myself without promising anything to anyone. I might simply say to myself or to someone else, "I am going to get a bottle of beer from the refrigerator," and then do that. Searle's two other categories of illocutionary acts are *expressives* and *declarations*, but I won't consider them in this paper.

I use the word 'statement' for those locutionary acts that can be used to perform positive assertive illocutionary acts. Statements are typically performed with sentences or sentential clauses, and are appropriately evaluated in terms of truth and falsity. Someone can perform an assertion by making a statement and accepting it for presenting or representing things as they are—or, as John Searle puts it, for "fitting the world." To deny a statement, a person *rules out* asserting that statement, because the statement fails to fit the world. For example, a person might deny that Buffalo is in Canada by saying "I deny that Buffalo is in Canada," which sounds a little pompous, or, more simply, by saying "Buffalo is not in Canada."

Suppositions occur frequently in arguments by natural deduction. Supposing a statement to be true is like temporarily accepting that statement, and supposing one to be false is like temporarily blocking or impeding its positive supposition. In deductive arguments especially, it is common both to introduce and to discharge suppositions.

The locutionary acts that can be used with positive directive force are what I call (second-person) *plans*, like this:

"Mark, please get up from your seat and close the door."

The plan represents the intended addressee as carrying out, or performing, the directed action. The person who formulates the plan can think it, or say it, without addressing the intended addressee—-he can perform a directive locutionary act without performing a directive illocutionary act. But for the speaker to actually direct Mark to implement the plan, he must address Mark.

The intended addressee can implement the plan without knowing about the plan. For if Mark does get up from his seat and close the door (on the appropriate occasion), he has *implemented* the plan. But in order to implement the plan, the addressee must *intend* to perform the action involved. If Mark stumbles and accidentally knocks the door shut, he has not implemented the plan "Mark, please close the door."

A person can use a *first-person plan* like this one:

I will get a bottle of beer from the refrigerator.

to perform a commissive illocutionary act. He *implements* the plan by doing what he represents himself as doing.

Just as statements can be asserted or denied, so someone can be directed to implement a second-person plan or to refrain from implementing that plan, and a speaker/writer/thinker can commit himself to implement or refrain from implementing a first-person plan. While a statement is *satisfied* if it is true, a plan is *satisfied* if it is implemented. (Statements have truth conditions, while plans have implementation conditions.)

For a speaker's directive act "Mark, please close the door" to be successful, or *complied with*, several things need to happen: Mark must hear, and understand, the speaker, Mark must agree to implement the speaker's plan, and Mark must close the door *in order* to implement that plan. If Mark makes his agreement explicit, he performs a commissive illocutionary act. Even if Mark agrees without saying or thinking any words, Mark's agreement *commits* him to close the door—-in such a case we might regard him as tacitly, or "in effect," performing a commissive act.

2 Some puzzling statements and plans

The locutionary acts that might be performed with these sentences:

(1) This statement is true.
(2) This statement is not true.

have often been regarded as problematic. Let us call statements performed with sentences like (1) *Truth-Teller* statements, and those performed with sentences like (2) *Liar* statements.

The printed sentences are not statements, and are not true or false. But it seems that a person can use each of these sentences to make a statement, a statement that the person making it characterizes as being true in the first case, and not true in the second case.

It is sometimes claimed that there are rules governing our use of language so that one or both of the sentences cannot properly be used to make a statement, and so are not eligible to be evaluated in terms of truth and falsity. However, it seems clear that both sentences can be used to perform meaningful language acts which represent themselves as being true in the first case, and not true in the second. If there is some concern that the formulation of (1) and (2) illicitly presupposes that the language acts are statements, we could replace (1) and (2) by the following:

(1*) This language act is a true statement.
(2*) This language act is not a true statement.

These pose the same problems as the original sentences and the language acts performed with them seem to do. There is no question that someone can use these most recent sentences to perform meaningful language acts. All a person needs to do is say them or write them or think them (so long as she understands what she is saying, writing, or thinking). She can also read them. So we might as well stick with the original formulation, and call the language acts *statements*, even if they do turn out to be defective in one way or another.

In this paper, I will argue that these statements *are* defective. They cannot properly be accepted/asserted for fitting the world. They can't be "measured" against the actual world. We have no reason to be puzzled or concerned by the existence of Truth-Teller, Liar, and other similar statements. For such statements aren't *serious*—we can't take them seriously, because they can't be used to perform assertive illocutionary acts. We can safely ignore these statements in developing theories of logic and language.

To develop my argument that Truth-Teller, Liar, and other similar statements aren't serious, I need to make clear, and emphasize, that language acts are events in space and time that are performed with tokens of expressions. All actual language acts are token acts.

3   One sentence token, multiple statements

Imagine that on a Monday morning, Mark writes 'Today is Monday' on the board in room 321 Hamilton Hall. The sentence on the board is not a language act, and is not true or false. But when Mark wrote the sentence, he made a statement. In writing 'Today,' Mark attended to the day on which he was writing the sentence, and in writing 'is Monday,' he characterized that day as being a Monday. Since Mark knew what day it was, he also accepted, and so asserted, the statement that he made.

Later that day, when the sentence is still on the board, other people, and even Mark himself, can read the sentence and make new statements of their own. In reading 'Today,' it would be natural for someone to direct her attention to the day on which she was reading the sentence, and in reading 'is Monday,' it would be natural to characterize the day as being a Monday. Since she probably knew what day it was, it would also be natural for her to accept the statement she made. In talking about language acts, we commonly adopt the "perspective" of the persons who produce the expressions that are used to perform language acts, but readers and listeners also perform language acts of their own.

If the sentence is still on the board the following day, those who read the sentence are likely to perform false statements, which they are likely to reject. But if Barb knows that Mark wrote the sentence on the previous day, she might very well use the sentence to make/perform a true statement. She would do this by using the word 'Today' to direct her attention to the preceding day, and using the rest of the sentence to characterize the preceding day as a Monday. If she knows enough to do this, she will probably accept the statement that she makes.

Although the different statements that are made with the one sentence on Monday might be considered to be *essentially similar*, they are not different occurrences of a single statement. We might also say that the different statements *have* or *express* the *same content*. But this is a metaphorical way of describing a relation between different statements and a feature that they have in common. A statement is not literally a container holding a content which it can share with other statements. For them to be essentially similar or to have the same content, it is sufficient for the most recent factual statements to be focused on the same day, and to attribute the same feature to that day. But the ideas of different language acts being essentially similar, or having the same content, are somewhat informal. In different circumstances we might employ slightly different criteria for what it takes for language acts

Serious Statements and Plans

to be essentially similar—which criterion is employed will depend on the inquiry we are carrying out.

Someone who uses this sentence:

> This statement is true.

to perform a Truth-Teller statement is attending to the statement she is then making, and characterizing that statement is true. If the sentence is used on another occasion, by the same person or a different person, to perform a Truth-Teller statement, the speaker is not attending to the statement that was performed first. Each person who uses the sentence to perform a Truth-Teller statement is referring to the statement she is then making.

For continuing our discussion of Truth-Teller, Liar, and other statements, it is slightly more convenient to use letters to label statements rather than to use the word 'this.' So now we shall use 'a' for the statement that someone makes with this sentence:

> Statement a is true.

But each time someone uses this sentence to make a Truth-Teller statement, she is making/performing a different statement a, and she is using the component expression 'statement a' to refer to the statement she is making. If Janet uses this sentence on a particular occasion to make a Truth-Teller statement, Max could use the "same" sentence to characterize Janet's statement as being true. Max's statement would not be a Truth-Teller statement, for he would be referring to Janet's statement and not to his own statement. And while both Max and Janet are using the sentence to refer to the same statement and to characterize that statement as true, we don't consider their statements to be essentially similar. For one of them is self-referential and the other is not.

We will use 'b' for the Liar statements that are performed with this sentence:

> Statement b is not true.

When this sentence is used more than once to make a Liar statement, it is being used to make different Liar statements.

4 Some locutionary acts can't be serious

I understand a factual statement to be one that is concerned with ordinary objects or events, and that can be "measured" against the actual world.

John T. Kearns

Factual statements can be contrasted with statements about abstract entities, so while a statement that seven horses and five other horses make twelve horses is a factual statement, the statement that 7 plus 5 is twelve isn't factual. Statements about fictional characters and events also fail to be factual.

But Truth-teller and Liar statements occur in space and time, and either predicate being true of themselves, or predicate failing to be true of themselves. In order for these statements to pose serious intellectual problems, these statements must be factual.

It seems more-or-less evident that locutionary acts are important primarily for their role in constituting illocutionary acts. We make statements so that we can make assertions and denials, and so that we can suppose things to be or not to be this way or that. We make (second-person) plans so that we can direct people to implement or refrain from implementing these plans. Illocutionary acts are the units of speech, we perform these acts to interact with the world and with one another.

If there are locutionary acts that aren't suited for performing illocutionary acts, then those acts aren't serious. I am concerned in this paper to show that Truth-Teller and Liar statements, and other statements that resemble them, aren't serious. But I will begin by considering plans rather than statements. For we can formulate plans which pose problems like those associated with Truth-Teller and Liar statements, and a discussion of these plans illuminates and clarifies the problems posed by the statements.

Someone can perform a locutionary act by saying, writing, or thinking the sentence (or other expression) that is involved. Such an act doesn't need to be communicated or made known to someone else. This is true even for a plan like this one:

> Mark, please get up from your seat and close the door.

Janet might speak or think this sentence without addressing Mark, in doing this she would be performing a locutionary act. And Mark could implement Janet's plan without knowing about her plan, so long as he intentionally gets up and shuts the door on the occasion in question. But Janet can't use her plan to perform a directive act without addressing Mark.

I will begin by considering some puzzling (second-person) plans which Mark has formulated for his friend Grace. Just as we used letters for Truth-Teller and Liar statements, Mark is using letters of the alphabet to label his plans. His first plan is p:

> Grace, please see to it that plan p is implemented.

Serious Statements and Plans

I will call plans like this *Implement me* plans. Mark can formulate his plan, and speak it or think it without letting Grace know about the plan. But if he wants to direct Grace to implement this plan, he needs to address Grace, and explain to her what plan p is.

But why would he bother? This plan, which is (obviously, I hope) analogous to a Truth-Teller statement, is clearly defective. It wouldn't make any sense, it wouldn't be rational, for Mark to take this to be a serious plan, and ask or order Grace to implement it. Because what is Grace supposed to do? The plan provides Grace no guidance about what she should do. And Grace can't refrain from implementing plan p either, because the *Implement me* plan has no implementation conditions.

A sensible plan must describe or represent actions for the addressee to carry out. A sensible plan is one that someone might conceivably be able to carry out. A serious plan might not be sensible, but it must certainly represent actions whose implementation is *intelligible*. Such a plan can't simply (and only) represent the addressee as carrying out this very same plan.

An impossible plan like this one:

Shane, please open the door, but leave the door shut.

can't actually be implemented. But the two component plans can each be implemented. And we understand how two plans can be conjoined to constitute a compound plan. This is sufficient to make the plan for Shane intelligible, which is what enables us to recognize that the plan is impossible to implement. We know what Shane would have to do, and recognize that this can't be done.

Mark can also formulate a plan q which is analogous to a Liar statement:

Grace, I am asking you to not implement plan q.

I will call plans like this *Don't implement me* plans. In order to implement plan q, Grace must refrain from implementing the plan. And if she intentionally refrains from implementing the plan, she will have implemented it. Plan q is such that Grace can only implement the plan by intentionally failing to implement this plan. Plan q is a *pathological* plan which undermines itself. Success and failure would be the same thing for a pathological plan. But that makes no sense. Plan q also lacks implementation conditions.

Like plan p, pathological plans like q aren't serious. Mark can't really want Grace to implement plan q, and there is no way that Grace could either implement the plan or intentionally refrain from implementing it. With plans that are either "empty", like p, or pathological (and also empty), like q, what we should do is simply *ignore* them. We are rationally committed, for these plans, to refrain from using them, or trying to use them, to perform directive illocutionary acts.

Logical theories for illocutionary directive and commissive arguments may be interesting and even important, but these will be theories dealing with serious plans and with arguments that involve or employ those plans.

Much the same thing can be said about Truth-Teller and Liar-sentence language acts. If Janet uses this sentence:

Statement a is true.

to represent the language act she is then performing as being true, she is not performing a serious assertive locutionary act. For a serious statement to truly predicate an expression of an object or event, there must be a criterion associated with the predicate expression, which is met or satisfied by the object or event. The criterion associated with a given predicate might make sense for certain objects but not others.

For a statement which predicates an expression of some object to be true, the criterion which is associated with that predicate expression must be satisfied by the object in question. So in such a case, for the criterion associated with 'true' to be satisfied, the criterion associated with some other predicate expression must be satisfied. But it makes no sense that the criterion associated with 'true' for a given statement is, simply and only, that the criterion associated with 'true' is satisfied by that statement. The criterion associated with 'true' can't be employed for a Truth-Teller Statement. Truth-Teller statements have no truth conditions. In spite of this, and because of this, we can truly say that Truth-Teller statements aren't true.

If Janet uses this sentence to make a Liar statement:

Statement b isn't true.

she is making a pathological statement. Her act would succeed in fitting the world just in case it failed to fit. Success and failure would be the same thing for the statement she is making, for her statement undermines itself. We have agreed to call her language act a statement, but it isn't a serious factual statement. The criterion that must be satisfied for Janet's most recent language act to be a true factual statement is, simply and only, that this act fails to satisfy the criterion for being a true factual statement.

When Janet, or anyone, says, and understands, this sentence:

> Statement b is not true.

or writes it, or thinks it, using 'Statement b' to attend to the act she is performing, she is not making a serious statement. She can't consider what it would be like for this statement to fit the world, for there is nothing it would be like, not even something impossible. For Janet's language act to satisfy the criterion for not being a true statement, it must, simply and only, satisfy the criterion for being true. But that is no criterion at all.

Each time that someone uses this sentence:

> Statement b is not true.

to perform a language act, using 'Statement b' to attend to the language act she is then performing, she is predicating being a not true statement of the particular "token" act she is then performing. If Janet uses the sentence on one occasion to perform a pathological language act that is not a serious factual statement, I can use the "same" sentence to say something true. To do this, I use 'Statement b' to attend to Janet's language act, and truly say that her act is not a true statement. After all, it can't be true if it has no truth conditions. In saying that Janet's act isn't true, I will not have said anything about the act that, on that occasion, I am performing.

Plans like those which we considered earlier:

> Grace, please implement plan p.
> Grace, I am asking you to not implement plan q.

seem more like jokes than like reasonable plans. A reasonable plan is one that someone might really use to direct an addressee to perform the plan's action, while intending, and wanting, for the addressee to actually do this. A reasonable plan is one that is appropriate for performing a directive illocutionary act. Not all serious plans are reasonable, but they are intelligible, and can be investigated and explored by developing illocutionary logical theories.

In a similar way, a serious factual statement is one that is appropriate for performing an assertive illocutionary act. It can be accepted for fitting the world, ruled out for failing to fit the world, or supposed to fit or to fail to fit the world. Neither Truth-Teller nor Liar statements can be used to make serious factual statements.

Some time ago, Patrick Greenough (2001) argued (in effect) that Liar sentence language acts aren't what he called "supposition apt." Even supposing that one of these acts is true, or false, leads to incoherence. Greenough did not situate his discussion in a framework involving locutionary and illocutionary acts. But he did consider suppositions, which are assertive illocutionary acts that can be performed with statements. By insisting that Liar statements are not supposition-apt, he was in a sense, but indirectly, recognizing that these statements are not suited for performing illocutionary acts. If he had considered pathological plans rather than statements, this move wouldn't have worked, because there are no counterparts to suppositions that are performed with directive locutionary acts.

My claim that Truth-Teller and Liar statements, and *Implement Me* and *Don't Implement Me* plans, aren't serious locutionary acts, and so can't be used to perform illocutionary acts of kinds for which they might seem to be suited, agrees with, but enlarges on, Greenough's discussion of statements that aren't supposition-apt. However, it is important to understand, and to emphasize, the fundamental character of illocutionary acts.

References

Austin, J. (1965). *How to Do Things with Words*. New York: Oxford University Press.

Greenough, P. (2001). Free assumptions and the liar paradox. *The American Philosophical Quarterly, 38*, 115–135.

Kearns, J. T. (2006). Conditional assertion, denial, and supposition as illocutionary acts. *Linguistics and Philosophy, 29*, 455–485.

Kearns, J. T. (2017). "I asked you to mail that letter, not to burn it," An illocutionary logical analysis of directive acts and arguments. In P. Arazim & T. Lávička (Eds.), *The Logica Yearbook* (pp. 169–180). London: College Publications.

Searle, J. R. (1969). *Speech Acts: An Essay in the Philosophy of Language*. London: Cambridge University Press.

John T. Kearns
University at Buffalo, SUNY, Department of Philosophy
United States
E-mail: kearns@buffalo.edu

The Third and Fourth Stoic Account of Conditionals

WOLFGANG LENZEN

Abstract: In the 3rd century B.C., Stoic logicians fiercely discussed the correct truth conditions of conditionals. Besides (1) *material* implication and (2) *strict* implication, two further accounts had been suggested. The third one which most likely has to be attributed to Chrysippus is sometimes interpreted as a *connexive* implication, while the fourth account which was probably defended by the Peripatetics remains fairly unclear. In this paper it will be argued that Chrysippus' conception (3) amounts to a *hybrid* form of strict *and* connexive implication while the Peripatetic account (4) consists in restricting implications to the case where the consequent is *properly* contained in the antecedent.

Keywords: logic of conditionals, Stoic logic, connexive logic

1 Introduction

In an oft-quoted passage from the "Outlines of Scepticism", Sextus Empiricus reports about a controversy among Stoic logicians concerning the correct analysis of conditionals. According to Kneale and Kneale (1962):

> [1] Philo says that a sound conditional is one that does not begin with a truth and end with a falsehood [...]. [2] But Diodorus says it is one that neither could nor can begin with a truth and end with a falsehood. [...] [3] And those who introduce the notion of connexion say that a conditional is sound when the contradictory of its consequent is incompatible with the antecedent. [...] [4] And those who judge by implication say that a true conditional is one whose consequent is contained potentially in the antecedent. According to them the statement 'If it is day, it is day' and similarly every conditional which is repetitive will apparently be false; for it is impossible for a thing to be contained in itself. (pp. 128–129)

Wolfgang Lenzen

Among historians of logic there is a wide agreement that the first conception as favoured by Philo basically amounts to a *material implication*, while the second conception as favoured by Diodorus basically amounts to a *strict implication*.[1] According to Kneale and Kneale (1962, p. 129), the third view was most likely put forward by Chrysippus, while the fourth one "is probably Peripatetic". Thus these accounts shall here be referred to as the Chrysippian and the Peripatetic conception, respectively, no matter whether these ascriptions are historically correct or not.[2]

Already in antiquity, the dispute about the different conceptions became so well known that (as reported in Kneale & Kneale, 1962, p. 128) "Callimachus wrote an epigram saying 'Even the crows on the roofs caw about the nature of conditionals'". Also in modern times, this topic has been discussed quite extensively.[3] Yet there remain several important questions which shall be addressed in this paper. First of all, it has to be examined what the Chrysippian conception, according to which "a conditional is sound when the contradictory of its consequent is incompatible with its antecedent" *exactly* amounts to. From the perspective of modern logic, two propositions p, q are incompatible if and only if their conjunction, $(p \wedge q)$, is (logically) *impossible*. Accordingly, Chrysippus' condition of the incompatibility of the antecedent, p, with the "contradictory", i.e., the negation, of the consequent, q, appears to amount to the requirement that the conjunction $(p \wedge \neg q)$ is impossible. This very condition, however, is nothing else but the standard definition of a strict implication.[4] Thus it might seem that, like his opponent Diodorus, also Chrysippus was defending a *strict* implication.

---

[1] The word 'basically' is meant to indicate that there may be subtle differences between the Stoic conceptions of conditionals and their formal explications as attempted here. Thus a referee of this paper pointed out to the temporal element in the definition of conditionals and the ensuing temporal dependency of the modal operators. These issues are beyond the scope of this paper, however. For a reconstruction of Diodorus' conception of the modal operators the reader is referred to (von Kutschera, 1986).

[2] The referee mentioned in the previous footnote also pointed out that Sextus Empiricus' report contains another, fifth kind of Stoic conditionals "called *paraconditional* (subconditional, παρασυνημμένον)" which besides the requirement that "the consequent follows from the antecedent" requires that "the antecedent is confirmed as true". This appears to correspond to the distinction which modern logicians (in the Anglo-Saxon world) make between a *valid* and a *sound* argument. The latter not only has to be formally valid but also all its premises have to be true.

[3] An overview of the literature may be found in (O'Toole & Jennings, 2004).

[4] Cf. (Lewis & Langford, 1959, p. 124): "'p strictly implies q' is to mean 'It is false that it is possible that p should be true and q false' or 'The statement 'p is true and q false' is not self-consistent'".

On the other hand, the Chrysippian conception must *differ* from the Diodorian one since only Diodorus, but not Chrysippus, considered the following somewhat paradoxical conditional as sound:

ATOMS   If atomic elements do not exist, then atomic elements do exist.

This proposition has the logical structure 'If not-p, then p'. Propositions of this form had been classified by Aristotle as "absurd". In Prior Analytics II, 4, 57b3-14, Aristotle used the "absurdity" of 'If not-p, then p' to prove that one and the same proposition q cannot be implied both by a proposition p and by not-p. Therefore McCall (1975) introduced the notion of a *connexive* implication which can be characterized by satisfying "Aristotle's Theses":

ARIST 1   No proposition p is implied by its own negation.
ARIST 2   No proposition q is implied both by a proposition p and by $\neg p$.

Now it is an important open question whether the third and/or the fourth Stoic account amounts to a connexive implication or not. To answer this question we will proceed as follows:

- In Section 2 the so-called "paradoxes" of (material and strict) implication will be summarized.
- In Section 3 the notion of the connexivity of an implication operator will be analyzed.
- In Section 4 Chrysippian implication will be shown to represent a "hybrid" form of strict and connexive implication.
- In Section 5 it will be argued that Peripatetic implication differs from the other accounts by requiring that the consequent is *properly* contained in the antecedent, i.e., whenever p implies (or entails) q, q must be "weaker" than p.
- In two Appendices (Section 6 and 7) some formal properties of Chrysippian and Peripatetic implication will be elaborated.

2   The "paradoxes" of material and strict implication

Since, in what follows, several different conceptions of implication are considered, we use '$\Rightarrow$' to symbolize (any kind of) implication *in general* and we denote the four Stoic conceptions just by adding subscripts $_{\text{'Phi'}}$, $_{\text{'Dio'}}$, $_{\text{'Chr'}}$, and $_{\text{'Per'}}$. Furthermore, we adopt the usual symbol '$\supset$' and '$\rightarrow$' for *material* and *strict* implication. Thus Philonian and Diodorian implication—which basically coincide with material and strict implication, respectively—can be defined as follows:

DEF 1 $\quad (p \Rightarrow_{\text{Phi}} q) =_{\text{df}} (p \supset q)$.
DEF 2 $\quad (p \Rightarrow_{\text{Dio}} q) =_{\text{df}} (p \rightarrow q)$.

Of course, strict implication '$p \rightarrow q$' might in turn be defined with the help of the modal operators 'it is possible that', '$\Diamond$', or 'it is necessary that', '$\Box$', as follows:

DEF 3 $\quad (p \rightarrow q) =_{\text{df}} \neg\Diamond(p \wedge \neg q) \quad$ or $\quad (p \rightarrow q) =_{\text{df}} \Box(p \supset q)$.

DEF 1 gives rise to the "paradoxes" of material implication:

PMI 1 $\quad$ Whenever a proposition p is *false*, then, for any proposition q, $(p \supset q)$ is true.

PMI 2 $\quad$ Whenever a proposition q is *true*, then, for any proposition p, $(p \supset q)$ is true.

These "paradoxes" are no *real* paradoxes like, e.g., the Liar paradox. They only exhibit certain *oddities* of the '$\supset$'-operator and indicate that '$\supset$' has different logical properties than the ordinary language connective 'If ..., then ...'. One further oddity is expressed by the following corollary which might be paraphrased somewhat dramatically by saying that *every proposition either* (materially) *implies or is* (materially) *implied by its own negation*:

PMI 3 $\quad$ For every proposition p: Either $(p \supset \neg p)$ is true, or $(\neg p \supset p)$ is true.

PMI 3 shows that the operator of material implication is far from being connexive since it heavily violates principles ARIST 1, ARIST 2.

In contrast to Philonian implication, Diodorian implication validates an important Stoic inference schema, the so-called *dialectical argument*, which was used, e.g., by Zeno as a means of *indirect proof*: "If p then q, and if p then not-q; therefore it is impossible that p".[5]

DEF 2 gives rise to the "paradoxes" of strict implication:

---

[5] See (Kneale & Kneale, 1962, p. 128); Kneales' variables 'P' and 'Q' have been replaced by 'p','q'.

The Third and Fourth Stoic Account of Conditionals

PSI 1 Whenever a proposition p is *necessarily false*, or *impossible*, then, for any proposition q, $(p \to q)$ is true.

PSI 2 Whenever a proposition q is *necessarily true*, then, for any proposition p, $(p \to q)$ is true.

As an immediate corollary one obtains:

PSI 3 Any *necessarily false* proposition strictly implies its own negation; and any *necessarily true* proposition is strictly implied by its own negation.

Thus in particular:

PSI 4 $(p \wedge \neg p) \to \neg(p \wedge \neg p)$.

These "paradoxes" show that also the operator of strict implication, $\to$, violates ARIST 1, 2 and hence it is not (fully) connexive.

Diodorus' strange example ATOMS appears to belong to the category of propositions captured by PSI 3. In Diodorus' opinion, the antecedent 'Atomic elements do not exist' is "always false", i.e., *impossible*, while the consequent, 'Atomic elements do exist' is "always true", i.e., *necessary*. From a more modern perspective, however, such a view appears rather problematic. First, it may rightly be doubted whether the Stoics' atomistic theory of matter is *true* at all. Second, even if it were *true*, it would at best be *nomologically* true, but it would not be *logically* true! We will return to this issue in Section 4 below.

3 Some facts about connexive implication

The idea of a connexive implication is sometimes taken in a vague sense just to indicate that there must be some "connection" between antecedent and consequent. Thus O'Toole and Jennings (2004, p. 479) denoted Chrysippus' account as the "connexivist view" because this label accords "with its attribution by Sextus to 'those who introduce connexion'." In the wake of McCall (1975), however, the notion of connexivity is mostly used as a technical term. Pizzi and Williamson (1997, p. 569) describe the intuitive idea as follows:

(1) No proposition implies its own negation.
(2) No proposition implies each of two contradictory propositions.

(3) No proposition implies every proposition.

(4) No proposition is implied by every proposition.

Since (3) and (4) express one and the same requirement in an "active" way ('implies') and in a "passive" way ('is implied by'), it seems advisable to distinguish also "active" and "passive" variants of (1) and (2).[6] Using '$\exists p$' and '$\forall q$' as symbols for the quantifiers 'there exists at least one proposition p' and 'for every proposition q', respectively, the above conditions can be formalized as follows:

CONN 1 $\neg \exists p(p \Rightarrow \neg p)$
CONN 2 $\neg \exists q(\neg q \Rightarrow q)$
CONN 3 $\neg \exists pq((p \Rightarrow q) \wedge (p \Rightarrow \neg q))$
CONN 4 $\neg \exists pq((p \Rightarrow q) \wedge (\neg p \Rightarrow q))$
CONN 5 $\neg \exists p \forall q(p \Rightarrow q)$
CONN 6 $\neg \exists q \forall p(p \Rightarrow q)$.

Now, for "almost all" systems of modal logic, the following *restrictions* of the connexive principles become *provable*.[7] With respect to the "active" versions just restrict the principle to *self-consistent antecedents*:

CONN 1_{rest} $\neg \exists p(\Diamond p \wedge (p \Rightarrow \neg p))$
CONN 3_{rest} $\neg \exists pq(\Diamond p \wedge (p \Rightarrow q) \wedge (p \Rightarrow \neg q))$
CONN 5_{rest} $\neg \exists p(\Diamond p \wedge \forall q(p \Rightarrow q))$.

With respect to the "passive" versions, restrict the principle instead to *non-necessary consequents*:

CONN 2_{rest} $\neg \exists q(\neg \Box q \wedge (\neg q \Rightarrow q))$
CONN 4_{rest} $\neg \exists pq(\neg \Box q \wedge (\neg p \Rightarrow q) \wedge (p \Rightarrow q))$
CONN 6_{rest} $\neg \exists q(\neg \Box q \wedge \forall p(p \Rightarrow q))$.

---

[6] As was pointed out by another referee of this paper, conditions (3) and (4) express one and the same requirement only "under some non-trivial assumptions". This is true in so far as principles (3) and (4) do not necessarily entail each other. But this also holds of the corresponding "active" and "passive" versions of the remaining connexive principles.

[7] See (Lenzen, 2020b, Section 2), where the phrase 'almost all' is made more precise by stating the minimal set of principles which the respective system of modal logic has to satisfy.

The Third and Fourth Stoic Account of Conditionals

The fact that all these principles are *theorems* of "almost all" systems of modal logic means that *restricted* or "*humble*" connexivity is not an extra requirement which strict implication might either fulfill or not.[8] Rather, *strict implication is (almost always) "humbly" connexive!*

Let us therefore turn to "non-humble" or "hardcore" connexivity! Since no *self-consistent* proposition entails its own negation *anyway*, the first thesis of hardcore connexivism boils down to the claim that (even) if p is *self-inconsistent*, p doesn't entail its own negation:

HARD 1   If $\neg \Diamond p$, then $\neg(p \Rightarrow \neg p)$.

Given that $(p \wedge \neg p)$ is the paradigm of an impossible proposition, HARD 1 entails the corollary $\neg((p \wedge \neg p) \Rightarrow \neg(p \wedge \neg p))$. Hence, any hardcore connexivist has to accept that (*even*) the *contradictory* proposition $(p \wedge \neg p)$ does *not* entail the *tautological* proposition $\neg(p \wedge \neg p)$. Therefore hardcore connexivism also has to reject the usual laws of *conjunction*, i.e.:

CONJ 1 $(p \wedge q) \Rightarrow p$
CONJ 2 $(p \wedge q) \Rightarrow q$.[9]

In a similar way, the "passive" counterpart of HARD 1 says that (even) if q is necessary, q will not be entailed by its own negation:

HARD 2   If $\Box q$, then $\neg(\neg q \Rightarrow q)$.

In particular it follows that the paradigm of a necessary proposition, $(p \vee \neg p)$, cannot be entailed by its own negation. Therefore hardcore connexivism also has to reject the usual laws of *disjunction*:

DISJ 1 $p \Rightarrow (p \vee q)$
DISJ 2 $q \Rightarrow (p \vee q)$.

In his exposition of the history of connexivity, McCall (2012) claimed that Aristotle, Chrysippus, Boethius, and Abelard had defended (unrestricted) principles of connexivity. In a recent examination it was shown, however, that at least *Aristotle* and *Boethius* (and probably also Abelard) never would have subscribed to *hardcore* connexivism because for them the laws CONJ 1, 2

---

[8] The notion of "humble connexivity" is due to Kapsner (2019).
[9] See (Lenzen, 2019). The proof that "hardcore connexivism" is incompatible with CONJ 1, 2 presupposes some elementary laws such as *contraposition* and *transitivity* of strict implication.

and DISJ 1, 2 appear indispensable.[10] In the next section it remains to be examined whether at least *Chrysippus* may be considered as a proponent of hardcore connexivism.

4 Chrysippus' Janus-faced conception of implication

Sextus' description of the Chrysippian conception of conditionals suggests *two different* interpretations: that of a *strict*, and that of a *connexive* implication. Unfortunately it is not always recognized that these two interpretations are *incompatible*. For instance, the author of a Wikipedia entry on Chrysippus' theory of conditionals wrote:

> Chrysippus adopted a much stricter view [than Diodorus] regarding conditional propositions, which made such paradoxes impossible: to him, a conditional is true if denial of the consequent is logically incompatible with the antecedent. This corresponds to the modern-day strict conditional.[11]

However, if Chrysippus' conception really amounted to a (modern-day) *strict* implication, then the "paradoxes" PSI 1–4 would arise inevitably! In order to resolve this apparent conflict, let us consider again the relevant passage from Sextus Empiricus' report:

> And those who introduce the notion of connexion (συναρτησις) say that a conditional is sound when the contradictory of its consequent is incompatible with its antecedent. According to them the conditionals mentioned above [and thus in particular ATOMS] are unsound, but the following is true: 'If it is day, it is day'.

Three indubitable points can be extracted from this passage:

(i) Chrysippus defines a conditional as sound if and only if the contradictory of its consequent, i.e., $\neg q$, is "*incompatible*" with its antecedent, p.

---

[10] See (Lenzen, 2020a). Abelard's view of connexivity had already been discussed in (Lenzen, 2019, Section 5). A detailed exposition of Abelards's logic is given in (Lenzen, 2021). For a recent attempt to defend Abelard against the criticism of his contemporary Alberic of Paris see (Estrada-González & Ramírez-Cámara, 2020).

[11] https://en.wikipedia.org/wiki/Chrysippus#Conditional_propositions

(ii) Chrysippus rejects the strange Diodorian example ATOMS which has the logical structure 'If not-p, then p' as *not sound*.

(iii) In contrast, 'If p, then p' is a (trivial) example of a *sound* Chrysippian conditional.

Let us discuss these three points in reverse order!

Ad (iii): The soundness of 'If p, then p' certainly matters in connection with the Peripatetics who reject 'If p, then p'. This example is not, however, of any help for clarifying the difference between Diodorus' and Chrysippus' view.

Ad (ii): In a recent paper[12] it was speculated that Chrysippus might have been an adherent of strict implication after all. *Perhaps* Chrysippus was willing to accept the "paradoxes" of strict implication, including corollary PSI 4, according to which a self-inconsistent proposition entails its own negation. *Maybe* he just rejected the particular *example* because the antecedent of ATOMS is not really *impossible*, nor is the consequent therefore really *necessary*. As was explained at the end of Section 2, from the perspective of modern physics the proposition 'Atomic elements of things exist' doesn't constitute a *necessary truth* at all. Therefore it seems *possible* that Chrysippus rejected ATOMS only because he had doubts concerning the "necessity" of the Stoic's theory of matter. However, O'Toole and Jennings (2004) pointed out:

> [...] that the axiōma 'Atomic elements of existents are without parts' is conceptually or analytically true, and hence necessary [...] *according to the versions of necessity both of Diodorus and Chrysippus*. The definition of Diodorus is worded as follows: 'The necessary is that which being true, will not be false' [...]; whereas that of Chrysippus is worded thus: 'The necessary is that which being true does not admit of being false, or admits of being false but is prevented by external factors from being false'. (pp. 481–2, my emphasis)

Therefore the aforementioned speculations about Chrysippus' reasons for rejecting ATOMS appear rather implausible and one would better look for an alternative explanation.

O'Toole and Jennings (2004) think that Chrysippus' rejection of ATOMS is based on the "coherence view" according to which "a sound conditional requires a connexion or coherence between the antecedent and the consequent". For the sake of illustration, they consider the example: 'If triangles have

---

[12] See (Lenzen, 2020a), Section 3.

four sides, then Chrysippus is the greatest of Stoic logicians.' According to the *modern* understanding of incompatibility, the impossible antecedent 'Triangles have four sides' is incompatible with *any consequent*, even if both propositions are entirely unconnected. However, as O'Toole and Jennings (2004) put it, it is "difficult to see how Chrysippus, or any other Stoic logician, might have considered these propositions to be in any sense 'in conflict'" (p. 490).

As Mates (1953) remarked, modern logicians are *almost invariably* "led to suppose that 'incompatible' is used [by Chrysippus] in its ordinary sense, according to which incompatible propositions *cannot* both be true, i.e., their conjunction is *logically false*" (p. 48, my emphasis). In view of the foregoing discrepancies, however, one should better check whether Chrysippus' expression 'μαχεται' can be interpreted in another sense than that of *logical* incompatibility.

Ad (i): In their edition of the *Outlines of Scepticism*, Annas and Barnes (2000) paraphrase Chrysippus' definition by the requirement that "the opposite of its consequent *conflicts* with its antecedent" (p. 12). O'Toole and Jennings (2004) explain that the primary meaning of the Greek verb, from which 'μαχεται' is derived,

> [...] is *to fight* or *to battle* or *to war*. Now, one would hardly want to translate the term 'μαχεται' in a logical context as 'fights' or 'battles' or 'wars'. [...] Probably 'conflicts' is just the right compromise. It is bloodless enough for a logic book, yet it remains faithful to the etymological origins of the Greek term [...]. (p. 490)

More specifically Sanford (1989) suggested that proposition ATOMS fails to satisfy Chrysippus' definition of a sound conditional because, "the denial of the main clause is *not incompatible* with the if-clause. Indeed, it *is* the if-clause" (p. 24). Of course, from a *contemporary* point of view, every impossible proposition is incompatible with any proposition, hence also with itself! But according to *Chrysippus'* view, the incompatibility of two propositions p, q has to arise from a "conflict" between *p and q*: One proposition, say p, somehow has to *contradict* the *other* proposition, q; otherwise there can be no "war", no "battle" and no "conflict" between them.[13] Given this interpretation of the notion 'μαχεται', Chrysippus' idea of a sound conditional may be defined as follows:

---

[13] This intuitive idea has been revived by Nelson (1930) who maintained that no property "of a single proposition p is sufficient to determine whether it is consistent or inconsistent with just any other randomly selected proposition q" (p. 143).

DEF 4 $(p \Rightarrow_{\text{Chr}} q) := \neg\Diamond(p \land \neg q) \land \Diamond p \land \Diamond \neg q.$

On the one hand, the antecedent has to be incompatible with the negation of the consequent. On the other hand, this incompatibility must not be caused by one of the propositions alone, i.e., p must be possible, and $\neg q$ must be possible, too; otherwise there is no real "conflict" *between* them. DEF 4 thus reconciles the two faces of Janus' head in an elegant way. Because of condition $\neg\Diamond(p \land \neg q)$, Chrysippian implication is sort of a *strict* implication, while condition $\Diamond p \land \Diamond \neg q$ also makes it sort of *connexive*.

To conclude this section let it be pointed out that a closely related interpretation was suggested by Nasti De Vincentis (2006, p. 241) where the following formula serves as a definiens for Chrysippian implication:

NASTI $\neg\Diamond(p \land \neg q) \land (\Box q \supset \Box p) \land (\Diamond q \supset \Diamond p).$

It is easy to see that the definiens of our DEF 4 entails condition NASTI. According to the truth-conditions of material implication, $\Diamond p$ entails $(\Diamond q \supset \Diamond p)$, and $\Diamond \neg q$, i.e., $\neg\Box q$, entails $(\Box q \supset \Box p)$. As Nasti De Vincentis noted, his interpretation of Chrysippian implication satisfies the principle "From the impossible only the impossible, and not the possible, follows". According to DEF 4, in contrast, an impossible antecedent p doesn't entail *any proposition at all*! Further formal properties of the operator $\Rightarrow_{\text{Chr}}$ shall be investigated in Section 6.

5 Peripatetic implication

The *fourth* Stoic conception of implication is scarcely discussed in the literature. Mates (1953) pointed out that this view had not been "discussed by any other ancient sources" (p. 49). As O'Toole and Jennings (2004, p. 480) noted "it has been little discussed by modern commentators" either. Thus Kneale and Kneale (1962) only explained that their attribution of this view to the Peripatetics was based on "the rejection of the form 'If p then p', which the Peripatetics regarded as a piece of useless Stoic verbalism" (p. 128). Long and Sedley (1990) suggested "that [the fourth view] may not be significantly different from the connexion account" (p. 211), i.e., from Chrysippus' conception.

Although—as Bonevac and Dever (2012) stressed—"we do not have enough evidence concerning this fourth option to *know* [!] what its advocates had in mind" (p. 180, my emphasis), the brief description given by Sextus suggests a rather clear and unique interpretation:

[...] those who judge by implication [...] say that a true conditional is one whose consequent is contained potentially in its antecedent. According to them the statement 'If it is day, it is day' and similarly every conditional which is repetitive [...] will apparently be false; for it is impossible for a thing to be contained in itself. (Kneale & Kneale, 1962, p. 129)

On the one hand, for the Peripatetics, a *sound* implication is characterized by the condition that the consequent must be "potentially contained" in the antecedent. On the other hand, merely "repetitive" conditionals of the structure 'If p, then p' are considered as *unsound*.

Now in general the relation 'A is contained in C' admits of at least two different interpretations. First, it may mean that A is a *subset* of C, $A \subseteq C$, which leaves open the possibility that, conversely, also $C \subseteq A$ so that A and C coincide. Secondly, 'containment' may be interpreted in the sense that A is a *proper subset* of C: $A \subset C =_{df} A \subseteq C \wedge A \neq C$. Evidently it is this latter interpretation which the Peripatetics must have had in mind when they argued that no "thing" can "be contained in itself".[14] From a purely linguistic point of view, it might be objected that the Greek 'δυνάμει' is far from expressing the meaning we are suggesting here. The translations offered by the Kneales ('contains *potentially*') or by Mates ('is *in effect* included') certainly are not synonymous with '*contains properly*' or 'is *properly* included in'. However, the entire context strongly suggests that the Peripatetics wanted to stress that no "thing" is *properly* contained in itself!

While "repetitive" propositions of the form 'If p, then p' are *sound* conditionals according to the conceptions of Philo and of Diodorus, the Peripatetics opposed this view because the containment of the consequent, p, in the antecedent, p, is not a *proper* one.[15] In Sextus' report this view was formulated somewhat irritatingly by saying that repetitive propositions are "apparently" or "probably" false. The latter expression made Bonevac and Dever (2012) wonder: "[...] if it is impossible for a thing to be included in itself, why are repeated conditionals 'probably' false rather than necessarily false?" (p. 179). Well, the answer is quite obvious. Sextus only wanted to

---

[14] As Ulrich Nortmann pointed out to me, the Stoics may have had another—more natural—interpretation of the containment relation in mind which can be illustrated, for example, by an apple lying in a basket while the basket itself can't lie (or be "contained") in itself!

[15] This analysis conforms with the view of Sanford (1989): "On the fourth and final view listed by Sextus, the second conditional here [which has the structure 'If p, then p'] would still not be true. Like the example Sextus gives, 'If it is day, it is day', the if-clause, being identical to the main clause, *does not properly contain* [!] the main clause" (p. 25, my emphasis).

point out that it is *his own interpretation* that repetitive propositions were considered by the Peripatetics as false. This becomes evident in Mates' translation of the passage:

> [...] those who judge by 'force' declare that a conditional is true if its consequent is in effect included in its antecedent. According to them, *I suppose* [!], 'If it is day, then it is day' and every repeated conditional will be false, for there is no way of a thing itself to be included in itself. (Mates, 1996, p. 143, my emphasis)[16]

Altogether, then, it seems safe to assume that Peripatetic implication differs from Diodorian implication or from Chrysippian implication *only* by the *extra requirement* that the consequent q must not be identical with—but instead has to be *properly contained* in—the antecedent p. Since the proper containment of q in p can be expressed by the condition that $p \Rightarrow q$ but not $q \Rightarrow p$, one obtains two possible explications:

DEF 5.1 $\quad (p \Rightarrow_{Per} q) =_{df} (p \Rightarrow_{Dio} q) \land \neg(q \Rightarrow_{Dio} p)$
DEF 5.2 $\quad (p \Rightarrow_{Per} q) =_{df} (p \Rightarrow_{Chr} q) \land \neg(q \Rightarrow_{Chr} p)$.

In a footnote to the relevant passage of Sextus' report, G. Bury, the editor of an older, bilingual edition of the *Outlines of Pyrrhonism*, explained:

> 'Implication' (*emphasis*) is power of signifying more than is explicitly expressed. An example of the 'potential inclusion' is 'If a man exists, a beast exists'. (Bury, 1933, p. 222/3, fn. *b*)

As Franz von Kutschera pointed out to me, the Greek expression 'periechetai dynamei' may also be translated as 'to contain according to the sense'.[17] This *philological* explanation is apt to support our *logical* interpretation in two respects. First, if the "emphasis" of an implication is the "power of signifying more than is explicitly expressed", then the antecedent of a sound

---

[16]This point has already been stressed by Frede (1974): "S[extus] E[mpiricus] bemerkt, daß nach dieser Definition eine Aussage der Form 'wenn p, dann p' wohl nicht wahr sein könne. Zu dieser Bemerkung ist zunächst einmal festzustellen, daß S. E. selbst deutlich macht, daß es sich bei seiner Bemerkung lediglich um eine eigene Interpretation der Definition handelt" (pp. 90/91).

[17]As a matter of fact, the editor of a German edition of Sextus Empiricus' work translated the passage as follows: "Die nach dem 'impliziten Sinn' Urteilenden schließlich sagen, eine Implikation sei wahr, deren Nachsatz im Vordersatz dem Sinn nach enthalten sei." (see Hossenfelder, 1968, p. 182).

conditional has to express *more* than is expressed in the consequent. But in a "*repetitive*" conditional, the consequent *coincides* with the antecedent p; hence it doesn't express more, but exactly as much as p!

Second, Bury's somewhat obscure remark, that the conditional 'If a man exists, a beast exists' exemplifies a "potential inclusion" and hence a sound Peripatetic conditional, probably has to be understood as follows. 'Beast' is a literal, but somewhat infelicitous translation of the Greek word for 'animal'. Similarly, 'exists' is just synonymous with 'is',[18] so that 'If a man is, an animal is' most likely has to be understood as saying that whenever something, x, is a man, x is an animal.[19] Hence Bury's 'If a man exists, a beast exists' may be considered as an awkward formulation of the logicians' pet example: 'Every man is an animal'. This example of a *true*—indeed *analytically* true—proposition satisfies the Peripatetics' requirement that the consequent, i.e., the predicate-term 'animal', is "potentially included" and "properly included" in the antecedent, i.e., in the subject-term 'man'. On the one hand, 'animal' is *properly* included in 'man', because the latter concept *also* contains the concept 'rational' which is not contained in 'animal'. On the other hand, 'animal' is included only "*potentially*" or *implicitly* in 'man', because the concept 'man' first has to be *analyzed* into the complex concept 'rational animal', before it becomes evident that it *contains* 'animal'!

Our interpretation of Peripatetic implication as a *restriction* of strict implication to the case where the consequent does not conversely entail the antecedent not only accords with the assessment of Mates (1953, p. 49) for whom "the fourth type of implication seems to be a restricted type of Chrysippean implication". It also accords with the more recent views of Anne Wersinger and Yale Weiss. Thus Wersinger (2008, p. 59) righly emphasized: „Pour être une vraie conséquence, la conséquence doit dire autre chose que les prémisses et pas la même chose". And Weiss (2019, p. 321) characterized the fourth account of conditionals as a "precursor of analytic entailment", where the "central idea behind analytic entailment is that the meaning of the consequent of a conditional should be contained in the meaning of the antecedent."

Although, then, the very *idea* of Peripatetic implication is entirely clear, the corresponding *logic* of this operator displays certain oddities which shall be examined in Section 7. Before turning to this rather technical appendix, let us summarize the upshot of the foregoing sections:

---

[18] Note also that what the Kneales and others translate as 'If it is day, it is day', Bury formulates as 'If day exists, day exists'.

[19] Thus also in Boethius' theory of hypothetical syllogisms a universal affirmative proposition is typically paraphrased as 'Si est A, est B' (see Prantl, 1927, p. 703, fn. 147).

- *Philonian* implication is just *material* implication.
- *Diodorian* implication is just *strict* implication and hence a strengthened form of Philonian implication.
- *Chrysippian* implication is a special case of Diodorian implication which requires that the antecedent, p, must not be impossible, and that the consequent, q, must not be necessary.
- *Peripatetic* implication is a restriction of Diodorian or Chrysippian implication which requires that the antecedent p is "stronger" than the consequent q, i.e., p entails q but q does *not conversely* entail p.

This confirms the wide-spread opinion formulated, e.g., by O'Toole and Jennings (2004) that Sextus ordered the Stoic conceptions "from the weakest to the strongest, in each case citing an example which is allowed by the next weaker interpretation, but which is rejected by the one under discussion" (p. 479).

6 Some formal properties of Chrysippian implication

(6.1) Condition '$\Diamond p$' in DEF 4 entails that a conditional with an impossible antecedent will *never* be sound:

CHRYS 1 $\forall p(\text{If } \neg \Diamond p, \text{ then } \neg \exists q(p \Rightarrow_{\text{Chr}} q))$.

(6.2) This principle entails that an impossible antecedent doesn't imply its own negation:

CHRYS 2 $\forall p(\text{If } \neg \Diamond p, \text{ then } \neg (p \Rightarrow_{\text{Chr}} \neg p))$.

CHRYS 2 is apt to explain why Chrysippus considered Diodorus' "anti-connexive" example ATOMS as unsound, namely simply because the antecedent 'Atomic elements of matter do not exist' is "impossible".

(6.3) CHRYS 1 further entails that an *impossible* proposition doesn't entail *itself*:

CHRYS 3 $\forall p(\text{If } \neg \Diamond p, \text{ then } \neg (p \Rightarrow_{\text{Chr}} p))$.

Hence the law of the *reflexivity* of implication is no longer (unrestrictedly) valid; it only holds for *contingent* propositions p:

CHRYS 4 $\forall p(\text{If } \Diamond p \wedge \Diamond \neg p, \text{ then } (p \Rightarrow_{\text{Chr}} p))$.

CHRYS 4 suffices to support the conclusion that for Chrysippus repetitive conditionals like 'If it is day, it is day' or 'If it is night, it is night' would count as sound. However, the *weakened* conditionals 'If it is day, it is day or it is night' and 'If it is night, it is day or it is night' would *not* count as sound because the *consequent* is necessary, hence its *negation* is impossible, so that there is no real Chrysippian—i.e., so to speak, no *mutual*—"conflict" between the antecedent and the negation of the consequent!

(6.4) The impossible conjunction $p \wedge \neg p$ implies none of its conjuncts:

CHRYS 5 $\neg((p \wedge \neg p) \Rightarrow_{\text{Chr}} p)$ and $\neg((p \wedge \neg p) \Rightarrow_{\text{Chr}} \neg p)$.

Hence Chrysippian conditionals fail to satisfy the (unrestricted) laws of *conjunction*, CONJ 1, 2.

(6.5) No necessary, and in particular no tautological, proposition can ever be the consequent of a sound Chrysippian conditional:

CHRYS 6 $\forall q(\text{If } \Box q, \text{ then } \neg \exists p(p \Rightarrow_{\text{Chr}} q))$.

Thus in particular $p \vee \neg p$ is not entailed by either of its disjuncts:

CHRYS 7 $\neg(p \Rightarrow_{\text{Chr}} (p \vee \neg p))$ and $\neg(\neg p \Rightarrow_{\text{Chr}} (p \vee \neg p))$.

Hence Chrysippian conditionals also fail to satisfy the (unrestricted) laws of *disjunction*, DISJ 1, 2.

7   Some formal properties of Peripatetic implication

The basic idea of Peripatetic implication consists in the requirement that the antecedent has to contain the consequent *properly*. This admits of two different explications as captured in DEF 5.1, DEF 5.2. The subsequent observations apply to both variants.[20]

(7.1) Peripatetic implication must be assumed to be *transitive*:

PERI 1   If $(p \Rightarrow_{\text{Per}} q)$ and $(q \Rightarrow_{\text{Per}} r)$ then $(p \Rightarrow_{\text{Per}} r)$.

---

[20] A closer examination of the differences between the "Diodorian variant" and the "Chrysippian variant" of Peripatetic implication must be left for future investigations. Suffice it to mention here that only the latter but not the former represents a *connexive* implication!

The Third and Fourth Stoic Account of Conditionals

For if p properly contains q which in turn properly contains r, then p properly contains r.

(7.2) On the other hand, by its very definition, Peripatetic implication is not reflexive, but *antireflexive*, i.e.:

PERI 2 $\neg \exists p : (p \Rightarrow_{\text{Per}} p)$.

(7.3) From this it follows that Peripatetic implication must be *antisymmetric*:

PERI 3   If $(p \Rightarrow_{\text{Per}} q)$, then not $(q \Rightarrow_{\text{Per}} p)$.

For if there would exist propositions p, q such that $(p \Rightarrow_{\text{Per}} q)$ and $(q \Rightarrow_{\text{Per}} p)$, one might infer by PERI 1 $(p \Rightarrow_{\text{Per}} p)$ which contradicts PERI 2.

(7.4) The (material or strict) *equivalence* of two propositions p, q is normally defined by the condition of *mutual implication*:

$$(p \equiv q) =_{\text{df}} (p \supset q) \wedge (q \supset p) \qquad (p \leftrightarrow q) =_{\text{df}} (p \rightarrow q) \wedge (q \rightarrow p).$$

In the case of Peripatetic implication, however, such a definition would *make no sense*. In view of PERI 3 there cannot exist propositions p, q which would be "Peripatetically equivalent" in the sense of $(p \Rightarrow_{\text{Per}} q) \wedge (q \Rightarrow_{\text{Per}} p)$. Hence familiar laws such as, e.g., the law of *double negation*, which in a logic of Philonian or Diodorian implication is expressed by $(\neg\neg p \equiv p)$ or $(\neg\neg p \leftrightarrow p)$, cannot, as such, be expressed in the logic of Peripatetic implication.

(7.5) The same holds true for laws which are otherwise used to express familiar properties of conjunction as, e.g., symmetry and idempotence:

$$(p \wedge q) \equiv (q \wedge p) \quad \text{or} \quad (p \wedge q) \leftrightarrow (q \wedge p)$$
$$(p \wedge p) \equiv p \quad \text{or} \quad (p \wedge p) \leftrightarrow p$$

None of these principles can be formulated as such, i.e., as equivalences, in the language of Peripatetic implication. This does not, however, mean that Peripatetic logicians would have to make a *substantial difference* between, say, $(p \wedge q)$ and $(q \wedge p)$—they just don't have the conceptual means to express their *equivalence*!

(7.6) Peripatetic logicians would instead have to resort to inference schemata like the following ones which correspond to the familiar rules of conjunction-elimination and conjunction-introduction in systems of natural deduction:

(*) $(p \wedge q) \Rightarrow_{\text{Per}} p$

(**) $(p \wedge q) \Rightarrow_{\text{Per}} q$

(***) If $r \Rightarrow_{\text{Per}} p$ and $r \Rightarrow_{\text{Per}} q$ then $r \Rightarrow_{\text{Per}} (p \wedge q)$.

However, a Peripatetic logician cannot even accept these principles all *together* because otherwise, after substituting '$(p \wedge q)$' for 'r' in (***), the premises (*) and (**) would allow to derive $(p \wedge q) \Rightarrow_{\text{Per}} (p \wedge q)$ in contradiction to PERI 2!

(7.7) Moreover, Peripatetic logicians would probably want to reject the idempotence of conjunction in either of the forms $(p \wedge p) \Rightarrow_{\text{Per}} p$ or $p \Rightarrow_{\text{Per}} (p \wedge p)$. After all, the propositions p and $(p \wedge p)$ appear to have *"the same content"* so that neither of them is *properly* contained in the other! But then Peripatetic logic can no longer postulate the basic laws of conjunction in the unrestricted form stated above. A general schema like (*) is normally taken to hold for *any* propositions p, q! Thus, in particular, one may substitute, e.g., 'p' for 'q' so that one would obtain $(p \wedge p) \Rightarrow_{\text{Per}} p$!

References

Annas, J., & Barnes, J. (Eds.). (2000). *Sextus Empiricus: Outlines of Scepticism*. Cambridge: Cambridge University Press.

Bonevac, D., & Dever, J. (2012). A history of the connectives. In D. M. Gabbay, F. J. Pelletier, & J. Woods (Eds.), *Handbook of the History of Logic: Vol. 11. Logic: A History of its Central Concepts* (pp. 175–233). Amsterdam: Elsevier.

Bury, R. G. (Ed.). (1933). *Sextus Empiricus: Outlines of Pyrrhonism (With an English Translation)*. Cambridge, MA: Harvard University Press.

Estrada-González, L., & Ramírez-Cámara, E. (2020). A Nelsonian response to 'the most embarrassing of all twelfth-century arguments'. *History and Philosophy of Logic, 41*(2), 101–113.

Frede, M. (1974). *Die stoische Logik*. Göttingen: Vandenhoeck & Ruprecht.

Hossenfelder, M. (Ed.). (1968). *Sextus Empiricus: Grundriss der pyrrhonischen Skepsis*. Frankfurt am Main: Suhrkamp.

Kapsner, A. (2019). Humble connexivity. *Logic and Logical Philosophy, 28*(3), 513–536.

Kneale, W., & Kneale, M. (1962). *The Development of Logic*. Clarendon Press.

Lenzen, W. (2019). Leibniz's laws of consistency and the philosophical foundations of connexive logic. *Logic and Logical Philosophy, 28*(3), 537–551.

Lenzen, W. (2020a). A critical examination of the historical origins of connexive logic. *History and Philosophy of Logic, 41*(1), 16–35.

Lenzen, W. (2020b). Kilwardby's 55th lesson. *Logic and Logical Philosophy, 29*(4), 485–504.

Lenzen, W. (2021). *Abaelards Logik*. Paderborn: mentis. (forthcoming.)

Lewis, C. I., & Langford, C. H. (1959). *Symbolic Logic*. New York: Dover Publications.

Long, A. A., & Sedley, D. N. (1990). *The Hellenistic Philosophers* (Vol. 1). Cambridge: Cambridge University Press.

Mates, B. (1953). *Stoic Logic*. Berkeley: University of California Press.

Mates, B. (1996). *The Skeptic Way: Sextus Empiricus' Outlines of Pyrrhonism*. New York/Oxford: Oxford University Press.

McCall, S. (1975). Connexive implication. In A. R. Anderson & N. D. Belnap (Eds.), *Entailment: The Logic of Relevance and Necessity* (pp. 434–452). Princeton: Princeton University Press.

McCall, S. (2012). A history of connexivity. In D. M. Gabbay, F. J. Pelletier, & J. Woods (Eds.), *Handbook of the History of Logic: Vol. 11. Logic: A History of its Central Concepts* (pp. 415–449). Amsterdam: Elsevier.

Nasti De Vincentis, M. (2006). Conflict and connectedness: Between modern logic and history of ancient logic. In E. Ballo & M. Franchella (Eds.), *Logic and Philosophy in Italy* (pp. 229–251). Monza: Polimetrica International Scientific Publisher.

Nelson, E. J. (1930). Intensional relations. *Mind, 39*(156), 440–453.

O'Toole, R. R., & Jennings, R. E. (2004). The Megarians and the Stoics. In D. M. Gabbay & J. Woods (Eds.), *Handbook of the History of Logic: Vol. 1. Greek, Indian and Arabic Logic* (pp. 397–522). Amsterdam: Elsevier.

Pizzi, C., & Williamson, T. (1997). Strong Boethius' thesis and consequential implication. *Journal of Philosophical Logic, 26*(5), 569–588.

Prantl, C. (1927). *Geschichte der Logik im Abendlande* (Vol. 1). Leipzig: G. Fock.

Sanford, D. H. (1989). *If P, then Q: Conditionals and the Foundations of Reasoning*. London/New York: Routledge.

von Kutschera, F. (1986). Zwei modallogische Argumente für den Determinismus: Aristoteles und Diodor. *Erkenntnis, 24*(2), 203–217.

Weiss, Y. (2019). Sextus empiricus' fourth conditional and containment logic. *History and Philosophy of Logic*, *40*(4), 307–322.

Wersinger, A. G. (2008). Sextus Empiricus et la »conséquence« inassignable: le scepticisme à l'épreuve de la logique. *Cahiers philosophiques*, *115*(3), 46–62.

Wolfgang Lenzen
University of Osnabrück, Department of Philosophy
Germany
E-mail: `lenzen@uos.de`

Normative Parties in Subject Position and in Object Position

TEREZA NOVOTNÁ AND MATTEO PASCUCCI[1]

Abstract: We analyze some normative relations as instances of a general schema of relations among a finite number of parties; in this schema parties can play various roles grouped into two main conceptual layers, called 'subject position' and 'object position'. Relying on the theoretical apparatus introduced, we develop a new symbolic representation for normative reasoning which constitutes an alternative to approaches available in the literature. Our contribution includes a semantic characterization for a series of logical systems built over the proposed framework.

Keywords: Hohfeldian theory, modal logic, normative relations, subject and object position

1 Introduction

In legal theory many examples of normative relations can be described in terms of *three main components*: a list of subjects, a content and an object. This is a standard classification that can be found, for instance, in Knapp (1995). The *subjects* of a normative relation are those parties that establish a relation (via a decision, an agreement, etc.) or that satisfy sufficient conditions for a relation to arise (e.g., being a member of a corporation is sufficient for having some duties). Each party can be thought of as a natural person, a collective body, a corporation, a state, etc.[2] The *content* of a normative relation is the sort of duty, right, etc., it expresses (such as a right of one person

---

[1] The authors thank Daniela Glavaničová, Amalia Haro Marchal, Miloš Kosterec, Mirco Sambrotta and Martin Vacek for useful comments during a presentation of this work. Matteo Pascucci was supported by the *Štefan Schwarz Fund* for the project "A fine-grained analysis of Hohfeldian concepts". Tereza Novotná gratefully acknowledges the support of Masaryk University - project "Právo a technologie IX" number MUNI/A/1292/2020 with the support of the Specific University Research Grant, as provided by the Ministry of Education, Youth and Sports of the Czech Republic.

[2] Sometimes in the legal literature a subject is strictly equated with a natural person. The argument is that complex legal relations can be reduced to very basic ones, involving natural persons only. For example, a labour-law relation between an employee and a company can be always decomposed into singular relations between two persons—the employee and the owner

against the entire world or a duty of one person towards another person). The object of a normative relation is the human behaviour or specific thing that the relation pertains to. More precisely, human behaviours can be regarded as *primary* objects of a relation, whereas specific things are sometimes mentioned as secondary objects. To give a very simple example: a subject x owes money to a subject y due to a contract they signed; thus, (i) there is a normative relation stipulated by the subjects x and y, (ii) its content is a duty of one person towards another, and (iii) its (primary) object is x's payment of a certain amount of money to y (while its secondary object is money *simpliciter*).

As this example shows, in addition to the possibility of being a subject in a relation, a normative party can be also mentioned in the object of a relation. We will say that all parties that act as subjects in stipulating a normative relation are in the *'subject position'*, whereas all parties that are mentioned in the object of a normative relation are in the *'object position'*.[3] In general, parties mentioned in these two components of a relation need not be the same. For instance, consider a slightly more complex scenario: x stipulates a contract with y that z has to give money to w: here x and y are in the subject position, while z and w are in the object position. In both positions parties may act according to a certain *direction*: we can imagine that in some cases a contract stipulated between x and y is a form of commitment of x *towards* y (in other cases a form of reciprocal commitment); furthermore, if a contract concerns the payment of an amount of money by z to w, we can speak of a duty that z has *towards* w. Not only this, but we can also have, in both positions, parties that act in representation of others or to the advantage/disadvantage of others. For instance: x stipulated a contract with y on behalf of v and to the advantage of t (subject position), that z has to give money to w on behalf of u and to the advantage of s (object position).[4] Thus, it is in principle possible to have an arbitrarily large number of parties in both positions.

In the present article we view normative relations as instances of a general schema of relations among a finite number of parties. The main novelty is

---

of the company. For the purpose of this article, we do not put any specific restrictions on the kind of entity taken as a subject (and, more generally, as a party) in a normative relation. Indeed, our aim is developing a formal framework that deals primarily with relations of a certain kind, rather than with entities of a certain kind.

[3] The subject position may be also called 'agreement position', as long as the analysis concerns relations that arise from the stipulation of an agreement. However, sometimes normative relations arise simply by virtue of the fact that certain conditions are satisfied, without any agreement being in place. For instance, when a party satisfies the condition of being a citizen of a state.

[4] Going a bit deeper into the details, we can imagine a scenario in which t benefits from the existence and validity of the contract, while s benefits from the fact that w, after receiving money, becomes able to solve some debts.

that these parties can play roles distributed over the two conceptual layers mentioned, i.e., the subject position and the object position. The article is arranged as follows. In Section 2 we review some fundamental approaches to the formal representation of normative relations. In Section 3 we discuss examples of relations among many parties taken from the normative domain, and explain how they could be analyzed in terms of the opposition between subject position and object position. In Section 4 we introduce a new symbolic language based on the theoretical framework proposed, and build a series of logical systems having an increasing deductive power. In Section 5 we provide characterization results for these systems in terms of possible worlds semantics and discuss, in passing, foreseen future developments of the present framework.

2 An overview of related approaches

Formal analyses of normative relations have been proposed for several decades and particular attention has been paid to the so-named *Hohfeldian squares of opposition*. These are reproduced here in Figure 1 and Figure 2, where dotted arrows connect opposites and solid arrows connect correlatives, according to the terminology employed by Markovich (2019). For the original theory behind the two squares, see Hohfeld (1913, 1917); for an extended discussion of the meta-relations between the two squares, see, e.g., Sileno (2016). The role played by a party in a normative relation can be defined in terms of either a *deontic concept* (duty, claim-right, privilege, no-claim) or a *potestative concept* (liability, power, immunity, disability). Roles defined by the second square may affect those defined by the first square (e.g., when a party has the power to create a duty of another party towards a third one).

Symbolic languages used to capture the Hohfeldian theory often represent concepts in the two squares via combinations of modal and boolean operators at a propositional level. The first systematic proposal is due to Kanger (see, e.g., Kanger, 1970). His framework can be presented in terms of syntactical constructions of the form $\pm \mathcal{O} \pm E_i \pm \phi$, where ϕ is a formula denoting a proposition, $\mathcal{O}$ a modal operator meaning 'it is obligatory that', E_i a modal operator meaning 'agent i sees to it that' and $\pm$ are alternative options for affirmative or negative propositions.[5] For instance, $-\mathcal{O}+E_i+\phi$ says that it is not obligatory for agent i to bring about ϕ.

---

[5]Alternatively, one could take ϕ to denote an action-type and E_i as a predicate meaning 'agent i performs'. In both approaches, the idea is that agent i contributes to ϕ.

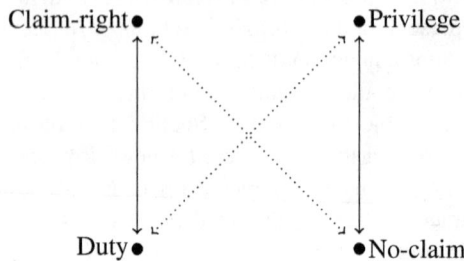

Figure 1: Hohfeldian square of deontic opposition.

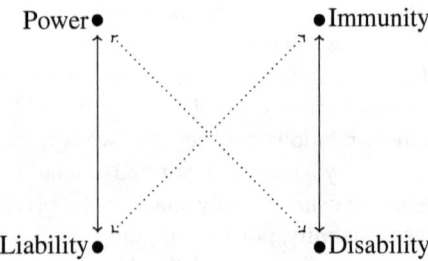

Figure 2: Hohfeldian square of potestative opposition.

Generalizations of Kanger's framework are proposed, e.g., by Lindahl (1977) and Sergot (2013). The idea is that a wide logical space of normative relations can be captured via maximal consistent conjunctions of formulas built over variations of Kanger's symbolic language. Maximal consistency is defined with reference to an underlying logical system, and authors usually opt for systems whose axiomatic basis includes at least the following principles for obligation (where $\neg$ stands for negation, $\equiv$ for material equivalence, $\wedge$ for conjunction and $\rightarrow$ for material implication):

Subject Position and Object Position

(E) $\quad \dfrac{\phi \equiv \psi}{\mathcal{O}\phi \equiv \mathcal{O}\psi}$

(M) $\quad \mathcal{O}(\phi \wedge \psi) \to (\mathcal{O}\phi \wedge \mathcal{O}\psi)$

(C) $\quad (\mathcal{O}\phi \wedge \mathcal{O}\psi) \to \mathcal{O}(\phi \wedge \psi)$

(P) $\quad \neg \mathcal{O}(\phi \wedge \neg \phi)$

Hansson (1970), Makinson (1986) and Lindahl (2001) observe that Kanger's framework focuses on the role played by only one party, whereas in many cases more parties are needed for a proper account of a normative relation. To address this problem, Herrestad and Krogh (1995) define a language including a modal operator $_i\mathcal{O}_j$, which expresses a duty *directed* from the party i (called 'bearer') towards the party j (called 'counterparty').

However, speaking of bearers and counterparties *simpliciter* is still not sufficient to distinguish all possible roles involved in a normative relation. In this regard, Duarte d'Almeida (2016), for instance, observes that the relational character of duties and rights is independent from their object. Consider the following two sentences:

A) y is under a duty towards x to stay off x's place;

B) y is under a duty to stay off x's place.

Both sentences express a normative relation between x and y; however, while in sentence A x plays the role of a contracting party in an implicit normative agreement (y's duty is 'towards x'), this is not the case in sentence B, where the mentioned duty might be a consequence of a normative agreement between y and some third party (e.g., a duty towards z). A broader notion of correlation among duties and rights can be formulated under the assumption that some of the parties involved are not always explicitly mentioned. Additional critical remarks on the thesis that normative relations can be reduced to an opposition between a bearer and a counterparty are provided, for instance, by Lyons (1970). All this suggests that normative relations can be more generally thought of as relations among an arbitrary number of parties playing different roles. We will here explore this idea, distributing possible roles over a 'subject position' and an 'object position'.[6]

---

[6] We refer the reader to Sergot (2013) for further discussion of related approaches available in the literature.

3 Examples of normative relations

In this section we illustrate the use of the notions of 'subject position' and 'object position' in the analysis of two groups of examples of normative relations: Group I includes situations involving many subjects that can be regarded as civil persons or private parties; Group II includes situations in which at least one of the parties involved is a state whose authority is represented by a certain institution (for instance, a court).

The first two examples that we are going to analyze belong to Group I.

Example 1 In civil law, the *institute of representation* is often used in normative relations involving many parties. When the representation is based on a contract, law or court order, it pertains to a situation where a third party enters into an originally two-sided legal problem, agreeing to act on behalf of one of the other two. Rights and duties are divided among the three parties. Clearly, representation can take place also for the other side originally involved, so as to give rise to a normative relation among four parties. Further generalization to a finite number of parties and their representatives is possible. As we said above, the primary object of a relation is a human behaviour that shall be affected by the content of the relation. In the case of representation, it is a third party's behaviour that turns out to be affected: when individual x will stand against individual y during the court procedure and individual z is representing individual x, the procedural duties of x shall be executed by z as a legal representative. Thus, x and z agree that z will act on behalf of x in the court: z plays a role both in the subject position (as a party stipulating the agreement) and in the object position (since her behaviour is the object of the agreement).

Example 2 The *institute of guarantee* is another typical example of multilateral normative relation. There are various kinds of guarantee; however, their conceptual foundations are similar. In the case of a bank guarantee, for instance, a party x owes a certain amount of money to a party y and this duty will affect the behaviour of a bank z (involved as a third party), with which x stipulated an agreement. The party y has a right to demand money from the bank. Such a right is usually related to a condition that needs to be fulfilled and, as soon as this happens, the bank must give money to y. Therefore, rights and duties of the original parties (i.e., x and y) directly affect the third party's behaviour, and the latter (i.e., z) is both in the subject position (due to the stipulation of an agreement with x) and in the object position (given that a certain behaviour from z is required).

Subject Position and Object Position

The next two examples that we will analyze belong to Group II.

Example 3 Consider the case in which a state is responsible for some *fundamental human rights*, such as the right of its citizens to a favorable environment. Every person has the right to a good environment and every person has to follow legal rules that prohibit environmental pollution. However, in specific cases, the state itself is responsible for environmental pollution of another (private) party. We assume, for instance, that party x is a power plant polluting the air that party y, an individual living nearby, breathes. Under certain circumstances, y can seek her right not to breathe polluted air against the state, whence the latter is involved as a third party z. The responsibility of z towards y that no party will pollute the air y has to breathe constitutes an *implicit normative convention*; therefore, the power plant x founds herself in the object position with respect to such a convention (her behaviour is affected by it).

Example 4 Consider a *criminal procedure*: we have a victim of a crime, the person who is accused of committing a crime and a state represented by different legal entities. In the perspective of restorative justice, the aim of the criminal procedure and of punishment is to restore the relationships between different parties (generally, between the victim and the convicted, but also between the convicted and the society as a whole). We can say that there are some rights and duties between the first two parties—victim and convicted—for example a duty not to steal another one's property. The behaviour of the convicted party interfering with this duty causes many new duties that the state shall fulfill towards the victim. The state is represented by different public authorities in different phases of the procedure—the police, a prosecutor, a court or a prison. Due to what the constitution of a state guarantees to its citizens, these public authorities have a duty to behave in such a way as to punish the convicted. Therefore, here parties qualifying as citizens can be said to play a role both in the subject position, since they satisfy a sufficient condition (i.e., citizenship) for a normative relation to arise (i.e., the duty of public authorities to ensure a criminal procedure), and in the object position, since they may benefit from the result of the criminal procedure (e.g., by receiving a compensation). Furthermore, according to this analysis, public authorities can be said to play (at least) a role in the object position, since the normative relation at issue establishes that they need to ensure that justice will be restored.

4 Formal framework

Relying on the theoretical apparatus described so far, we introduce below a new symbolic language $\mathcal{L}$ for normative reasoning. The peculiar feature of $\mathcal{L}$ is the presence, among its primitive symbols, of a family of modal operators to express the opposition between normative parties in the subject position and normative parties in the object position. These operators do not make reference to any specific role played by normative parties; therefore, they can be used to formalize heterogeneous scenarios of normative reasoning in which multiple roles are involved.[7] However, possible roles are arranged according to a *'direction'*: both in the subject position and in the object position we have roles involving a certain form of commitment (or *active roles*) and roles involving a certain form of reception (or *passive roles*). The direction is graphically represented by the arrow $\Rightarrow$: symbols listed on its left indicate active roles, those on its right passive roles. For instance, an active role in the subject position can be played by a party that stipulates a contract in order to make a concession to another party (the latter playing therefore a passive role in the same stipulation); an active role in the object position can be played by a party that brings about—or asks a delegate to bring about—a state-of-affairs to the advantage/disadvantage of another party (the latter playing therefore a passive role in the same scenario).

We start by specifying the vocabulary of our language.

Definition 1 (Vocabulary) *The language $\mathcal{L}$ includes the following primitive symbols:*

- *a countable set of normative parties* PAR, *denoted by* x, y, z, *etc.;*

- *a countable set of propositional letters* PRO, *denoted by* p, q, r, *etc.;*

- *for any finite (and possibly empty) lists of normative parties* σ_1, σ_2, σ_3 *and* σ_4, *modal operators of the types* $\Delta_{(\sigma_3 \Rightarrow \sigma_4)}^{(\sigma_1 \Rightarrow \sigma_2)}$ *and* $\Delta_{(\sigma_3 \not\Rightarrow \sigma_4)}^{(\sigma_1 \Rightarrow \sigma_2)}$;

- *the boolean connectives* $\neg$ *(negation) and* $\to$ *(material implication);*

- *round brackets.*

---

[7]Moreover, the language here introduced is in principle suitable to formalize also non-normative contexts involving interactions among parties, as long as one can group these parties into the two mentioned layers.

Subject Position and Object Position

The reading of the two types of modal operator will be clarified below, after introducing the notion of a well-formed formula of $\mathcal{L}$.[8]

Definition 2 (Well-formed formulas) *The set* WFF *of well-formed formulas of $\mathcal{L}$ is defined by the grammar below, where $p \in$ PRO:*

$$\phi ::= p \mid \neg\phi \mid \phi \to \phi \mid \Delta^{(\sigma_1 \Rightarrow \sigma_2)}_{(\sigma_3 \Rightarrow \sigma_4)} \phi \mid \Delta^{(\sigma_1 \Rightarrow \sigma_2)}_{(\sigma_3 \not\Rightarrow \sigma_4)} \phi$$

Additional boolean operators (including the 0-ary operators $\top$ and $\bot$) can be defined in the usual way in terms of the primitive ones. The notion of *main operator* in a formula is also standard. We will use the label $\mathcal{L}^\Delta$ to represent the set of formulas of $\mathcal{L}$ whose main operator is a modal one.

Let us now examine the meaning of modal formulas in detail. In the case in which none of the lists of parties referred to in a modal operator is empty, we can read $\Delta^{(\sigma_1 \Rightarrow \sigma_2)}_{(\sigma_3 \Rightarrow \sigma_4)} \phi$ as "on the basis of a stipulation between σ_1 and σ_2, σ_3 has to ensure ϕ for σ_4", and we can read $\Delta^{(\sigma_1 \Rightarrow \sigma_2)}_{(\sigma_3 \not\Rightarrow \sigma_4)} \phi$ as "on the basis of a stipulation between σ_1 and σ_2, σ_3 does not have to ensure ϕ for σ_4". It is intended that σ_1 and σ_3 play active roles, while σ_2 and σ_4 play passive roles; notice, however, that the same party x can occur both in the list on the left and in the list on the right of $\Rightarrow$ (as well as both in the subject position and in the object position): the formula $\Delta^{(x,y \Rightarrow x,y,z)}_{(x,w \Rightarrow z)} p$ is an example of this kind.

When some of the lists of parties are empty, the reading can be changed accordingly. In particular, an empty list of parties in the subject position suggests that we are representing either the *fulfillment of a condition* (involving passive roles only in that position) or a *decision* (involving active roles only in that position), rather than a two-sided stipulation. If both lists in the subject position are empty, then the subject position is vacuously mentioned, and we are actually representing a normative relation without any reference to its source; then, if the two lists in the object position are non-empty, what we get is a simple command like "John ought to give Mary 10 €"—and we basically avoid saying which is the source of the command. Furthermore, the absence of a list of normative parties in the object position suggests that we are formalizing either a relation with no target groups, such as beneficiaries, etc. (when only active roles are mentioned in that position) or a norm that is just intended to constitute an advantage/disadvantage for some parties (when only passive roles are mentioned in that position). An example of

---

[8] For the sake of brevity, hereafter we will omit saying that modal operators can be construed for an arbitrary choice of four (and possibly empty) lists of normative parties.

the former is "John ought to wear a uniform (possibly on the basis of a certain convention)", an example of the latter is "Mary ought to be treated in a respectful way". If both lists in the object position are empty, then we are actually representing a pure *ought-to-be* statement (which, again, may depend on a stipulation). For instance: "there ought to be three chairs in the prison cell (due to an internal regulation)". Finally, if all lists in the subject and in the object position are empty, then a formula can be taken to express a sort of *universal norm*, with no reference to particular categories of individuals/entities. This discussion already witnesses that the range of possible relations expressed by Δ-operators is quite broad.

Empty lists of parties in the subject or the object position will be hereafter denoted also by a blank space, *whenever it will be important to emphasize this aspect*, so as to distinguish them from non-empty lists. By contrast, when no blank space is used and no further restriction is mentioned, *all lists* of normative parties in a formula have to be regarded as *possibly empty*. As examples of formulas where the empty/non-empty distinction is emphasized, consider $\Delta\genfrac{(}{)}{0pt}{}{\Rightarrow \sigma_2}{\Rightarrow \sigma_4}\phi$ and $\Delta\genfrac{(}{)}{0pt}{}{\sigma_1 \Rightarrow}{\sigma_3 \Rightarrow}\phi$. The first can be read "due to a condition satisfied by σ_2, ϕ has to be the case to the advantage/disadvantage of σ_4"; the second can be read "due to a decision of σ_1, σ_3 has to ensure ϕ".[9]

We will see that in some logical systems over $\mathcal{L}$ it is possible to establish formal connections between empty and non-empty occurrences of lists of normative parties in formulas of $\mathcal{L}^\Delta$. To this aim, the following notion will be fundamental, where the symbol $*$ denotes either $\Rightarrow$ or $\not\Rightarrow$:

Definition 3 (Reference-abstraction) *Whenever a formula ξ_1 is obtained from a formula $\xi_2 = \Delta^{(\sigma_1 \Rightarrow \sigma_2)}_{(\sigma_3 * \sigma_4)}\phi$ by possibly removing some occurrence of a normative party in the main operator, we say that ξ_1 is a reference-abstraction of ξ_2.*

The notion of reference-abstraction applies both to single parties and to entire lists. For instance, $\Delta\genfrac{(}{)}{0pt}{}{\Rightarrow \sigma_2}{\sigma_3 \Rightarrow}\phi$ is a reference-abstraction of $\Delta\genfrac{(}{)}{0pt}{}{\Rightarrow \sigma_2}{\sigma_3 \Rightarrow \sigma_4}\phi$, whereas $\Delta^{(x \Rightarrow y,z)}_{(z \Rightarrow w)}\phi$ is a reference-abstraction of $\Delta^{(x,y \Rightarrow y,z)}_{(x,z \Rightarrow w)}\phi$. Notice also that, according to Definition 3, any formula whose main operator is a modal one is a reference-abstraction of itself.

We will now introduce a series of systems over $\mathcal{L}$. As a preliminary system, we can mention the linguistic extension of the Propositional Cal-

---

[9] Whether ϕ is brought about to the advantage or disadvantage of σ_4 depends on the context; in both cases we can say that the list of parties σ_4 is the main 'target' or 'recipient' of ϕ.

culus (**PC**) obtained via our modal operators. Its axiomatic basis corresponds to the set of principles $\{A0, R0\}$ below, and can be strengthened with specific axioms for normative discourse, such as $\neg \Delta_{(\sigma_3 * \sigma_4)}^{(\sigma_1 \Rightarrow \sigma_2)} \bot$ or $\Delta_{(\sigma_3 * \sigma_4)}^{(\sigma_1 \Rightarrow \sigma_2)} \phi \rightarrow \neg \Delta_{(\sigma_3 * \sigma_4)}^{(\sigma_1 \Rightarrow \sigma_2)} \neg \phi$, which would block the possibility of inconsistent or conflicting objects of agreements/stipulations. These extensions would not impose any peculiar property on the parties involved in normative relations.

System S_1 is obtained from **PC** by adding an axiom (*A1*) governing reference-abstraction. The idea is that, in most cases, we can remove reference to some normative parties in a formula ϕ representing a normative relation and get a formula ϕ' representing a more general normative relation that is entailed by the original one. However, S_1 poses a restriction in this regard: that the symbol $\not\Rightarrow$ does not occur in the main operator (and we know, by Definition 2, that it can occur only in the object position). For instance, S_1 does not allow one to infer "it is not the case that ϕ has to be brought about to the advantage/disadvantage of σ_4" from "it is not the case that ϕ has to be brought about *by* σ_3 to the advantage/disadvantage of σ_4". The intuition behind is that sometimes a list of parties *different from* σ_3 might be required to bring about ϕ to the advantage/disadvantage of σ_4. Therefore, we will say that reference-abstractions not involving parties on the left or on the right of $\not\Rightarrow$ are *safe*. We also observe that in S_1 the substitution of logically equivalent formulas in the scope of a modal operator does not preserve derivability. For this reason, S_1 can be used for reasoning with hyperintensional modalities.

Definition 4 (System S_1) *System S_1 is defined by the set of deductive principles $\{A0, R0, A1\}$, where:*

A0 All substitution instances of **PC**-*tautologies;*

R0 if $\vdash \phi$ and $\vdash \phi \rightarrow \psi$, then $\vdash \psi$;

A1 $\phi_1 \rightarrow \phi_2$, whenever ϕ_2 is a safe reference-abstraction of ϕ_1.

We then consider a system, called S_2, which strengthens the deductive power of S_1 in two senses. First, it becomes possible to transform certain combinations of modal operators into a single modal operator. The idea is the following. Suppose that there is a stipulation between two lists of normative parties σ_1 and σ_2 that ϕ has to be brought about by σ_3 to the advantage/disadvantage of σ_4. Furthermore, suppose that there is a finite chain of other stipulations which step-by-step transfer the commitment to bring about ϕ to new parties, so that the chain ends with a commitment of a list of parties σ_m to bring about ϕ to the advantage/disadvantage of a list of parties σ_{m+1}. In similar cases one could argue that a new stipulation emerges from the combination of all those forming the chain: it is a stipulation between σ_1 and σ_2 that σ_m brings about ϕ to the advantage/disadvantage of σ_{m+1}. Thus, according to this view, we can have a domino effect on stipulations.[10] Second, in S_2 it is possible to replace logically equivalent formulas in the scope of a modal operator.[11]

Definition 5 (System S_2) *System S_2 is defined by the set of deductive principles $\{A0, R0, A1, A2, R2\}$, where:*

A2 $\left(\Delta^{(\sigma_{1.1} \Rightarrow \sigma_{1.2})}_{(\sigma_{1.3} \Rightarrow \sigma_{1.4})} \phi \land \ldots \land \Delta^{(\sigma_{n.1} \Rightarrow \sigma_{n.2})}_{(\sigma_{n.3} \Rightarrow \sigma_{n.4})} \phi \right) \to \Delta^{(\sigma_{1.1} \Rightarrow \sigma_{1.2})}_{(\sigma_{n.3} \Rightarrow \sigma_{n.4})} \phi$,

provided that, for $1 \leq i \leq (n-1)$, $\sigma_{i.3} = \sigma_{(i+1).1}$ and $\sigma_{i.4} = \sigma_{(i+1).2}$;

R2 *if* $\vdash \phi \equiv \psi$, *then* $\vdash \Delta^{(\sigma_1 \Rightarrow \sigma_2)}_{(\sigma_3 * \sigma_4)} \phi \equiv \Delta^{(\sigma_1 \Rightarrow \sigma_2)}_{(\sigma_3 * \sigma_4)} \psi$.

The last system we will describe, called S_3, is the smallest regular system for operators of type $\Delta^{(\sigma_1 \Rightarrow \sigma_2)}_{(\sigma_3 \Rightarrow \sigma_4)}$ that extends S_2.[12]

Definition 6 (System S_3) *System S_3 is defined by the set of deductive principles $\{A0, R0, A1, A2, R2, A3\}$, where:*

A3 $(\Delta^{(\sigma_1 \Rightarrow \sigma_2)}_{(\sigma_3 \Rightarrow \sigma_4)} \phi \land \Delta^{(\sigma_1 \Rightarrow \sigma_2)}_{(\sigma_3 \Rightarrow \sigma_4)} \psi) \equiv \Delta^{(\sigma_1 \Rightarrow \sigma_2)}_{(\sigma_3 \Rightarrow \sigma_4)} (\phi \land \psi)$.

We finally observe that the principle $\Delta^{(\sigma_1 \Rightarrow \sigma_2)}_{(\sigma_3 * \sigma_4)} \top$, which would be needed to get normal modal systems, does not seem to have a plausible reading in this context (and this is especially true for those instances in which $\not\Rightarrow$ is present). Therefore, one should carefully check which instances of such a schema could be added as axioms in order to have *at least some* normal operators in extensions of S_3 (e.g., the operator in which all lists of normative parties are empty). This aspect will be investigated in future work.

---

[10]We would like to stress the difference between the domino effect on a chain of distinct stipulations conveyed by A2 and the simplification conveyed by the principle $\Delta^{(\sigma_{1.1} \Rightarrow \sigma_{1.2})}_{(\sigma_{1.3} \Rightarrow \sigma_{1.4})} \ldots \Delta^{(\sigma_{n.1} \Rightarrow \sigma_{n.2})}_{(\sigma_{n.3} \Rightarrow \sigma_{n.4})} \phi \to \Delta^{(\sigma_{1.1} \Rightarrow \sigma_{1.2})}_{(\sigma_{n.3} \Rightarrow \sigma_{n.4})} \phi$, which ultimately concerns a *single* stipulation.

[11]The combination of these two properties in a single system is just an option for the sake of illustration: in principle, they are reciprocally independent.

[12]For axiomatizations of regular systems of modal logic, see, e.g., Segerberg (1971).

Subject Position and Object Position

5 Semantic characterization

For the semantic analysis of systems built over $\mathcal{L}$ we will rely on a very simple approach that can be extended in a modular way. The idea is to use a set of possible worlds and to specify, for each world w, a set of formulas of $\mathcal{L}^\Delta$ that constitute the norms in effect at w. This set will satisfy certain properties, depending on the logical system considered.

Definition 7 (Frames) *The language $\mathcal{L}$ is interpreted on normative frames, i.e., structures of kind $\mathfrak{F} = \langle W, N, \mathfrak{g} \rangle$, where:*

- W *is a set of possible worlds (or states), denoted by w_1, w_2, w_3, etc.;*

- $N = \{n(\phi) : \phi \in \mathcal{L}^\Delta\}$ *is a set of possible norms;*

- $\mathfrak{g} : W \longrightarrow \wp(N)$ *is a function determining the set of norms that are in effect at each possible world.*

According to this definition, N contains a norm for each proposition in $\mathcal{L}^\Delta$. However, it is in principle possible to have $n(\phi) = n(\psi)$ for two distinct propositions $\phi, \psi \in \mathcal{L}^\Delta$. In a trivial case, N is a singleton (i.e., all $\mathcal{L}^\Delta$-formulas express the same norm).

Definition 8 (Models) *A model over a normative frame is a structure of kind $\mathfrak{M} = \langle \mathfrak{F}, V \rangle$, where:*

- $\mathfrak{F}$ *is the underlying frame;*

- $V : \text{PRO} \longrightarrow \wp(W)$ *is a valuation function.*

Definition 9 (Truth-conditions) *The truth of a formula with reference to a state w in a model $\mathfrak{M}$ is defined below, where $p \in \text{PRO}$:*

- $\mathfrak{M}, w \vDash p$ *iff* $w \in V(p)$;

- $\mathfrak{M}, w \vDash \neg \phi$ *iff* $\mathfrak{M}, w \nvDash \phi$;

- $\mathfrak{M}, w \vDash \phi \to \psi$ *iff either* $\mathfrak{M}, w \nvDash \phi$ *or* $\mathfrak{M}, w \vDash \psi$;

- $\mathfrak{M}, w \vDash \Delta^{(\sigma_1 \Rightarrow \sigma_2)}_{(\sigma_3 * \sigma_4)} \phi$ *iff* $n(\Delta^{(\sigma_1 \Rightarrow \sigma_2)}_{(\sigma_3 * \sigma_4)} \phi) \in \mathfrak{g}(w)$;

Definition 10 (Validity) *A formula ϕ is valid in a model $\mathfrak{M}$ iff $\mathfrak{M}, w \vDash \phi$ for every $w \in W$; ϕ is valid in a frame $\mathfrak{F}$ iff it is valid in all models over $\mathfrak{F}$; finally, ϕ is valid in a class of models/frames iff it is valid in all models/frames of the class.*

Notice that, as a consequence of Definition 9 and Definition 10, in the case of a formula $\phi \in \mathcal{L}^\Delta$, we have that ϕ is valid in a model $\mathfrak{M}$ iff ϕ is valid in the frame of $\mathfrak{M}$, whereas this does not hold, in general, for formulas of $\mathcal{L}$. We describe below properties of frames to interpret systems S_1, S_2 and S_3. The class of all normative frames can be used to interpret a mere linguistic extension of **PC**.

Definition 11 (S_1-frame) *An S_1-frame satisfies the following property, for every $w \in W$:*

P1   if $n(\phi_1) \in \mathfrak{g}(w)$ and ϕ_2 is a safe *reference-abstraction of ϕ_1, then $n(\phi_2) \in \mathfrak{g}(w)$.*

The validity of a formula ϕ in the class of S_1-frames will be denoted as $\models_{S_1} \phi$. An analogous notation will be used for the other classes of frames below.

Definition 12 (S_2-frame) *An S_2-frame satisfies P1 and the following properties, for every $w \in W$ and every modal operator $\Delta_{(\sigma_3 * \sigma_4)}^{(\sigma_1 \Rightarrow \sigma_2)}$:*

P2.1   if $n(\Delta_{(\sigma_{1.3} \Rightarrow \sigma_{1.4})}^{(\sigma_{1.1} \Rightarrow \sigma_{1.2})}\phi), \ldots, n(\Delta_{(\sigma_{n.3} \Rightarrow \sigma_{n.4})}^{(\sigma_{n.1} \Rightarrow \sigma_{n.2})}\phi) \in \mathfrak{g}(w)$ and, for $1 \leq i \leq (n-1)$, $\sigma_{i.3} = \sigma_{(i+1).1}$ and $\sigma_{i.4} = \sigma_{(i+1).2}$, then $n(\Delta_{(\sigma_{n.3} \Rightarrow \sigma_{n.4})}^{(\sigma_{1.1} \Rightarrow \sigma_{1.2})}\phi) \in \mathfrak{g}(w)$;

P2.2   if $n(\Delta_{(\sigma_3 * \sigma_4)}^{(\sigma_1 \Rightarrow \sigma_2)}\phi) \in \mathfrak{g}(w)$ and $\models_{S_2} \phi \equiv \psi$, then $n(\Delta_{(\sigma_3 * \sigma_4)}^{(\sigma_1 \Rightarrow \sigma_2)}\psi) = n(\Delta_{(\sigma_3 * \sigma_4)}^{(\sigma_1 \Rightarrow \sigma_2)}\phi)$.

Definition 13 (S_3-frame) *An S_3-frame satisfies P1, P2.1, P2.2 and the following property, for every $w \in W$ and every modal operator $\Delta_{(\sigma_3 \Rightarrow \sigma_4)}^{(\sigma_1 \Rightarrow \sigma_2)}$:*

P3 $n(\Delta_{(\sigma_3 \Rightarrow \sigma_4)}^{(\sigma_1 \Rightarrow \sigma_2)}\phi), n(\Delta_{(\sigma_3 \Rightarrow \sigma_4)}^{(\sigma_1 \Rightarrow \sigma_2)}\psi) \in \mathfrak{g}(w)$ iff $n(\Delta_{(\sigma_3 \Rightarrow \sigma_4)}^{(\sigma_1 \Rightarrow \sigma_2)}(\phi \wedge \psi)) \in \mathfrak{g}(w)$.

We sketch below a proof of the semantic characterization of S_1 in terms of the class of S_1-frames. The symbol $\vdash_{S_1}$ will denote derivability in this system.

Theorem 1 (Soundness of S_1)   *If $\vdash_{S_1} \phi$, then $\models_{S_1} \phi$.*

Proof. An induction on the length of derivations. The validity of *A0* and the fact that *R0* preserves validity are straightforward consequences of the definition of a model over an S_1-frame. In the case of *A1*, the result follows from the fact that, if ϕ_2 is a *safe* reference-abstraction of ϕ_1 and, for some $w \in W$ and model $\mathfrak{M}$, we have $\mathfrak{M}, w \vDash \phi_1$ and $\mathfrak{M}, w \nvDash \phi_2$, then we must have, due to the truth-conditions associated with modal operators, $n(\phi_1) \in \mathfrak{g}(w)$ and $n(\phi_2) \notin \mathfrak{g}(w)$, which contradicts *P1*. □

Theorem 2 (Completeness of S_1) *If* $\vDash_{S_1} \phi$*, then* $\vdash_{S_1} \phi$.

Proof. We can rely on the method of canonical models. The peculiar aspect of the canonical model for our systems is the definition of the set N. Let $\asymp$ be the (metalinguistic) relation of provable equivalence among formulas in $\mathcal{L}^\Delta$, that is: for any $\phi, \psi \in \mathcal{L}^\Delta$, $\phi \asymp \psi$ iff $\vdash_{S_1} \phi \equiv \psi$. Clearly, $\asymp$ is an equivalence relation, and we can take $[\phi]_\asymp$ to be the equivalence class of a formula ϕ. Then, $N = \{\underline{n}([\phi]_\asymp) : \phi \in \mathcal{L}^\Delta\}$. In the case of S_1 each equivalence class is a singleton, while it is an infinite class already in S_2. For a given state (i.e., maximal S_1-consistent set of formulas) w, let $\mathfrak{g}(w) = \{\underline{n}([\psi]_\asymp) : \phi \in w\}$. Moreover, let $n(\phi) = \underline{n}([\phi]_\asymp)$, for any $\phi \in \mathcal{L}^\Delta$. In order to prove that the canonical model satisfies property *P1*, it is sufficient to rely on the standard truth-lemma (where, for any $\phi \in \mathcal{L}^\Delta$, we have $\mathfrak{M}, w \vDash \phi$ iff $\underline{n}([\phi]_\asymp) \in \mathfrak{g}(w)$) and observe that maximal S_1-consistent sets of formulas are closed under *A1*. □

Theorem 3 (Semantic characterization of S_2 and S_3) $\vDash_{S_i} \phi$ *iff* $\vdash_{S_i} \phi$*, for* $i \in \{2, 3\}$.

Proof. The soundness part can be easily adapted. In the completeness part, the canonical model for S_2 satisfies *P2.1* and *P2.2*. due to the fact that every maximal S_2-consistent set of formulas contains all instances of axiom *A2* and is closed under the rule *R2*. Furthermore, the canonical model for S_3 satisfies *P3* due to the fact that every maximal S_3-consistent set of formulas contains all instances of *A3*. □

We finally observe that frames for S_2 can be transformed into *neighborhood frames* and frames for S_3 can be transformed into *relational frames* with a set of 'queer'-worlds. Similar transformation procedures are described by Segerberg (1971). In future work we plan to provide translational embeddings of other formal frameworks into ours and to unveil the whole range of possible relations that can be expressed in terms of the Δ-operators; we will focus on those relations that are relevant for legal reasoning. Additionally, we plan to develop a first-order extension of the present framework with a more sophisticated semantics that is expected to shed further light on the entities that can play a role in subject and in object position. For instance, one could associate distinct categories of entities to parties in the two positions, in order to represent distinct types of stipulations, as the two Groups mentioned in Section 3 suggest.

Contributions

The contents of this work are the result of a joint research of the two authors. Sections 1 and 2 were jointly written. Section 3 is mainly due to Tereza Novotná. Sections 4 and 5 are mainly due to Matteo Pascucci.

References

Duarte d'Almeida, L. (2016). Fundamental legal concepts. The Hohfeldian framework. *Philosophy Compass, 11,* 554–569.
Hansson, B. (1970). Deontic logic and different levels of generality. *Theoria, 36,* 241–248.
Herrestad, H., & Krogh, C. (1995). Obligations directed from bearers to counterparties. In *Proceedings of ICAIL 1995* (pp. 210–218).
Hohfeld, W. N. (1913). Some fundamental legal conceptions as applied in legal reasoning. *Yale Law Journal, 23,* 16–59.
Hohfeld, W. N. (1917). Fundamental legal conceptions as applied in judicial reasoning. *Yale Law Journal, 26,* 710–770.
Kanger, S. (1970). New foundations of ethical theory. In R. Hilpinen (Ed.), *Deontic Logic: Introductory and Systematic Readings* (pp. 36–58). Springer.
Knapp, V. (1995). *Teorie Práva*. Praha: C.H. Beck.
Lindahl, L. (1977). *Position and Change. A Study in Law and Logic*. Dordrecht: Springer.
Lindahl, L. (2001). Stig Kanger's theory of rights. In G. Holmström-Hintikka, S. Lindström, & R. Sliwinski (Eds.), *Collected Papers of Stig Kanger with Essays on His Life and Work* (pp. 151–171). Springer.
Lyons, D. (1970). The correlativity of rights and duties. *Noûs, 4,* 45–55.

Makinson, D. (1986). On the formal representation of rights relations: Remarks on the work of Stig Kanger and Lars Lindahl. *Journal of Philosophical Logic, 15*, 403–425.

Markovich, R. (2019). Understanding Hohfeld and formalizing legal rights: the Hohfeldian conceptions and their conditional consequences. *Studia Logica, 108*, 129–158.

Segerberg, K. (1971). *An Essay in Classical Modal Logic*. Uppsala: Filosofiska Studier.

Sergot, M. (2013). Normative positions. In D. Gabbay, J. Horty, X. Parent, R. van der Meyden, & L. van der Torre (Eds.), *Handbook of Deontic Logic and Normative Systems* (pp. 353–406). College Publications.

Sileno, G. (2016). *Aligning law and action. A conceptual and computational inquiry*. PhD Dissertation. University of Amsterdam.

Tereza Novotná
Masaryk University, Faculty of Law, Institute of Law and Technology
The Czech Republic
E-mail: tereza.novotna@mail.muni.cz

Matteo Pascucci
Slovak Academy of Sciences, Institute of Philosophy
The Slovak Republic
E-mail: matteopascucci.academia@gmail.com

Logic and Human Practices

JAROSLAV PEREGRIN[1]

Abstract: What, outside of our logical theories, makes us believe that the theories are reliable, and what is it that warrants them? What I propose is that it is just our argumentative practices; that logic is a theory of the practices in a sense similar to (though not the same as) that in which physics is a theory of the antics of spatio-temporal objects. Critics object that this approach would degrade logic to something on the level of etiquette, insisting that the laws of logic are absolute and hence independent of any parochial human practices. This paper argues that once we understand the true nature of our *practices* (such as that of argumentation or drawing inferences), our suggestion becomes feasible. What we must understand is that the practices consist not only of moves (like giving reasons or drawing inferences), but also of consonant assessments of (or the assuming of "normative attitudes" toward) such moves.

Keywords: normativity of logic, practices, normative attitudes, reliability of logic

1 What are logical theories about?

Pursuing logic, we produce various theories; and it seems that in this enterprise we have been quite successful. To date, our logical theories are plentiful; we have revealed various logical laws (such as, e.g., *modus ponens*, or *disjunctive syllogism*) and we know a lot about deduction, proofs, models, etc. Not everything our theories state (and especially not all the laws stated by them) is universally accepted and some theories are hotly debated, but nevertheless we usually do not doubt that our logical pursuit has substance.

One of the main tasks of logic is to help us ensure the reliability of our theories concerning the world around us; yet how can we be sure that the theories of logic themselves (and the laws they pinpoint) are reliable? This is a question of what Schechter (2013, 2018) aptly calls the "reliability challenge": what, outside of our logical theories, makes us believe that these

---

[1] Work on this paper was supported by the grant No. 20-18675S of the Czech Science Foundation.

theories are reliable, and how do we confront them with it? When we produce theories of nature, we also cannot be always sure that we are not mistaken and that the theories are reliable, but we know how to confront them (and the putative laws of nature incorporated in them) with reality, and thereby confirm or disconfirm them. Thus, if our theories of the world are misguided, we are likely to notice a discrepancy which will reveal our mistake. Do we have some similar "reality check" for the theories of logic?

We may think that as logic cannot just "float free in the void", there *must* be an anchor and hence a "reality check"—and if it cannot be found in the real world, it must be sought somewhere beyond it, perhaps in the transcendental depths of our minds, or in a supernatural realm accessible only via some peculiar ability of our minds (such as the *intellectus*, which, as Aquinas put it, "reads inside" things[2]). Just as what we see in the real world may confirm/disconfirm our physical theories, so our logical theories may be confirmed/disconfirmed by what we see, though now not in the real world, but rather in some peculiar world which we are able to see with our "inner eyes".

Schechter (2013) himself concludes that we need an *objective* domain with which the laws of logic are to be confronted. The objectivity, according to him, manifests itself in certain principles, involving the claim that "the truth of logical truths and the falsity of logical falsehoods do not depend ... on our thoughts, language, or social practices" (pp. 214–5). Hence we are to pursue a domain which is independent of what we, reasoners and players of the game of giving and asking for reasons, do, and which thus grounds logical truths in an absolute sense.

This picture is seductive and difficult to subvert (though it is often disregarded that it is equally difficult to underpin). But we must remember that logic is also closely connected to certain forms of our discursive practices, to overt reasoning and argumentation. Hence, cannot these mundane phenomena provide a reality check for logic less esoteric than the "facts" to be found beyond the real world?[3] I am convinced they can, and hence that the theories taking the reality check for logic to consist in a confrontation with some unworldly reality are unnecessary;[4] we can make do with a thoroughly naturalistic explanation.

---

[2] Aquinas (1882, VI lect. 5 n. 1179).

[3] This would help us classify logic as one of the ordinary sciences and vindicate its "non-exceptionality" in the sense of Hjortland (2017).

[4] This view of the subject matter of logic reveals some surprising parallels between logic and natural sciences (see Peregrin, 2019).

What I think stands in the way of this is our frequent misunderstanding of the nature of these human practices. What we must appreciate is that our discursive practices are essentially *rule-governed* in the sense that they *incorporate* their rules; and that what can serve as the reality check for us are not the data detailing which moves people, as a matter of fact, carry out, but rather which moves they *hold for correct*. In the case of the inferential practices targeted by logic our reality check is then constituted not by the inferences people, as a matter of fact, draw, but by the inferences they take to be correct.

2   The normativity of practices

That there are senses in which some human practices, such as reasoning or argumentation, are "rule-governed" is assumed to go almost without saying, at least since the "rule-following discussion" fanned especially by Kripke (1982).[5] The problem, however, is that the senses of this rule-governedness are numerous and some of them are quite esoteric as the conceptions of logic questioned above. In contrast to this, I want to present a very down-to-earth and transparent account of the rule-governedness of our practices, which, I am convinced, we must take into account to understand what our theories of logic must rest on.

The rule-governedness I have in mind consists in the fact that such practices constitutively involve certain consonant assessments, by their practitioners, of the actions which constitute their substrate. Thus the practice of drawing inferences[6] consists not only of the moves from premises to conclusions, but rather also from the ever-present attitude of taking some of such moves for correct and others for incorrect—just like the practice of playing chess consists not only of moving pieces across the chessboard, but also of holding the ever-present attitudes of taking some of such moves for correct and others for incorrect.

The fact that the assessments, which we may call, borrowing a term from Brandom (1994), *normative attitudes*[7], are consonant, i.e., that different people tend to take the same kinds of moves for correct, may be seen to constitute an implicit (social) rule. As Wittgenstein taught us, not all rules

---

[5] See McDowell (1984); Goldfarb (1985); Boghossian (1989); Haugeland (2000).

[6] Aka giving reasons, as part of the public practice of argumentation.

[7] Brandom presented an extensive theory of normative attitudes, but he was not the one to coin the term. It is used, for example, by Hart (1961).

can be explicit, the explicit ones must ultimately rest on implicit ones;[8] and the situation when the assessment of correctness by individual members of a society comes to resonate with most of the other members provides for the existence of precisely such an implicit rule, at least in a rudimentary form. Hence the most basic level of rule-governedness of human practices consists precisely in the presence of such consonant assessments.[9]

This is to say that the rules which govern practices such as argumentation are not simply in the eye of a beholder, but are part of the practices themselves (though often just an implicit part). Thus, they are not merely items of the toolbox of the theoretician, they are already part of the subject matter of their theories. Our inferential practices consist not only of drawing inferences, but also of evaluating the correctness of others' (and one's own) inferences. (These two components may come apart—we may realize that an inference we have drawn was not correct.) More generally, distinctively human kinds of practice encompass rules as their integral parts. These practices consist not only of "doing things", but also of monitoring and regulating how the things are done, i.e., also of "doings targeting the first-order doings".

3 Absorbing "the meta"

To elucidate this peculiar feature of human practices, let us look at them from a different angle. Imagine a behavioral pattern displayed by some animals; e.g., a flock of hens rushing out of a henhouse looking for food. From our viewpoint (though, presumably, not from the hens') there is a "metalevel" to this behavior. On the "metalevel" *we* (though not *they*) can describe their behavior, we can take it for "correct" (measured by our aims)

---

[8] See Brandom (1994, §I.2.4); see also Peregrin (2014, §4.1).

[9] It is worth noting that this, as documented by the increasing number of reports of the scientists targeting human ontogeny, is not just a philosophical speculation. The fact that "a person establishes a social reference group [who] evaluate and demonstrate approval or disapproval, even if the behavior in question does not affect them directly" (Castro, Castro-Nogueira, Castro-Nogueira, & Toro, 2010, p. 353) is becoming a common observation. As Schmidt and Rakoczy (2019) summarize their long-term research, "young children develop normative attitudes toward a variety of different acts in different contexts. They enforce social norms as unaffected third parties, suggesting that they take an impersonal perspective regarding norms and understand something about the normative force and generality of norms. ... Hence, early in ontogeny human beings start developing into normative beings and care about upholding shared standards, which suggests some attachment to their cultural group beyond strategic motives". Findings concerning the crucial role of normativity for human ontogeny are reported also by other empirical studies (but to analyze this in detail is a topic for a different paper.)

or "incorrect", and we can attempt to regulate it. (We can open the doors of the henhouse at certain hours, prepare food for the hens at certain places, etc.) The existence of the "metalevel" is given by the fact that we humans can assume certain attitudes towards the behavior of the hens, we can report it or try to influence it.

Of course, the same happens if the animals displaying the basic level pattern are us, humans. However, in this case it can happen that those who display the attitudes on the "metalevel", who display a "second-order" behavior targeting the "first-order" one, are *the same* humans as those who display the "first-order" behavior. In this case the whole pattern, consisting of the two levels, becomes what we can call *self-reflective*. And the thesis which I defend here—and the appreciation of which I am convinced is crucial for understanding both reasoning and our theories thereof, including logic—is that human practices are characterized by being self-reflective in this sense.

In other words, while any behavior of animals (or, for that matter, "behavior" of inanimate things) can be described on a meta-level and regulated from without, what we call human practices already *incorporates* the meta-level, they are regulated from within (hence: self-regulated).[10] To become a competent practitioner of the human language games, *viz.* a speaker of language, for instance, the speaker, apart from becoming able to produce appropriate "languagings", must also, as Sellars (1974, p. 424) puts it, "acquire the ability to language about languagings, to criticize languagings, including his own". Similarly, Brandom (2000, p. 20ff.) stresses that it is the self-reflective quality of our human conceptual activities that enables us to put the rules that regulate them into words, and thereby become "semantically self-conscious".

4 What is a practice?

In her attempt to solve the challenge posed by Kripke (1982), Ginsborg (2011) invokes what she calls "primitive normativity", which amounts to the fact that we hold some acts for correct as a matter of primitive fact, and not as a result of any conscious application of a rule. "Thus," she writes, "your disposition is not just to say '125' in answer to '68 plus 57,' '126' in answer to '68 plus 58,' and so on; it is also, in each case, to take what you

---

[10]The term "self-regulation" is used, e.g., by Tomasello (2019). An interesting problem, which, however, goes beyond the scope of the present paper, would be to research a possible connection between this kind of self-regulation and what some authors call the "self-domestication" of *Homo sapiens* (Wrangham, 2019).

are saying to be the appropriate response to the query. You are disposed not only to respond with a number which is in fact the sum, but to consider that particular response appropriate."

I think that the "primitive normativity" Ginsborg is urging here is the kind manifested by the "normative attitudes" urged above. When we carry out the additions, there are *two* things in play: not only the disposition to produce the results, but also the disposition to take the results as adequate or correct. It is crucial to stress that the normative attitudes' being "primitive" involves their being nothing like propositional attitudes, and not being based on an appreciation of rules or meanings. On the contrary, these attitudes *underlie* all rules in general, and the linguistic rules that underlie meanings in particular. It is a feature of us humans that we have developed these idiosyncratic behavioral patterns (which I tend to call practices) involving these pro- and con- attitudes to the behavior.

Hence in this sense, Ginsborg's proposal is wholly in the spirit of our approach. But there is an important difference: it renders the responses of a person correct or incorrect because of the existence of the corresponding normative attitudes of the *very same person*. This, I think, stems from the conviction of the author that counting is primarily an individual, mental activity. The same would hold, according to many authors, about reasoning. But this assumption, I am convinced, is mistaken, and it blocks us from arriving at an adequate understanding of human practices.

The point is that practices such as counting and reasoning cannot evolve as purely private, because a public dimension is in their very essence. This is not to say that an individual, independently of a society, cannot have evolved some technique of classifying groups of objects according to their numerosity, or a technique to estimate what will be the case if something else is the case; but it is to say that to make this into fully-fledged counting or reasoning the individual needs a society, because it is only within the context of a society that the practices can acquire the normative dimension which qualifies them as being the distinctively human ("self-conscious") practices.

It may seem strange that something as essentially mental as reasoning would have evolved not in the mind but in the arena of the intersection of many minds; however, current research is bearing this out. Most forcefully it is put forward by Mercier and Sperber (2011, 2017), who argue, I think rightly, that private reasoning is secondary to public argumentation, rather than the other way around. This is of a piece with the recent trend to see the human mind as much more a social product than used to be usual (see, e.g., Tomasello, 2014).

Hence I think that to understand the practice of reasoning, which is the subject matter of our logical theories, we must accept that it involves normative attitudes, but not merely the normative attitudes of a subject to her own inferences (as Ginsborg insists). Rather, we must accept that the very practice, along with so many of other distinctively human practices, presupposes assuming such normative attitudes *to each other*. It is this kind of mutual assessment that constitutes the practices as such and gives them their essence.

From this viewpoint, the conception of practices presented here is close to that of Rouse (2007), according to which "a performance belongs to a practice if it is appropriate to hold it accountable as a correct or incorrect performance of that practice." This, I am convinced, is the key; and of course it follows that each of the participants also assesses their own performances as correct/incorrect, as assumed by Ginsborg. Thus the essence of practices of this kind consists in what Rouse calls "the mutual accountability of their constitutive performances".

5 Wright on inferential practices

I think that the full appreciation of the nature of human practices lets us resolve some vexing problems concerning the nature of logical theories, especially concerning their descriptive versus normative character. Let us consider how the question concerning the relationship between our logical theories and our argumentative practices was posed by Wright (2018). He asks: "What is the relationship between our basic inferential competences and logic as an explicit scientific-theoretical subject?" And his answers runs as follows:

> There is a possible, perfectly reputable scientific project which would consist in the attempt to codify and systematize our actual deductive inferential habits. This would be an empirical sociological project. It would stand comparison with empirical linguistics or the attempt to write up the rules of Chess, say, in a scenario where the game continued to be widely played in a community—perhaps among the descendants of a small number of survivors after a nuclear holocaust—but where no explicit statement of the rules and object of the game had survived. But to think of logical theory on that model ignores the point that logic, as usually conceived, is a normative science. Its project is

> not, or not merely, the systematic general description of actual inferential practices but the development of theory that is apt for the evaluation of those practices, a theory at least part of whose brief is to constrain our judgements about what follows from what, about which are good inferences and which are bad, and why. (pp. 426–7)

The picture Wright presents to reject it, the picture of logic as "codifying and systematizing our actual deductive inferential habits" is quite similar to the one we are proposing and defending here. Hence can our proposal, stating that what makes logical theories correct/incorrect are our argumentative practices, be defended against Wright's criticism? Can we contravene the objection that it rids logic of its normativity, which is its *sine qua non*?

I think that in the light of what has been presented above the criticism can be shown to miss its point. The first thing to note is that the project as Wright describes it, the project "of codifying and systematizing of our actual deductive inferential habits" (just like the project of codifying post-nuclear chess) *does* have a "normative dimension". The point is that our inferential practice (just like chess and many other things we humans do) is rule-governed in the sense discussed above; and the core of "codifying and systematizing it" is capturing its *rules*.

This is quite obvious in the case of chess (be it "post-nuclear" or whatever). There is no chess without rules. It is not that without rules it would be an incomplete or impoverished or rudimentary chess—moving pieces of wood over a chessboard in the utter absence of rules would have nothing to do with chess at all. Note that this is not to say that we need *explicit* rules—what we need is that moves are consonantly assessed as correct or incorrect. And what I claim is that in this respect, our deducing and inferencing, our practices of reasoning and argumentation, are like chess: they are "rule-governed games", not necessarily in the sense that they would be governed by *explicit* rules, but in that their integral part are ongoing consonant assessments of their moves as right or wrong. No moves from thoughts to thoughts, or from sentences to sentences, would be derivations or inferences if they could not be carried out correctly or incorrectly. And no comprehensive "codifying and systematizing" can leave out this part of the practice.

True, the fact that rules are already a part of the *subject matter* of logic (rather than a merely a matter of its outcomes), does not yet make it normative. Admittedly, it might be that it yields us a pure *description* of the rules, disengaged statements that certain communities, as a matter of fact, accept/uphold/follow certain rules. But this is not what logic, as it is standardly pursued, does: logic aims at rules that are binding (also) *for us*. Therefore, logical theories are not construed as merely descriptions of rules, they are construed as their "explicitations"—they do not only state that the rules hold for somebody, they present, in an explicit form, rules which are to be followed.

Wright stresses that logical theory should be "apt for the evaluation of those practices" (p. 427). In one sense, then, the descriptive enterprise already is—in so far as what it captures are rules, it can be seen as an explicit articulation of the means of such evaluation, of criticism of individual inferential moves people make—it can classify them as correct or incorrect in the sense of respecting or violating the rules.

To be sure, there is a secondary level of normativity, the level which allows us to criticize not only the inferential moves of the practitioners, but to a certain extent even their rules. This happens when we identify the rules implicit to their practices, make them explicit, streamline and polish them (usually during a process of zooming in on a reflective equilibrium; see Peregrin & Svoboda, 2017) and use the result as an explicit norm which allows us to correct and rectify individual normative attitudes and hence the implicit rules. Giving the project this, second, normative dimension makes logic into something ultimately more than purely a descriptive project.

6 Logic and etiquette

But there is one more objection that follows from Wright's considerations, the objection that logic understood in the way we propose degrades logic to something like etiquette. Again, as Wright puts it, commenting on such a view: "If there is normativity involved, it is a normativity broadly comparable to that of rules of etiquette. 'That's not how it is *done.*' It is possible, but intellectually hugely unattractive, to take such a view of logic. The normativity of logic, we think, is an altogether more substantial matter..." (p. 427). Is it? There is no doubt that there are serious differences between logic and etiquette and that there is a sense in which logic is much more important than etiquette. However, does the difference consist in the baseline normativity involved?

Compare the rule that we should greet each other in the morning with the rule of *modus ponens* (hereafter MP). It seems that whereas violating the former rule just means breaking with some local customs, in the latter case a violation is much more substantial: it has nothing to do with any local habits, it seems to be breaking with something indisputably objective and crucially important, perhaps rationality. In short, rules of logic, unlike rules of etiquette, seem to be *absolute*—and nothing short of the absoluteness seems to be able to assume their role. But I think this is disputable.

What exactly is a rule like MP? It tells us that we may derive a consequent of a conditional from the conditional itself plus its antecedent. This much is quite clear; but what, exactly, is a conditional? A conditional, in a typical case, is a thought, a proposition or a sentence, consisting of an antecedent, a connective that we can call *implication*, and a consequent (where the antecedent and the consequent are of the same kind as the whole conditional, i.e., a thought, a proposition or a sentence). But what makes a connective into an implication? How do we identify it? Not, it would seem, by its look: in the case of linguistic expressions, we know their look cannot tell us anything about their meaning nor of their functioning; and in the case of a part of a proposition or a thought, it is not even clear what it would mean to talk about their "look".

Therefore, it would seem that the only way to identify an implication is in terms of its *function*, on the basis of what it *does*. How can we specify its function? Hardly without mentioning MP or something very close to it. For example, if we characterize implication in terms of the usual truth table, then we say, *inter alia*, that a conditional is false if its antecedent is true and its consequent is false; i.e., that if the conditional and its antecedent are true, its consequent is bound to be true too.

From this viewpoint, MP would seem to say that if we connect two thoughts or propositions by a connective that produces MP-obeying conditionals, then the result will obey MP. And this triviality, of course, cannot be the important law of logic we all cherish! So if we do not want to accept that we are all under a mere illusion that there is such a law as MP, we must construe it in a different way. Since identifying implication in a functional way appears to render MP trivial, we must present it as something identified not in terms of its function. But it seems that if we consider something as "Implication" (note the capital) as an abstract object, then it will be *incurably functional*—the only thing all items classified as implications have in common is their function, so the corresponding abstract cannot but just consist of the function. And the claim that *this kind of entity* has the function of implication is thus bound to be trivial.[11]

---

[11] See Peregrin (2010).

What, however, may be a non-trivial fact is that a specific item, like the horseshoe of classical logic, or the *if-then* of English, does, as a matter of fact, function like implication, i.e., that it, *inter alia*, obeys MP. And if MP is construed in this way, the barrier between the rules of logic and those of etiquette breaks down. In both cases, we have 'That's not how it is *done*' ('That's not how *if-then* is used in English.')

Let me stress, once more, that this is not to say that there is no substantial difference between etiquette and logic. The roles of the two enterprises within our coping with the world may be very different, and likewise their levels of importance for us. The point is only that the difference is not a matter of the former enterprise being local and human-made, while the latter is global and human-independent. Once we realize that MP may have non-trivial content only with a *specific* sign in place of implication, we can see that it too is bound to be local and human-made. It is a rule for the usage of a cultural tool.

One way to describe the situation that MP is, essentially, a prescription for handling an item (which we call implication, like the horseshoe or *if-then*), is to compare it with the rule of chess stipulating that the bishop moves only diagonally. This is also prescribing us how to handle an item (a bishop). And just as the chess rule is nontrivial because it does not tell us merely that an item that obeys it, i.e., one that moves only diagonally, moves only diagonally, but rather that it is a certain specific item (perhaps a piece of wood at which I point) that moves only diagonally, so MP tells us that a certain specific item (the horseshoe, *if-then*, ...) behaves in a certain way.

But does not MP come out of these considerations as all too cheap? It does not make us do anything, it only lets us make an item obey it. Is this not a far cry from the absoluteness that logical rules are to display—from "the hardness of the logical *must*" (as Wittgenstein (1956) would put it)? We must realize that MP as well as other logical rules are *constitutive*; and their importance does not consist in the fact that it would show our thinking a definite direction—rather it provides us with certain (extremely important) vessels that can take our thinking into spheres which were hitherto inaccessible.

MP *does* make us do something, it makes us handle an item in a specific way—but only after the item is subordinated to the corresponding rules—after it is "constituted" as an item governed by the rules.

7 Rule of logic as constitutive

So the picture to which we are converging, in outline form, is the following: we humans have developed certain "rule-governed" practices. These practices are characterized by having absorbed their "metalevel", on which they are assessed and deemed correct or incorrect. This is the result of the fact that within these practices we tend to assume normative attitudes to each other's doings. One of such practices is meaningful talk and its sub-practice is argumentation—the rules of this particular practice being what is studied by logic. No doubt all such practices have developed within the framework of evolution and, as everything in the biological world, they exist because they either proved to be adaptive, or to be piggybacking on something that is.[12]

The practices of argumentation make room for *justifying* those claims that can be involved in the game. It is clear that rules which govern the game are not something which can be subject to this kind of justification. Thus, primarily the rules are "justified" in that here they turned out to be useful in the course of evolution, they are not justified in the sense that there would be *reasons* for them.

This brings us to the important point stressed above. Rules of language in general, and those of logic in particular, cannot be seen as instrumental rules, which direct us how to use concepts. Hence they are *not* like instructions how to employ a spear to kill a hare. They are more like *constitutive* rules that produce certain "cognitive gadgets"[13]. These rules constitute concepts like conjunction, negation or implication with which we can then reason. (Here any attempted continuation of the analogy with a spear would break down, for a spear cannot be produced by rules, but only by material workings.) There may be other rules instructing us how to use the logical concepts, but these are applicable only once we have already helped ourselves to the concepts, i.e., once the constitutive rules are in place.

Let us return to Schechter (2013). As we have already noted, he claims that "the truth of logical truths and the falsity of logical falsehoods do not depend ... on our thoughts, language, or social practices" (ibid.). But here it is extremely important to clarify what is meant by "depending on our thoughts, language, or social practices". Does the truth of *That walrus over*

---

[12] As for the question why we have developed them, there is an extensive literature on the evolutionary origins of language; and less extensive on those of argumentation. But see the works of Mercier and Sperber mentioned above.

[13] I borrow this term from Heyes (2018), who uses it in a slightly different, though not utterly unrelated context.

there is hairy depend on such things? Well, in a sense it surely does. It is true also because *walrus* means what it does in English, and what it means in English is a matter of the social practices of its speakers. But this is most probably not what Schechter wants to take into account—so perhaps what he means is "only on social facts" or "given the meanings are fixed". Well and good. What about *We tend to greet each other in the morning*? Obviously, the truth of *this* statement *does* depend on our social practices, in particular it is true because the practices are what they are. What about *We tend to infer B from A and ⌜If A then B⌝*? It depends on our practices in a similar way. But it seems to be irrelevant for logic, and in particular it is a far cry from MP. Now what about: *We tend to hold it for correct to infer B from A and ⌜If A then B⌝*? Again, it is true thanks to our linguistic practices, but is this relevant for logic?

If we hold fast to the absolutistic conception of logic, then it would seem that not, for our tendencies are irrelevant for what *really* holds. If we were to tend to infer *A* from *B* and ⌜*If A then B*⌝, this would not shatter the validity of MP. However, as we saw, MP is a directive for handling (an) *implication*, and our normative attitudes determine which expressions of our language (if any) are implications. An item of our language is an implication iff it obeys MP (or something very close to it) plus certain other rules. So insofar as truths of logic are to be found in natural languages, the facts about our linguistic behavior *are* relevant.

But is this not a *reductio ad absurdum* of the fact that we should see the logical truths as sentences of natural languages? Are these truths not something much more abstract? But here we face the problem we encountered above: "Implication", *qua* an abstract item, cannot but be a purely *functional* entity, and formulating MP for such an entity renders it trivial. The only way to make it nontrivial is to see it as a prescription for a *specific* item, such as a concrete specific expression of a language (perhaps a language *of thought*, but this does not rid us of the problem).

8 Laws of logic and correctness

The view that the laws of logic cannot be correct/incorrect—that they are "like etiquette" in that they can be at most useful/useless—may seem to be a non-starter. Is it not obvious that, for example, MP *is* correct, while affirming the consequent (AC) is *not*? Is it not obvious that she who argues *It rains and if it rains, the streets are wet; hence the streets are wet* argues correctly, while he who argues *The streets are wet and if it rains, the streets are wet; hence it rains* argues incorrectly?

Yes, this much indeed is clear. However, what does this show? Well, we assume that MP, but not AC, is a rule governing the English *if-then*. Given this, the move from *It rains* and *If it rains, the streets are wet* to *The streets are wet* is correct (for it is an instance of MP), while that from *The streets are wet* and *If it rains, the streets are wet* to *It rains* is not correct (for it is an instance of AC). But this concerns the correctness of the individual moves given the rules, not the correctness of the rules.

But is it not obvious that MP is correct for the English *if-then*, while AC is not? It is obvious in the sense that MP holds for *if-then* and AC does not (minor objections, that are raised from time to time, aside). However, insofar as this is so, it is because the rule has been, as a matter of fact, associated with this English expression, not because this association would be itself correct. It is a matter of a historical contingency; we could easily imagine that this kind of sound might have to come to be used as a connective obeying AC, and not MP. In no sense is it *correct* that *if-then* has come to obey MP, rather than AC.

The fact that a connective obeys MP, but not AC, would be a necessity (rather than a historical contingency) only if it were something as an "Implication" (not just an arbitrary item which would become an implication by a deliberate stipulation or historical development, but one that is such inherently). But we saw that an inherent implication could only be an essentially functional object, an object which would already incorporate obedience to MP (rather than AC).

There is no sense in which the rules of chess are correct as they are. We know that some of them can be changed in ways that would lead us to alternative, perhaps more or perhaps less interesting games; and there are, beyond doubt, many changes that would lead to the entire disruption of the whole game, so it would not be a game at all. We might, perhaps, call those alternative rules that would lead to such disruption *incorrect*, but thus we would only use *incorrect* in the sense of *useless* (or *harmful* or *devastating* ...).

9 Normativity of logic

There are various classifications of the ways in which logic can be considered normative. Russell (2020), for example, distinguishes three such ways:

One is that logic directly tells us how to reason (hence that logical theory is normative by its nature). The second way is that logical theory is not normative in itself, but has normative consequences. The third is that logic does not even have, by itself, normative consequences, but can help us derive normative consequences from some normative premises. Russell argues that as a matter of fact, logic only is normative in the third, weakest sense.

Taking a different visual angle Steinberger (2019) concludes that logic can be seen as normative in three different senses, namely as articulating directives, articulating evaluations, and articulating appraisals. Leaving aside the third of the senses, Steinberger comes to the conclusion that logic is normative both in the first and the second sense.

Despite the opposing results these two studies reach, there is something that is common to them (and to a lot of other contemporary work on the normativity of logic). They consider normativity as an attribute of logical *theory*; they see the situation so that there is a domain of human activities (drawing inferences, arguing, proving, ...) and there is a theory (logic) which may or may not be telling us how we should carry out these activities. Then, of course, there is a question on the basis of which logicians can issue such prescriptions, what is the source of authority of logic. And this may lead us to the search for a domain, urged by Schechter, underlying logical claims in a similar sense in which the real world underlies empirical claims.

In contrast to this, what I argue is that rules—and hence normativity—is already inherent to the practices, and they can make the theory normative by permeating into it. This has tremendous consequences especially for understanding the source of authority of logical theories. According to this view, the authority does not come from any supernatural domain which logical laws and rules would bring to light; it comes from the (proto)rules which are already implicitly present within the practices. True, the (proto)rules are not quite definite and unambiguous, and there is some work for logical theories to make them such, and there are alternative ways to do this work, so that we can have a plurality of logics;[14] the practices, however, constitute as much of the "reality check" for our theories as needed.

10 Conclusion

The reality check and the warrant of reliability of our logical theories can be provided by our mundane argumentative practices; it is not necessary to

---
[14]To discuss details of this process is another story—see Peregrin and Svoboda (2017).

search for them in any unworldly spheres. The point is that our distinctively human practices are characterized by "mutual accountability", *viz.* by the fact that they consist not only of some "first-order" performances, but also of "second-order" *normative attitudes* taking the performances as correct or incorrect. Hence insofar as we identify rules with consonant normative attitudes, human practices incorporate the rules that govern them.

Argumentation and reasoning, which are the subject matter of logic, are such practices; and the practices incorporate the normative attitudes and hence implicit rules. And what logic is after are not merely regularities of the practices, but precisely the *rules* inherent in them. As the rules incorporated in the practices are not always quite determinate, the job of logic, along with making them explicit, is also to make them more determinate.

The basic rules logic captures are not correct/incorrect; they can at most be useful/useless (and this is usually not a property of individual rules, but rather of their systems). Also the most basic rules of logic (such as those spelled out by Gentzenian natural deduction) cannot but be constitutive; they do not tell us how to reason, they equip us with gadgets with which—or in terms of which—to reason. Just as the rules of etiquette take their part in constituting a niche in which we can feel comfortable and safe, so the rules of logic help constitute a space in which we can talk meaningfully, reason and argue.

References

Aquinas, T. (1882). Sententia libri Ethicorum. In *Opera Omnia*. Roma: Vatican Polyglot Press.

Boghossian, P. A. (1989). The rule-following considerations. *Mind*, *98*(392), 507–549.

Brandom, R. (1994). *Making it Explicit: Reasoning, Representing, and Discursive Commitment*. Cambridge (Mass.): Harvard University Press.

Brandom, R. (2000). *Articulating Reasons*. Cambridge (Mass.): Harvard University Press.

Castro, L., Castro-Nogueira, L., Castro-Nogueira, M. A., & Toro, M. A. (2010). Cultural transmission and social control of human behavior. *Biology & Philosophy*, *25*(3), 347–360.

Ginsborg, H. (2011). Primitive normativity and skepticism about rules. *The Journal of Philosophy*, *108*(5), 227–254.

Goldfarb, W. (1985). Kripke on Wittgenstein on rules. *The Journal of Philosophy*, *82*(9), 471–488.
Hart, H. L. A. (1961). *The Concept of Law*. Oxford: Oxford University Press.
Haugeland, J. (2000). Truth and rule-following. In *Having Thought* (pp. 305–362). Cambridge (Mass.): Harvard University Press.
Heyes, C. (2018). *Cognitive Gadgets: The Cultural Evolution of Thinking*. Cambridge (Mass.): Harvard University Press.
Hjortland, O. T. (2017). Anti-exceptionalism about logic. *Philosophical Studies*, *174*(3), 631–658.
Kripke, S. A. (1982). *Wittgenstein on Rules and Private Language: An Elementary Exposition*. Cambridge (Mass.): Harvard University Press.
McDowell, J. (1984). Wittgenstein on following a rule. *Synthese*, *58*(3), 325–363.
Mercier, H., & Sperber, D. (2011). Why do humans reason? Arguments for an argumentative theory. *Behavioral and Brain Sciences*, *34*(2), 57–111.
Mercier, H., & Sperber, D. (2017). *The Enigma of Reason*. Cambridge (Mass.): Harvard University Press.
Peregrin, J. (2010). Logic and natural selection. *Logica Universalis*, *4*(2), 207–223.
Peregrin, J. (2014). *Inferentialism: Why Rules Matter*. Basingstoke: Palgrave.
Peregrin, J. (2019). Logic as a (natural) science. In I. Sedlár & M. Blicha (Eds.), *The Logica Yearbook 2018* (pp. 177–196). London: College Publications.
Peregrin, J., & Svoboda, V. (2017). *Reflective Equilibrium and the Principles of Logical Analysis: Understanding the Laws of Logic*. New York: Routledge.
Rouse, J. (2007). Social practices and normativity. *Philosophy of the Social Sciences*, *37*(1), 46–56.
Russell, G. (2020). Logic isn't normative. *Inquiry*, *63*(3–4), 371–388.
Schechter, J. (2013). Could evolution explain our reliability about logic? *Oxford Studies in Epistemology*, *4*, 214–239.
Schechter, J. (2018). Is there a reliability challenge for logic? *Philosophical Issues*, *28*, 325–347.
Schmidt, M. F., & Rakoczy, H. (2019). On the uniqueness of human normative attitudes. In N. Roughley & K. Bayertz (Eds.), *The Normative Animal* (pp. 121–135). Oxford: Oxford University Press.

Sellars, W. (1974). Meaning as functional classification. *Synthese*, *27*(3–4), 417–437.
Steinberger, F. (2019). Three ways in which logic might be normative. *The Journal of Philosophy*, *116*(1), 5–31.
Tomasello, M. (2014). *A Natural History of Human Thinking*. Cambridge (Mass.): Harvard University Press.
Tomasello, M. (2019). *Becoming Human: A Theory of Ontogeny*. Cambridge (Mass.): Belknap Press.
Wittgenstein, L. (1956). *Bemerkungen über die Grundlagen der Mathematik* [Remarks on the Foundations of Mathematics] (G. H. Von Wright & R. Rhees, Eds. & G. E. M. Anscombe, Trans.). Oxford: Blackwell.
Wrangham, R. (2019). *The Goodness Paradox: The Strange Relationship Between Virtue and Violence in Human Evolution*. New York: Pantheon.
Wright, C. (2018). Logical non-cognitivism. *Philosophical Issues*, *28*, 425–450.

Jaroslav Peregrin
Czech Academy of Sciences, Institute of Philosophy
The Czech Republic
E-mail: `peregrin@flu.cas.cz`

A Note on Paradoxical Propositions from an Inferential Point of View

IVO PEZLAR[1]

Abstract: In a recent paper by Tranchini (2019), an introduction rule for the paradoxical proposition ρ^* that can be simultaneously proven and disproven is discussed. This rule is formalized in Martin-Löf's constructive type theory (CTT) and supplemented with an inferential explanation in the style of Brouwer-Heyting-Kolmogorov semantics. I will, however, argue that the provided formalization is problematic because what is paradoxical about ρ^* from the viewpoint of CTT is not its provability, but whether it is a proposition at all.

Keywords: proof-theoretic semantics, constructive type theory, paradox, inductive definitions, Martin-Löf

1 Introduction

How do we define the meaning of logical constants? What does, e.g., the conjunction $\wedge$ mean? The standard answer put forward by the inferentialist (proof-theoretic) tradition is relatively simple: the meaning of logical constants within a certain natural deduction system is specified by introduction rules. These rules should effectively work as " 'definitions' of the symbols concerned" (Gentzen, 1969, p. 80, English translation). For example, the meaning constituting introduction rule for the conjunction $\wedge$ can be schematized as follows:

$$\frac{A \quad B}{A \wedge B} \wedge \text{I}$$

with the assumption that A and B are true, i.e., proven, propositions. It tells us that if we want to prove the proposition $A \wedge B$, first we have to prove the propositions A and B. In other words, the proof of $A \wedge B$ consists of a pair

---

[1] Work on this paper was supported by Grant no. 19-12420S from the Czech Science Foundation, GA ČR. I would also like to thank the anonymous reviewers for their most helpful comments.

of proofs of its conjuncts. Thus, the general idea is that we understand the meaning of $\wedge$ when we can properly use it, which in this case corresponds to the ability to prove the proposition of the form $A \wedge B$.

Where do introduction rules come from? The original Gentzen's set of introduction rules arose from analysing the structure of actual mathematical proofs. He wanted to capture "the forms of deduction used in practice in mathematical proofs" and develop a formal system that would come "as close as possible to actual reasoning" (Gentzen, 1969, p. 68). From this perspective, the origins of introduction rules were purely empirical. Hence, there were no general restrictions on them, they were—or rather should have been—just codifying the actual mathematical practice.

However, considering introduction rules should act as "definitions" of the symbols appearing in their conclusions, it seems reasonable to assume that the symbols to be defined should not appear among the premises of the corresponding rules to avoid circularity.[2] For example, assume that the introduction rule for conjunction would look as follows:

$$\frac{A \wedge B}{A \wedge B} \wedge I'$$

It would be difficult to see in what sense it constitutes or illuminates the meaning of $\wedge$.

Furthermore, if these "definitions" provided by introduction rules are to be of any practical value, we have to know how to use them. In natural deduction systems, this is a task for elimination rules, whose role is, simply put, to enact those definitions. In Gentzen's words: "[elimination rules] are no more, in the final analysis, than the consequences of these definitions [i.e., of introduction rules]" (Gentzen, 1969, p. 80). For example, the elimination rules for conjunction are as follows:

$$\frac{A \wedge B}{A} \wedge E_1 \qquad \frac{A \wedge B}{B} \wedge E_2$$

It seems unproblematic that if $A \wedge B$ was "defined" using A and B, we should be able to unpack this definition and get back its constituents, i.e., A and B in this case.

The observation that we should not be able to infer from a derived proposition more (or less) than what went into its derivation is crucial. It was this

---

[2]However, as we will see later, this is not always so straightforward, especially in the case of inductive definitions.

general concept—that later become known as the inversion principle (see Lorenzen, 1955; Prawitz, 1965) or harmony (see Dummett, 1991; Tennant, 1978)—that was violated by the famous counterexample by Prior (1960) to the idea that introduction and elimination rules alone can determine the meaning of logical constants. He proposed a new logical constant tonk governed by the following introduction and elimination rules:

$$\frac{A}{A \text{ tonk } B} \text{tonkI} \qquad \frac{A \text{ tonk } B}{B} \text{tonkE}$$

With these rules it is easy to derive the paradoxical conclusion that $\neg A$ is true assuming that A is true:

$$\frac{\dfrac{A}{A \text{ tonk } \neg A} \text{tonkI}}{\neg A} \text{tonkE}$$

What went wrong? It is the elimination rule that causes the paradoxical behaviour. Specifically, the elimination rule is not sanctioned by the corresponding introduction rule. As was said, elimination rules should not go beyond what introduction rules stipulate. In this case, it is the derivation of $A \text{ tonk } B$ from A. And since no B went into deriving $A \text{ tonk } B$, we should not be able to derive B back from it.[3]

However, as was already observed by Prawitz (1965, Appendix B), there are scenarios in which introduction and elimination rules are harmonious, yet paradoxical behaviour still arises. In these cases, the culprit is not the elimination rules as was the case with tonk, but the introduction rules.[4]

It is these problematic introduction rules, namely those that exhibit paradoxical behaviour due to some form of circularity, that will be the main topic of this paper. Specifically, I will examine an introduction rule discussed by Tranchini (2019) determining the meaning of the paradoxical proposition ρ^* that can be simultaneously proven and disproven, i.e., we can have proofs for both ρ^* and its negation. In the same paper, this rule is then formalized in the framework of Martin-Löf's constructive type theory (CTT) and supplied with a corresponding clause to the Brouwer-Heyting-Kolmogorov (BHK)

---

[3] See (Tranchini, 2014) for a more thorough discussion of tonk and its difference from other paradoxical connectives.

[4] I would like to thank the reviewer for this remark. See also (Schroeder-Heister, 2012) for a discussion of paradoxical behaviour in the sequent calculus setting, i.e., in an environment without introduction and elimination rules but with left and right rules.

semantics. I will, however, argue that the provided formalization is problematic because what is paradoxical about ρ^* from the viewpoint of CTT is not its provability, but whether it is a proposition at all.

The introduction and elimination rules and BHK clause for the paradoxical proposition ρ^* proposed by Tranchini (2019) are based on the introduction and elimination rules and BHK clause for implication $\supset$. So we begin by examining the latter, then we discuss the former.

2 Implication

According to the BHK semantics, the proof, and hence the meaning of the proposition $A \supset B$ consists of a method (procedure, function, program) which takes any proof of A and returns a proof of B. The standard introduction and elimination rules are as follows:

$$\frac{\begin{array}{c}[A]\\ B\end{array}}{A \supset B} \supset I \qquad \frac{A \supset B \quad A}{B} \supset E$$

The introduction rule tells us that if we want to prove proposition $A \supset B$, we first have to be able to derive proposition B from assumption A which should then be discharged.[5] In other words, it tells us how to construct (canonical) proofs of the proposition $A \supset B$. The elimination rule then tells us how we can use this proposition in proofs: if we derive $A \supset B$ together with A, we can then proceed to B alone.

Note that if we apply the $\supset$E rule immediately after the $\supset$I rule, i.e., construct $A \supset B$ and then remove it right away, we are making an unnecessary detour in a derivation. To get rid of these detours we use the following reduction meta rule (see Prawitz, 1965):

$$\frac{\dfrac{\begin{array}{c}[A]^n\\ \mathcal{D}\\ B\end{array}}{A \supset B}\supset I^n \quad \begin{array}{c}\mathcal{D}'\\ A\end{array}}{B} \quad \Rightarrow \quad \begin{array}{c}\mathcal{D}'\\ [A]\\ \mathcal{D}\\ B\end{array}$$

If a closed derivation, i.e., a derivation with no open assumptions, contains no detours, it is said to be in normal form.

---

[5]Hence, the assumption A is essentially just a placeholder to be withdrawn. For an alternative approach to assumptions, see (Pezlar, 2020).

A Note on Paradoxical Propositions

Utilizing the propositions-as-types principle[6], which is fully adopted by Martin-Löf's constructive type theory (CTT), we can make the BHK clause as well as the rules for implication more explicit and precise:

$$\frac{[x:A] \\ b(x):B}{\lambda x.b(x):A \supset B} \supset I \qquad \frac{c:A \supset B \quad a:A}{\mathrm{app}(c,a):B} \supset E$$

assuming that $A : prop$, $B : prop$, and $A \supset B : prop$, i.e., that A, B, and $A \supset B$ are propositions. The meta rule for detour reduction can then be captured as the following computation rule (also known as reduction rule or equality rule):

$$\frac{[x:A] \\ t(x):B \quad s:A}{\mathrm{app}(\lambda x.t(x),s) = t(s/x):B} \supset C$$

where $t(s/x)$ is the result of substituting s for x in t. Informally, the rule states that a derivation with a detour is equal to the derivation we obtain by removing this detour.

In what sense are these rules more explicit and precise? Regarding the explicitness, note that the premises and conclusions of these rules are no longer propositions but judgments of the form $a : A$ which can be read as "a is a proof of A". Hence, the proofs themselves are internalized in the object language and coded as terms. As for the precision, note that the informal statement of the corresponding BHK clause "a proof of $A \supset B$ consists of a method that takes any proof of A and returns a proof of B" is made more exact by the judgment $\lambda x.b(x) : A \supset B$ where the unspecified notion of a method is replaced by a specific lambda term, namely abstraction.

It is important to mention that from the perspective of CTT, the function $\lambda x.b(x)$ appearing in the conclusion of $\supset I$ is a function in a secondary sense, the more basic notion of a function appears in the premise of this rule, i.e., it is captured by the hypothetical derivation of $b(x) : B$ under the assumption $x : A$ (see, e.g., Klev, 2019a). We can liken this difference to Frege's distinction between functions as course-of-values and functions as unsaturated entities, respectively (see Frege, 1893).

---

[6] Also known as the Curry-Howard correspondence (see Curry & Feys, 1958; Howard, 1980).

Ivo Pezlar

One last note. So far we have presupposed that $A \supset B$ is a proposition, assuming A and B are propositions. In CTT, however, this is a judgment that can and should be demonstrated as well by using a special kind of rules called formation rules. These rules tell us how to form new propositions from other propositions. For example, $A \supset B$ receives the following formation rule:[7]

$$\frac{A : prop \qquad B : prop}{A \supset B : prop} \supset F$$

Note that this rule tells us how to form the proposition $A \supset B$, i.e., how to derive the judgment $A \supset B : prop$. However, the rule itself requires further justification. Generally, in CTT, we can judge that A is a proposition if we know what counts as a canonical proof of A, i.e., if we can recognize a canonical proof of A when we are presented with one (the same goes for equal canonical proofs). And to tell us what counts as canonical proofs is the purpose of the introduction rules. Therefore, formation rules are justified by the corresponding introduction rules. Thus, we can judge that $A \supset B : prop$ since we know (via the $\supset I$ rule) what should the canonical proofs of $A \supset B$ look like.

Consequently, this means that the rule $\supset I$ takes, if we want to be fully explicit, three premises, including those of the corresponding formation rules. The rule then should look as follows:

$$\frac{A : prop \qquad B : prop \qquad \begin{array}{c}[x : A]\\ b(x) : B\end{array}}{\lambda x.b(x) : A \supset B} \supset I'$$

Note In CTT, there are four basic kinds of rules: introduction rules, elimination rules, formation rules, and computation rules. Introduction rules are considered self-justifying, elimination rules correspond to introduction rules and are justified by computation rules (analogous to Prawitz's reduction rules, see Prawitz, 1965), formation rules are justified by introduction rules, and computation rules relate elimination rules to introduction rules. For more, see (Martin-Löf, 1984).

Now, let us finally proceed to the paradoxical proposition ρ^*.

---

[7] I am omitting the variant for showing how to form equal implicational propositions.

3 Paradoxical proposition ρ^*

To incorporate the paradoxical proposition ρ^* that can be simultaneously proven and disproven, Tranchini (2019) suggests the following extension to the BHK semantics. The informal clause explaining the corresponding proof condition for ρ^* goes as follows, where $\bot$ denotes absurdity:

> a proof of ρ^* is the result of applying a self-referential abstraction-like operation to a function (as an unsaturated entity) from proofs of ρ^* to proofs of $\bot$. The result of this operation are objects whose nature is similar to that of the functions as courses-of-value that constitute proofs of sentences of the form $A \supset B$, with the crucial difference that proofs of ρ^* take proofs of ρ^* as arguments and yield proofs of $\bot$ as values. (Tranchini, 2019, p. 601)

The clause for the paradoxical proposition ρ^* is given formalization in the framework of Martin-Löf's constructive type theory (CTT). Specifically, its inferential behaviour is specified by the following introduction and elimination rules (see also Read, 2010):

$$\dfrac{[x:\rho^*] \atop t(x):\bot}{\curlyvee x.t(x):\rho^*}\,\rho^*\mathrm{I} \qquad \dfrac{s:\rho^* \quad t:\rho^*}{\mathfrak{sbb}(s,t):\bot}\,\rho^*\mathrm{E}$$

which are then related by the following computation rule:

$$\dfrac{[x:\rho^*] \atop t(x):\bot \quad s:\rho^*}{\mathfrak{sbb}(\curlyvee x.t(x),s)=t(s/x):\bot}\,\rho^*\mathrm{C}$$

The ρ^*C rule shows how the function $\mathfrak{sbb}$, defined by the rule ρ^*E, operates on the canonical proofs of the proposition ρ^* generated by the rule ρ^*I, and thus in turn justifies the ρ^*E rule.

With these rules we can both prove ρ^* and disprove ρ^*, i.e., prove $\neg\rho^*$ understood as $\rho^* \supset \bot$:

$$\dfrac{\dfrac{[x:\rho^*]^1 \quad [x:\rho^*]^1}{\mathfrak{sbb}(x,x):\bot}\,\rho^*\mathrm{E}}{\curlyvee x.\mathfrak{sbb}(x,x):\rho^*}\,\rho^*\mathrm{I}^1 \qquad \dfrac{\dfrac{[x:\rho^*]^1 \quad [x:\rho^*]^1}{\mathfrak{sbb}(x,x):\bot}\,\rho^*\mathrm{E}}{\lambda x.\mathfrak{sbb}(x,x):\rho^*\supset\bot}\,\supset\mathrm{I}^1$$

If we combine these two derivations via the $\supset$E rule, we obtain the following derivation:

$$\cfrac{\cfrac{\cfrac{[x:\rho^*]^1 \quad [x:\rho^*]^1}{\mathfrak{sbb}(x,x):\bot}\rho^*E}{\mathcal{Y}x.\mathfrak{sbb}(x,x):\rho^*}\rho^*I^1 \quad \cfrac{\cfrac{[x:\rho^*]^1 \quad [x:\rho^*]^1}{\mathfrak{sbb}(x,x):\bot}\rho^*E}{\lambda x.\mathfrak{sbb}(x,x):\rho^*\supset\bot}\supset I^1}{\mathrm{app}(\lambda x.\mathfrak{sbb}(x,x),\mathcal{Y}x.\mathfrak{sbb}(x,x)):\bot}\supset E$$

Note that there is a redundancy on the right side of the tree: we derived $\rho^* \supset \bot$ via the $\supset$I rule and then immediately eliminated it by an application of the $\supset$E rule.

If we remove this detour using the $\supset$C rule (i.e., essentially compute the term $\mathrm{app}(\lambda x.\mathfrak{sbb}(x,x),\mathcal{Y}x.\mathfrak{sbb}(x,x))$), we get the following derivation:[8]

$$\cfrac{\cfrac{\cfrac{[x:\rho^*]^1 \quad [x:\rho^*]^1}{\mathfrak{sbb}(x,x):\bot}\rho^*E}{\mathcal{Y}x.\mathfrak{sbb}(x,x):\rho^*}\rho^*I^1 \quad \cfrac{\cfrac{[x:\rho^*]^1 \quad [x:\rho^*]^1}{\mathfrak{sbb}(x,x):\bot}\rho^*E}{\mathcal{Y}x.\mathfrak{sbb}(x,x):\rho^*}\rho^*I^1}{\mathfrak{sbb}(\mathcal{Y}x.\mathfrak{sbb}(x,x),\mathcal{Y}x.\mathfrak{sbb}(x,x)):\bot}\rho^*E$$

with yet another detour. But if we try to remove it, we discover that this derivation reduces to itself via the ρ^*C rule, and thus we get caught in what Neil Tennant called a loop: "the normalisation sequence never terminating with a proof in normal form" (Tennant, 1982, p. 270).

As was already mentioned before, note that the rule ρ^*I is essentially assembled as a self-referential variant of the implication introduction rule $\supset$I, which, as we discussed above, takes, in its fully explicit version, three premises, not just one. So, analogously, the fully revealed version of the ρ^*I rule for the proposition ρ^* should be:

$$\cfrac{\rho^*: prop \quad \bot: prop \quad \cfrac{[x:\rho^*]}{t(x):\bot}}{\mathcal{Y}x.t(x):\rho^*}\rho^*I'$$

Now, let us examine more closely the object $\mathcal{Y}x.t(x)$ of type ρ^*, i.e., the object constructed by the ρ^*I rule. Analogously to $\lambda x.b(x)$, it is supposed to be a coding (a name) of the function $t(x)$. Note, however, that the domain of $t(x)$ is also ρ^*, i.e., $\mathcal{Y}x.t(x)$ itself belongs to the domain ρ^* of the function

---

[8] Alternatively, we could construct this derivation directly from the two copies of the initial closed derivation of ρ^*, as does Tranchini (2019).

A Note on Paradoxical Propositions

$t(x)$ it is supposed to be coding. But that is a problem: we are forming a canonical object of type ρ^* from a function whose domain is ρ^* itself. In other words, we are generating an object of type ρ^* from itself (see Dyckhoff, 2016 for analogous observations).[9]

But is this really problematic in general? For example, the type $\mathbb{N}$ of natural numbers seems to be defined also with some degree of circularity but everything works just fine. Specifically, it has the following two introduction rules:

$$\frac{}{0 : \mathbb{N}} \text{NI}_1 \qquad \frac{n : \mathbb{N}}{succ(n) : \mathbb{N}} \text{NI}_2$$

The first rule simply stipulates that 0 is a natural number, so there is no issue. However, in the case of the second rule it seems like we are trying to generate an object of type $\mathbb{N}$ from itself: note that the premise of the successor rule NI_2 seems to presuppose that we already understand what it means for some n to be a natural number.

Why is this case unproblematic as opposed to $\rho^*\text{I}$? Is it perhaps because with the type $\mathbb{N}$ we have the base object $0 : \mathbb{N}$ which is missing in the case of the type ρ^*? Unfortunately no, because then we would have to conclude that the type of well-founded trees (W-types) has also problematic introduction rules, since they also do not have such base cases.[10]

The answer to this question can be found, if we examine the premises of the involved rules. Specifically, note that the rule ρ^* tells us that we can construct a canonical proof of ρ^* assuming we have a function that takes an arbitrary proof of ρ^* and returns a proof of $\bot$ (recall the BHK explanation of implication).

In contrast, the premise of the successor rule for $\mathbb{N}$ does require us to be in a possession of an arbitrary object of this type, we just need an object of this type. Or as Dyckhoff (2016, p. 82) put it: "we don't need to grasp *all* elements of $\mathbb{N}$ to construct a canonical element by means of the rule, just one of them, namely n."

To make these observations more general and precise, we can borrow a few notions from the literature on inductive types and then carry them over to the logical side in accordance with the Curry-Howard correspondence.[11]

---

[9] I thank one of the reviewers for pointing this out to me and invite the reader to consult Dyckhoff's paper as well, specifically pp. 81–83. Furthermore, the elimination rule $\rho^*\text{E}$ is also of some interest, however, in this paper I will be focused primarily on introduction rules.

[10] I thank an anonymous reviewer for this remark. Although W-types would deserve a closer examination, they are beyond the scope of this paper.

[11] As noted, e.g., by Dyckhoff (2016), whose simplified presentation we follow below.

Inductive types are often formally treated as the least fixed point of an operator Φ defined via type variables $X, Y, \ldots$, constants, and type constructors $+$ (addition), $\times$ (product), and $\to$ (function space). For example, the type of natural numbers $\mathbb{N}$ can be generated by the definition $\Phi(X) = X + 1$.[12]

Furthermore, a definition of a unary operator Φ is said to be *positive* if and only if only occurrences of X in it are positive.

1. An occurrence of X in X is *positive*.

2. An occurrence of X in $A \to B$ is *positive* if and only if it is (i) a positive occurrence in B or (ii) a negative occurrence in A

3. An occurrence of X in $A \to B$ is *negative* if and only if it is (i) a negative occurrence in B or (ii) a positive occurrence in A

4. An occurrence of X in $A + B$ and $A \times B$ is *positive* if and only if it is a positive occurrence in A or B

5. An occurrence of X in $A + B$ and $A \times B$ is *negative* if and only if it is a negative occurrence in A or B

A definition of a type as the least fixed point of an operator can then be said to be *positive* if and only if the operator definition is positive. Furthermore, a definition can be said to be *strictly positive*, if only occurrences of the type variable X in the definition are strictly positive.

6. An occurrence of X in X is *strictly positive*.

7. An occurrence of X in $A \to B$ is *strictly positive* if and only if it is a strictly positive occurrence in B

8. An occurrence of X in $A + B$ and $A \times B$ is *strictly positive* if and only if it is a strictly positive occurrence in A or B

For example, in $A \to B$, B occurs positively and A occurs negatively. Also note that if X does not occur positively in A then either X does not occur in A or X occurs negatively in A.

Now, the type ρ^* can be then defined as the least fixed point of the operator $\Phi(X) = X \to \bot$ (see Dyckhoff, 2016, p. 83). Note that this is not a positive definition, since X has a negative occurrence in the definition (see the clause 3

---

[12] See, e.g., (Dybjer, 1997), (Mendler, 1987).

above). Definitions with negative occurrences such as this are generally avoided (so called "strict positivity condition") for the unsurprising reason that they allow us to construct looping terms, analogous to our non-normalizable looping proofs (see, e.g., Bertot & Castéran, 2004; Chlipala, 2013).

Intuitively, we can think of these type definitions as corresponding to introduction rules under the Curry-Howard correspondence.[13] Analogously, we can also carry over the notions of positive/negative occurrence. For example, in the case of ρ^*, the negative occurrence of ρ^* in the corresponding type definition then coincides with the fact that on the logical side ρ^* appeared as an assumption. Similarly, we can also adopt the notions of (strictly) positive/negation definitions, e.g., we can say that an introduction rule for a proposition A is strictly positive if and only if A does not appear among its premises as an assumption/antecedent.[14]

Now, if we return to the difference between ρ^*I and NI_2, we can then say that the reason why the former is problematic, but the latter is not, is because ρ^* occurs negatively in a premise of ρ^*I, which is not the case for N in NI_2. Thus, we can say that ρ^*I is a negative introduction rule.

So, to conclude, the problem with ρ^*I is not just that ρ^* itself appears in the premise (as we have seen, e.g., N also appears in the premise of the corresponding rule NI_2 but causes no issues) but that it occurs negatively in the premise. Why is this problematic from the viewpoint of CTT? Recall that the premise of ρ^*I is a hypothetical judgment stating that $t(x)$ is a function from ρ^* to $\bot$. In order to fully understand this function, however, we have to understand what an arbitrary proof of ρ^* is. But to achieve this, we first need to understand what a canonical proof of ρ^* is. And to understand this, we need to understand the corresponding introduction rule. Thus, understanding the premise of this rule presupposes that we already understand the rule as a whole.[15] More generally put, the rule invites us to assume we know something (what does the canonical proof object of ρ^* look like) which is unknowable at that point. For these reasons, the rule ρ^*I cannot be considered as properly constituting the meaning of ρ^*.

Furthermore, recall that in CTT, formation rules are justified by introduction rules. In practice, this means that to be able to make the judgment

---

[13] The type constructor $\to$ can be roughly understood as corresponding to the implication operator $\supset$ on the logical side.

[14] As observed by Klev (2019b), negative occurrences of propositions in their own introduction rules are already banned implicitly by Martin-Löf (1971), explicitly by Dybjer (1994).

[15] For an analogous observation, see also (Klev, 2019b). I thank one of the reviewers for pointing this out.

that ρ^* is a proposition, i.e., $\rho^* : prop$, we first have to be able to show how its canonical proofs are constructed. But, as we just showed, this cannot be done, since the rule ρ^*I fails in its task to explain properly the meaning of ρ^*, i.e., of its canonical proofs. Consequently, we are not in a position to justifiably make the judgment $\rho^* : prop$.

Therefore, if there is a paradox from the viewpoint of CTT, it is rather about the formability of ρ^*. In other words, what can perhaps be viewed as paradoxical is the justifiability and explainability of the judgment $\rho^* : prop$. Utilizing the rules above we can provide some form of an "explanation" but simultaneously it cannot really be considered as a proper explanation due to its circular nature and the negative occurrence of ρ^* in the premise of its introduction rule can be understood as an indication of this. Thus, what seems to be in question is not the provability of ρ^*, but whether it is a proposition at all.[16]

4 Variants of ρ^*

The other variants of ρ^* discussed by Tranchini (2019) seem to suffer from analogous issues and, in the remaining place, I will try to briefly sketch why. These variants are: 1) a paradoxical proposition ρ with a negative self-reference operator ! and its inverse i and 2) semi-paradoxical propositions σ and τ whose paradoxical nature does not come from self-reference, or negative self-reference, but from their circular meaning-dependencies.

First, we consider the paradoxical proposition ρ with a negative self-reference operator. It is governed by the following introduction and elimination rules:

$$\frac{t : \neg \rho}{!t : \rho} \rho\text{I} \qquad \frac{t : \rho}{\text{i}t : \neg \rho} \rho\text{E}$$

The corresponding computation rule is as follows:

$$\frac{t : \neg \rho}{\text{i}!t = t : \neg \rho} \rho\text{C}$$

Analogously to ρ^*, with these rules we can construct proofs for ρ as well as $\neg \rho$, i.e., $\rho \supset \bot$, and combine them into a looping derivation:

---

[16]That is, of course, not to say that Russell-like self-referential paradoxes cannot be recreated in CTT at all (see, e.g., Coquand, 1992).

A Note on Paradoxical Propositions

$$\cfrac{\cfrac{\cfrac{[x:\rho]^1}{i x: \neg \rho}\rho E}{app(ix, x): \bot}}{\lambda x.app(ix, x): \neg \rho}\supset I^1 \qquad \cfrac{\cfrac{\cfrac{[x:\rho]^1}{i x: \neg \rho}\rho E \quad [x:\rho]^1}{app(ix, x): \bot}\supset E}{\cfrac{\lambda x.app(ix, x): \neg \rho}{!\lambda x.app(ix, x): \rho}\rho I}\supset I^1$$
$$\cfrac{}{app(\lambda x.app(ix, x), !\lambda x.app(ix, x)): \bot}\supset E$$

Now, returning to the rule ρI, if we make it fully explicit, we obtain:

$$\cfrac{\rho : prop \qquad t : \rho \supset \bot}{!t : \rho}\rho I'$$

which commits the same violation as the rule ρ^*I. Simply put, the rule ρI', which should act as a meaning explanation for ρ, presupposes that we already understand ρ. Here again it is indicated by the negative occurrence of ρ in the premise of the ρI rule, analogously to ρ^*I. Consequently, we are not justified in making the judgment $\rho : prop$.

Now, let us consider the case involving the propositions σ and τ with circular meaning-dependencies. They are specified by the following introduction and elimination rules:[17]

$$\cfrac{\cfrac{[x:\tau]}{t(x) : \bot}}{\mathcal{Y}'x.t(x) : \sigma}\sigma I \qquad \cfrac{s : \sigma \quad t : \tau}{\mathfrak{sbb}'(s, t) : \bot}\sigma E \qquad \cfrac{s : \sigma}{!'s : \tau}\tau I \qquad \cfrac{t : \tau}{i't : \sigma}\tau E$$

Although Tranchini does not supplant them with computation rules, I assume they might look as follows:

$$\cfrac{\cfrac{[x:\tau]}{t(x) : \bot} \quad s : \sigma}{\mathfrak{sbb}'(\mathcal{Y}'x.t(x), s) = t(s/x) : \bot}\sigma C \qquad \cfrac{t : \tau}{i'!'t = t : \tau}\tau C$$

Finally, Tranchini hints at the end of his paper that with these rules we can recreate Jourdain's paradox. To test it out, consider the following derivation:

---

[17]Tranchini (2019) presents these rules without the corresponding proof terms. Here we assume that the operators marked with "'" behave analogously to their unmarked variants with the exceptions generated by different typing (e.g., ! expects $\neg\rho$, while !' expects σ).

$$\cfrac{\cfrac{[x:\tau]^1}{\mathsf{i}'x:\sigma}\tau E \qquad [x:\tau]^1}{\cfrac{\mathsf{gbb}'(\mathsf{i}'x,x):\bot}{\cfrac{\mathcal{Y}'x.\mathsf{gbb}'(\mathsf{i}'x,x):\sigma}{\mathsf{gbb}'(\mathcal{Y}'x.\mathsf{gbb}'(\mathsf{i}'x,x),!'\mathcal{Y}'x.\mathsf{gbb}'(\mathsf{i}'x,x)):\bot}}\sigma I^1}\sigma E \qquad \cfrac{\cfrac{\cfrac{[x:\tau]^1}{\mathsf{i}'x:\sigma}\tau E \qquad [x:\tau]^1}{\cfrac{\mathsf{gbb}'(\mathsf{i}'x,x):\bot}{\cfrac{\mathcal{Y}'x.\mathsf{gbb}'(\mathsf{i}'x,x):\sigma}{!'\mathcal{Y}'x.\mathsf{gbb}'(\mathsf{i}'x,x):\tau}\tau I}\sigma I^1}}{}\sigma E$$

Again, if we try to reduce this derivation to a normal form, it enters into a loop (compare this with the derivation of $\mathtt{app}(\lambda x.\mathtt{app}(\mathtt{i}x,x),!\lambda x.\mathtt{app}(\mathtt{i}x,x)) : \bot$ from earlier).

Now, let us return to the introduction rules σI and τI. It is clear that the respective introduction rules are dependent on each other. Specifically, σI presupposes that we already understand what τ means and τI presupposes that we already understand what σ means. Again, let us begin by considering their fully explicit versions. We start with the rule σI:

$$\cfrac{\tau : prop \qquad \bot : prop \qquad \cfrac{[x:\tau]}{t(x):\bot}}{\mathcal{Y}'x.t(x):\sigma}\sigma I'$$

First, note that there is no apparent circularity, as the corresponding type definition $\Phi(X) = Y \to \bot$ suggests. But also note that $\sigma I'$ is in a way an incompletely specified rule since it refers to τ which we do not yet know to be a proposition. More specifically, we do not yet know how to prove it, i.e., how to construct its canonical proofs. For that, we need to consider an additional introduction rule τI (recall that the meanings of σ and τ should be interdependent). Once again, let us consider its explicit version:

$$\cfrac{\sigma : prop \qquad s : \sigma}{!'s:\tau}\tau I'$$

Again, at this stage, still no circularity appears, as the corresponding type definition $\Phi(Y) = X$ shows us. Note, however, that if we put these two definitions together, once again we obtain a negative definition of the form $\Phi(X) = X \to \bot$ (either if we substitute X for Y in the type definition corresponding to $\sigma I'$ or $Y \to \bot$ for X in the type definition corresponding to $\tau I'$). Thus, a complete specification of these introduction rules seems to come at a price of negativity.

A Note on Paradoxical Propositions

Note The Liar-like paradoxes considered in this paper can be generalized to a Curry-like paradox. For example, Read (2010) considers a proof-theoretic variant of a Curry paradox specified by the following introduction and elimination rules:

$$\dfrac{\begin{array}{c}[curry\,A]\\ A\end{array}}{curry\,A}\,curry\,\mathrm{I} \qquad \dfrac{curry\,A \quad curry\,A}{A}\,curry\,\mathrm{E}$$

where $curry$ is a unary logical connective. To obtain the original paradoxical rules $\rho^*\mathrm{I}$ and $\rho^*\mathrm{E}$, all we need to do is replace in these rules the arbitrary A with $\bot$ and $curry\,A$ with ρ^*.[18]

5 Conclusion

In this paper I tried to show that the analysis of the paradoxical proposition ρ^* utilizing constructive type theory (CTT) suggested by Tranchini (2019) is problematic because this proposition cannot be correctly formed in CTT, let alone proven. In other words, in CTT we are not able to properly justify the judgment $\rho^* : prop$. The same seems to apply to the other discussed variants of ρ^*, namely ρ, σ and τ.

The main issue with ρ^* lies in the circular nature of its introduction rule, more specifically, in the fact that there is a negative occurrence of ρ^* in its premise. This clashes with the general justification scheme of constructive type theory: formation rules, which tell us how to form new propositions, should be justified by the corresponding introduction rules, which tell us what these propositions mean, i.e., how to prove them. In the case of the proposition ρ^*, this justification requirement is, however, not met, since the introduction rule that should explain the meaning of ρ^* presupposes that we already understand it. Consequently, the formation rules cannot, strictly speaking, be understood as justified.

Therefore, we reached the conclusion that what is paradoxical about ρ^* from the perspective of CTT is rather whether or not it is a proposition at all, not that it can be proven and disproven at the same time.

Although some of the issues we have dealt with here might seem rather technical, from a more general point of view this paper is meant as a contribution to the general discussion concerning the nature of introduction rules. And even though at this time I am not yet ready to open the fundamental

---

[18] I thank one of the reviewers for this remark.

question "What is an introduction rule?", whatever the answer will be, it will have to take into account the matters discussed in this paper, i.e., how to deal with circularity, or more specifically, with negative occurrences within premises of introduction rules.

References

Bertot, Y., & Castéran, P. (2004). *Interactive Theorem Proving and Program Development*. Berlin: Springer.

Chlipala, A. (2013). *Certified Programming with Dependent Types*. Cambridge, MA: MIT Press.

Coquand, T. (1992). The paradox of trees in type theory. *BIT*, *32*(1), 10–14.

Curry, H. B., & Feys, R. (1958). *Combinatory Logic* (Vol. 1). Amsterdam: North-Holland Publishing Company.

Dummett, M. (1991). *The Logical Basis of Metaphysics*. London: Duckworth.

Dybjer, P. (1994). Inductive families. *Formal Aspects of Computing*, *6*(4), 440–465.

Dybjer, P. (1997). Representing inductively defined sets by wellorderings in Martin-Löf's type theory. *Theoretical Computer Science*, *175*, 329–335.

Dyckhoff, R. (2016). Some remarks on proof-theoretic semantics. In T. Piecha & P. Schroeder-Heister (Eds.), *Advances in Proof-Theoretic Semantics* (Vol. 43, pp. 79–93). Cham: Springer.

Frege, G. (1893). *Grundgesetze der Arithmetik I*. Jena: Hermann Pohle.

Gentzen, G. (1969). *The collected papers of Gerhard Gentzen*. Amsterdam: North-Holland Publishing Company.

Howard, W. A. (1980). The formulae-as-types notion of construction. In H. B. Curry, J. R. Hindley, & J. P. Seldin (Eds.), *To H. B. Curry: Essays on Combinatory Logic, Lambda Calculus, and Formalism*. London: Academic Press.

Klev, A. (2019a). A comparison of type theory with set theory. In S. Centrone, D. Kant, & D. Sarikaya (Eds.), *Reflections on the Foundations of Mathematics* (pp. 271–292). Cham: Springer.

Klev, A. (2019b). The justification of identity elimination in Martin-Löf's type theory. *Topoi*, *38*(3), 577–590.

Lorenzen, P. (1955). *Einführung in die operative Logik und Mathematik*. Berlin: Springer.
Martin-Löf, P. (1971). Hauptsatz for the intuitionistic theory of iterated inductive definitions. In J. E. Fenstad (Ed.), *Proceedings of the Second Scandinavian Logic Symposium (Oslo 1970)* (pp. 197–215). Amsterdam: North-Holland Publishing Company.
Martin-Löf, P. (1984). *Intuitionistic Type Theory*. Napoli: Bibliopolis.
Mendler, P. F. (1987). *Inductive definition in type theory* (Unpublished doctoral dissertation). Cornell University.
Pezlar, I. (2020). The placeholder view of assumptions and the Curry–Howard correspondence. *Synthese*.
Prawitz, D. (1965). *Natural Deduction: A Proof-theoretical Study*. New York: Dover Publications, Incorporated.
Prior, A. (1960). The runabout inference ticket. *Analysis*, 21(1), 38–39.
Read, S. (2010). General-elimination harmony and the meaning of the logical constants. *Journal of Philosophical Logic*, 39(5), 557–576.
Schroeder-Heister, P. (2012). Proof-theoretic semantics, self-contradiction, and the format of deductive reasoning. *Topoi*, 31(1), 77–85.
Tennant, N. (1978). *Natural Logic*. Edinburgh: Edinburgh University Press.
Tennant, N. (1982). Proof and paradox. *Dialectica*, 36(2–3), 265–296.
Tranchini, L. (2014). Proof-theoretic semantics, paradoxes and the distinction between sense and denotation. *Journal of Logic and Computation*, 26(2), 495–512.
Tranchini, L. (2019). Proof, meaning and paradox: Some remarks. *Topoi*, 38(3), 591–603.

Ivo Pezlar
Czech Academy of Sciences, Institute of Philosophy
The Czech Republic
E-mail: pezlar@flu.cas.cz

Three Conditionals: Contraposition, Difference-making and Dependency

ERIC RAIDL[1]

Abstract: This article investigates three conditionals for which Right Weakening fails: Rott's difference-making and dependency conditionals (aka Spohn's sufficient reason and Lewis' counterfactual dependency), and the contraposing conditional (aka Crupi and Iacona's evidential conditional). The article proves completeness results by conceiving of the new conditionals as defined from a basic conditional in a common semantics.

Keywords: difference making, sufficient reason, counterfactual dependency, evidential conditional, contraposition, Right Weakening, definable conditionals

1 Three strengthened conditionals

Conditionals $A > C$ are binary logical operators, where A is the antecedent and C the consequent. They are used to analyze natural language sentences of the form 'if A, [then] C', 'C, because/since A', 'A is a reason for C'. The standard account (Lewis, 1971; Stalnaker, 1968) suggests (roughly) that the conditional $A > C$ is true in world w if and only if the closest A-worlds to w are C-worlds. Recent reflections argue to strengthen the defining clause. I compare three such proposals:

$$\varphi \gg \psi := (\varphi > \psi) \wedge \neg(\neg\varphi > \psi) \qquad \text{sufficient reason}$$
$$\varphi \succ \psi := (\varphi > \psi) \wedge (\neg\varphi > \neg\psi) \qquad \text{dependency}$$
$$\varphi \triangleright \psi := (\varphi > \psi) \wedge (\neg\psi > \neg\varphi) \qquad \text{contraposition}$$

$\gg$ and $\succ$ are both based on the idea of difference making: the antecedent should make a difference to the consequent. But whereas $\gg$ says that the difference the antecedent makes is that its absence does not trigger the consequent, $\succ$ says that it triggers the negated consequent. The latter is a stronger difference than the former, if absence of the antecedent is possible.

---

[1] Funding for this research was provided under Germany's Excellence Strategy—EXC-Number 2064/1—Project number 390727645. I thank Jake Chandler, Hans Rott and Wolfgang Spohn for comments, and Vincenzo Crupi and Andrea Iacona for numerous discussions.

Thus $\varphi \succ \psi$ implies $\varphi \gg \psi$ if $\Diamond \neg \varphi$. On the other hand, $\rhd$ is motivated by having contraposition built into it.[2] Rott (1986) uses $\gg$ to analyze *because*, and in Spohn's (2015) ranking-theoretic account it represents the *sufficient reason relation* (Raidl, 2020a). Lewis (1973) suggested $\succ$ as *counterfactual dependency*. Crupi and Iacona (2020) use $\rhd$ to represent *evidential support*. It is also a special case of the support relation from Booth and Chandler (2020). Rott (2019, 2020) investigates $\gg$, $\succ$ and $\rhd$ under the name of difference-making, dependency, and contraposing conditional in belief-revision semantics. Raidl (2020a) proves completeness for $\gg$ in various semantics. Raidl, Iacona, and Crupi (2021) prove completeness for $\rhd$ in strongly centered Lewisean semantics. For comparison, I use one and the same underlying semantics—consistent Lewisean set-selection models. These are weaker than the strongly centered Lewisean models used by Crupi and Iacona (2020) and Lewis (1973), and are equivalent to Rott's semantics and the ranking semantics.

The three conditionals invalidate Right Weakening. For this, consider a Stalnakerian model where only one world is selected, if any, and the selection is based on a total well order: Let w_0 be an abc world, w_1 a $\neg a \neg bc$ world, and w_2 an $a \neg b \neg c$ world. Order the worlds according to their indicies, seen from w_0. Then, although we have $a \gg b$ in w_0, we don't have $a \gg (b \vee c)$, since $\neg a > (b \vee c)$. Similarly for $\succ$ and $\rhd$. Here is an example:

> If Ann gets really lucky, she will be accepted to Oxford, Cambridge, Yale, and Podunk state university
> ♯ If Ann gets really lucky, she will be accepted to Podunk state university

Although her scores are great, Ann needs a great deal of luck to get accepted to all of her preferred top universities. However, getting accepted to the insignificant Podunk state university, Ann needs no luck.[3]

Section 2 fixes the logic and semantics for $>$. Section 3 introduces the translation and backtranslation between conditional languages. Sections 4–6 prove completeness for the three conditionals, using a method from Raidl (2020a). The logic of $\rhd$ in Section 4 is weaker than in Raidl et al. (2021). The logic for $\gg$ in Section 5 is an alternative to Raidl (2020a, Corollary 2, last line). The logic of $\succ$ in Section 6 is new. If we restrict the language to non-nested conditionals which don't enter boolean combinations, the logics for $\gg$ and $\succ$ can also be compared to Rott's (2019) non-monotonic reasoning systems.

---

[2] Gomes (2019) argues that meaning preserving contraposition holds for so-called *implicative* natural language conditionals.

[3] Rott (2019) gives another example.

2 The Lewisean basic conditional

The alphabet is given by a set of propositional variables Var, the classical connectives $\neg, \wedge, \vee, \rightarrow$, the basic conditional $>$,) and (. The set of formulas $\mathcal{L}_>$ is defined inductively. $\top$ denotes any classical tautology, and $\bot = \neg\top$.

Definition 1 *The rules of* VN *are Modus Ponens for* $\rightarrow$ *(MoPo) and*

$$\frac{\vdash \varphi \leftrightarrow \varphi'}{\vdash (\varphi > \chi) \rightarrow (\varphi' > \chi)} \text{ LLE} \qquad \frac{\vdash \varphi \leftrightarrow \varphi'}{\vdash (\chi > \varphi) \rightarrow (\chi > \varphi')} \text{ RLE}$$

The axiom schemes are

All substitution instances of propositional tautologies	PT
$(\varphi > \psi) \rightarrow (\varphi > (\psi \vee \chi))$	DW
$((\varphi > \psi) \wedge (\varphi > \chi)) \rightarrow (\varphi > (\psi \wedge \chi))$	AND
$\varphi > \varphi$	ID
$((\varphi > \chi) \wedge (\psi > \chi)) \rightarrow ((\varphi \vee \psi) > \chi)$	OR
$((\varphi > \chi) \wedge (\varphi > \psi)) \rightarrow ((\varphi \wedge \psi) > \chi)$	CMon
$((\varphi > \chi) \wedge \neg(\varphi > \neg\psi)) \rightarrow ((\varphi \wedge \psi) > \chi)$	RMon
$\neg(\top > \bot)$	Cons

VN is in between Lewis' (1971) weakest conditional logic V (VN without Cons) and his logic VW = V + MP, where MP is $(\varphi > \psi) \rightarrow (\varphi \rightarrow \psi)$.[4] Thus VN is also weaker than the logic VC = VW + CS for strongly centered Lewisean models, where CS is $(\varphi \wedge \psi) \rightarrow (\varphi > \psi)$. $L \subseteq \mathcal{L}_>$ is a *classical conditional logic* iff it contains PT and is closed under MoPo, LLE and RLE. The smallest such logic is denoted CE.

Observation 1 *Given* MoPo + PT, RLE + DW *is equivalent to* Right Weakening

$$\frac{\vdash \varphi \rightarrow \varphi'}{\vdash (\chi > \varphi) \rightarrow (\chi > \varphi')} \text{ RW}$$

Thus VN is a normal conditional logic.[5] In VN, the *inner necessity* $\boxdot \varphi = \top > \varphi$ and the *outer necessity* $\Box \varphi = \neg\varphi > \bot$ are KD necessities, $\Box$ implies $\boxdot$, and object language versions of the belief-revision postulates of Inclusion, Preservation and Strong Preservation hold:

Observation 2 VN *derives*

$\Box \varphi \rightarrow \boxdot \varphi$	OI
$(\varphi > \psi) \rightarrow \boxdot(\varphi \rightarrow \psi)$	INC
$(\boxdot \psi \wedge \Diamond \varphi) \rightarrow (\varphi > \psi)$	PRES
$(\boxdot(\varphi \rightarrow \psi) \wedge \Diamond \varphi) \rightarrow (\varphi > \psi)$	SPRES

---
[4] Indeed, Cons follows from MP.
[5] The smallest such logic is CK = CE + RW + AND + $(\varphi > \top)$.

203

To model the logic VN, I will use a set-selection semantics (Chellas, 1975). $f\colon X \longrightarrow Y$ indicates that f is a total function from X to Y.[6]

Definition 2 $\mathfrak{M} = \langle W, f, V \rangle$ *is a* standard model *iff* $W \neq \emptyset$, $f\colon (W \times \wp(W)) \longrightarrow \wp(W)$ *and* $V\colon \mathrm{Var} \longrightarrow \wp(W)$. *It is* consistent Lewisean *iff f satisfies:*

$f(w, A) \subseteq A$	id
$f(w, A \cup B) \subseteq f(w, A) \cup f(w, B)$	or
If $f(w, A) \subseteq B$, then $f(w, A \cap B) \subseteq f(w, A)$	cmon
If $f(w, A) \not\subseteq \overline{B}$, then $f(w, A \cap B) \subseteq f(w, A)$	rmon
$f(w, W) \neq \emptyset$	cons

Points in W are *worlds*, subsets are *propositions*. The valuation V fixes truth of propositional variables in worlds. The *set-selection function f* associates to every world and every proposition A a proposition $f(w, A)$ – the closest A-worlds to w. The semantics is equivalent to Rott's (2019; 2020) full belief-revision semantics (Raidl, 2020a, Def. 12).[7] For this we need to understand $f(w, A)$ as belief core or the strongest believed proposition after a revision by A, and $f(w, W)$ as prior belief core, so that $\boxdot$ represents belief.

Definition 3 (Truth for $\mathcal{L}_>$) Truth *in $\mathfrak{M}$ for variables and classical connectives is defined inductively as usual, and*

$(>)$ $w \vDash_> \varphi > \psi$ iff $f(w, [\varphi]_>) \subseteq [\psi]_>$

where $[\varphi]_> := \{w \in W : w \vDash_> \varphi\}$ *is defined once* $w \vDash_> \varphi$ *is defined.*

We write $\mathfrak{M} \vDash_> \varphi$ iff for all $w \in W$, $w \vDash_> \varphi$. By slightly altering the canonical model construction from Lewis (1971) for V, one easily proves:

Theorem 1 VN *is sound and complete for consistent Lewisean models.*

VN is also the logic of Rott's full belief-revision semantics (Raidl, 2020a, Theorem 15) and of Spohn's ranking semantics (Raidl, 2019a, Theorem 3.6).

3 Translating back and forth

Let's consider another conditional language $\mathcal{L}_\beth$, like $\mathcal{L}_>$, except that $\beth$ replaces $>$ and $\mathcal{L}_\beth$ might contain $\boxdot$ as primitive (non-definable) symbol. The shared fragment with $\mathcal{L}_>$ is the classical propositional language $\mathcal{L}$.

---

[6] One could also use a sphere semantics or an ordering semantics.

[7] It has the belief-revision properties *1–*8,*8c, assuming consistent prior belief *0. *1, *6 correspond to a standard f, *2, *7, *8c, *8, *0 to (id), (or), (cmon), (rmon), (cons). Inclusion *3, Preservation *4 and Rott's *5a, b (the outer modality is normal) are redundant.

Three Conditionals

Consider $\mathcal{L}_\gg, \mathcal{L}_\succ$ and $\mathcal{L}_\triangleright$, where $\gg, \succ$ and $\triangleright$ are the difference-making, the dependency and the contraposition conditional. Only $\mathcal{L}_\triangleright$ contains $\square$ as primitive symbol. The models are the same, only the interpretation of the conditional changes:

Definition 4 (Truth for $\mathcal{L}_\gg, \mathcal{L}_\succ, \mathcal{L}_\triangleright$) *The truth clauses for propositional variables and classical connectives are as in $\mathcal{L}_>$, and*

$(\gg)$ $w \vDash_\gg \varphi \gg \psi$ iff $f(w, [\varphi]_\gg) \subseteq [\psi]_\gg$ & $f(w, [\neg\varphi]_\gg) \not\subseteq [\psi]_\gg$
$(\succ)$ $w \vDash_\succ \varphi \succ \psi$ iff $f(w, [\varphi]_\succ) \subseteq [\psi]_\succ$ & $f(w, [\neg\varphi]_\succ) \subseteq [\neg\psi]_\succ$
$(\triangleright)$ $w \vDash_\triangleright \varphi \triangleright \psi$ iff $f(w, [\varphi]_\triangleright) \subseteq [\psi]_\triangleright$ & $f(w, [\neg\psi]_\triangleright) \subseteq [\neg\varphi]_\triangleright$
 $w \vDash_\triangleright \square \varphi$ iff $f(w, W) \subseteq [\varphi]_\triangleright$

In $\mathcal{L}_>$, we have $\square \varphi \equiv \top > \varphi$. $\square$ is also definable in $\mathcal{L}_\gg$ and in $\mathcal{L}_\succ$, namely by $\square^S \varphi = \neg(\bot \gg \varphi)$ (Raidl, 2020a, Section 7) and $\square \varphi = \top \succ \varphi$. Since $\square$ is not definable by $\triangleright$ alone (Rott, 2020), we need it as a logical operator. If $>$ were in the languages $\mathcal{L}_\gg, \mathcal{L}_\succ$ and $\mathcal{L}_\triangleright$, we could define $\gg, \succ, \triangleright$ as in the Introduction. Since $>$ is not in these languages, we need a proxy. Let $\alpha \in \mathcal{L}_>$ and p, q variables. I write $\alpha = \alpha[p, q]$ iff α has its variables among $\{p, q\}$. $\alpha[p, q] \in \mathcal{L}_>$ abbreviates $\alpha \in \mathcal{L}_>$ and $\alpha = \alpha[p, q]$. $\alpha[\varphi/p, \psi/q]$ is the formula obtained by substituting φ for p and ψ for q in α.

Definition 5 $\circ: \mathcal{L}_\beth \longrightarrow \mathcal{L}_>$ *is a translation iff* $p^\circ = p$, $(\neg\varphi)^\circ = \neg\varphi^\circ$, $(\varphi * \psi)^\circ = (\varphi^\circ * \psi^\circ)$ *for* $* \in \{\wedge, \vee, \rightarrow\}$, *and there is* $\alpha[p, q] \in \mathcal{L}_>$ *such that for all* $\varphi, \psi \in \mathcal{L}_\beth$, $(\varphi \beth \psi)^\circ = \alpha[\varphi^\circ/p, \psi^\circ/q]$. *And if $\square$ is primitive in $\mathcal{L}_\beth$, there is* $\beta[p] \in \mathcal{L}_>$ *such that for all* $\varphi \in \mathcal{L}_\beth$, $(\square\varphi)^\circ = \beta[\varphi^\circ/p]$.

For $\varphi \in \mathcal{L}$, we obtain $\varphi^\circ = \varphi$, so that $\top^\circ = \top$ and $\bot^\circ = \bot$.

The translation essentially depends on the translation of $\beth$ (and eventually $\square$). α provides a *form* for the translate of $\varphi \beth \psi$. (Similarly for β.) One such form can be read off from the semantic definition of $\beth$ (and $\square$):

$(\circ_\gg)$ $(\varphi \gg \psi)^\circ := (\varphi^\circ > \psi^\circ) \wedge \neg(\neg\varphi^\circ > \psi^\circ)$
$(\circ_\succ)$ $(\varphi \succ \psi)^\circ := (\varphi^\circ > \psi^\circ) \wedge (\neg\varphi^\circ > \neg\psi^\circ)$
$(\circ_\triangleright)$ $(\varphi \triangleright \psi)^\circ := (\varphi^\circ > \psi^\circ) \wedge (\neg\psi^\circ > \neg\varphi^\circ)$
 $(\square\varphi)^\circ := (\top > \varphi^\circ)$

A translation should conserve meaning in the following sense:

Definition 6 *Let* $\circ: \mathcal{L}_\beth \longrightarrow \mathcal{L}_>$ *be a translation and M a model class in $\mathcal{L}_>$. We define the* automorphism modulo $\circ$, $M \stackrel{\circ}{\approx} M$, *iff for all* $\alpha \in \mathcal{L}_\beth$, *all* $\mathfrak{M} \in M$ *and all* $w \in W$, *we have*: $w \vDash_\beth \alpha$ *iff* $w \vDash_> \alpha^\circ$.

This builds a semantic bridge between $\vDash_{\sqsupset}$ and $\vDash_>$ which allows to understand the unknown ($\mathcal{L}_{\sqsupset}$) in terms of the known ($\mathcal{L}_>$). The translation $\circ$ is semantically well behaved in the sense that the translate φ° expresses the same proposition in $\mathcal{L}_>$ as the original formula φ in $\mathcal{L}_{\sqsupset}$.

Lemma 1 (Translation) *Let M be the class of standard models and $\circ = \circ_\triangleright, \circ_\gg,$ or $\circ_\succ$. Then $M \overset{\circ}{\approx} M$.*

Proof. $\circ_\triangleright$: By induction on the complexity of the formula. It suffices to verify $\triangleright$ and $\Box$. Let $\alpha = (\varphi \triangleright \psi)$ and assume the property for φ and ψ (our induction hypothesis IH). Thus $[\varphi]_\triangleright = [\varphi^\circ]_>$, and similarly for ψ. Hence:

$$
\begin{array}{lll}
w \vDash_\triangleright \varphi \triangleright \psi & \text{iff} & f(w, [\varphi]_\triangleright) \subseteq [\psi]_\triangleright \ \& \ f(w, [\neg\psi]_\triangleright) \subseteq [\neg\varphi]_\triangleright \quad \triangleright \\
 & \text{iff} & f(w, [\varphi^\circ]_>) \subseteq [\psi^\circ]_> \ \& \ f(w, [\neg\psi^\circ]_>) \subseteq [\neg\varphi^\circ]_> \quad \text{IH} \\
 & \text{iff} & w \vDash_> (\varphi^\circ > \psi^\circ) \wedge (\neg\psi^\circ > \neg\varphi^\circ) \quad > \\
 & \text{iff} & w \vDash_> (\varphi \triangleright \psi)^\circ \quad \circ
\end{array}
$$

Similarly for $\alpha = \Box \varphi$.

For $\circ_\gg$, see Raidl (2020a, Lemma 2). The argument is similar for $\circ_\succ$. $\square$

It is more difficult to find a *backtranslation* $\bullet$ of $>$ into $\sqsupset$:

$$
\begin{array}{lll}
(\bullet'_\gg) & (\varphi > \psi)^{\bullet'} & := \neg(\neg\varphi^{\bullet'} \gg (\neg\varphi^{\bullet'} \vee \psi^{\bullet'})) \\
(\bullet_\gg) & (\varphi > \psi)^\bullet & := \Box^S(\varphi^\bullet \wedge \psi^\bullet) \vee (\varphi^\bullet \gg (\varphi^\bullet \wedge \psi^\bullet)) \\
(\bullet_\succ) & (\varphi > \psi)^\bullet & := \varphi^\bullet \succ (\varphi^\bullet \wedge \psi^\bullet) \\
(\bullet_\triangleright) & (\varphi > \psi)^\bullet & := \Box(\varphi^\bullet \wedge \psi^\bullet) \vee (\varphi^\bullet \triangleright (\varphi^\bullet \wedge \psi^\bullet))
\end{array}
$$

$\bullet_\triangleright$ modifies the backtranslation from Crupi and Iacona (2020). The clauses for $\bullet'_\gg$ and $\bullet_\succ$ were introduced semantically by Rott (2019, Def2*, Ddef*). $\bullet_\gg$ is a variation of an alternative proposal (Rott, 2019, footnote 20). Raidl (2020a) uses $\bullet'_\gg$ to investigate $\gg$ in weaker semantics. For comparative reasons, I use $\bullet_\gg$. The backtranslations are also well behaved, but we won't need it. Translations are powerful tools. They allow deriving results for a defined conditional $\sqsupset$ from known results for $>$ (Raidl, 2020a). We will use this in what follows to obtain a complete axiomatization of $\triangleright, \gg$ and $\succ$.

Since the above languages $\mathcal{L}_\gg, \mathcal{L}_\succ, \mathcal{L}_\triangleright$ use the same models as $\mathcal{L}_>$, with a different interpretation of the conditional, we need a qualified notion of soundness and completeness: L is *sound for the class of models M (interpreted) in $\mathcal{L}_\sqsupset$* iff $\vdash_L \alpha$ implies $M \vDash_\sqsupset \alpha$. L is *complete for the class of models M (interpreted) in $\mathcal{L}_\sqsupset$* iff $M \vDash_\sqsupset \alpha$ implies $\vdash_L \alpha$.

4 Contraposition conditional

Definition 7 *The* contraposition logic C *is* CE + ID + AND + CMon *augmented by the Necessitation rule* N *for* $\Box$, *and*:

$(\varphi \triangleright \bot) \to (\varphi \triangleright \psi)$	IA
$(\varphi \triangleright \psi) \to (\neg\psi \triangleright \neg\varphi)$	C0
$((\varphi \triangleright (\varphi \land \psi)) \land ((\psi \lor \varphi) \triangleright \psi)) \to (\varphi \triangleright \psi)$	C1
$(\Diamond \varphi \land ((\psi \lor \varphi) \triangleright \psi)) \to (\varphi \triangleright (\varphi \land \psi))$	C2
$(\Box(\varphi \to \psi) \land \Diamond \varphi \land \Diamond \neg\psi) \to (\varphi \triangleright \psi)$	SPRES^C
$(\varphi \triangleright \psi) \to \Box(\varphi \to \psi)$	$\Box$MP
$\Box(\varphi \to \psi) \to (\Box\varphi \to \Box\psi)$	K
$\neg \Box \bot$	$\Box$Cons
$((\varphi \triangleright (\varphi \land \psi)) \land \Diamond \neg\varphi) \to (\varphi \triangleright (\varphi \land (\psi \lor \chi)))$	DW^C
$((\varphi \triangleright (\varphi \land \chi)) \land (\psi \triangleright (\psi \land \chi)) \land \Diamond \neg(\varphi \lor \psi)) \to ((\varphi \lor \psi) \triangleright ((\varphi \lor \psi) \land \chi))$	OR^C
$((\varphi \triangleright (\varphi \land \chi)) \land \neg(\varphi \triangleright (\varphi \land \neg\psi)) \land \Diamond \neg\varphi) \to ((\varphi \land \psi) \triangleright (\varphi \land \psi \land \chi))$	RMon^C

C is a classical conditional logic. But contrary to its basic companion VN, it is non-normal, since DW fails. However, it has the weak substitute DW^C. By N, K and $\Box$ Cons, $\Box$ is a KD necessity. C1 and C2 are needed for technical reasons. By C0, $\triangleright$ contraposes. Due to this, OR is derivable from AND. By ID, AND, OR and C0, we also have the reverse of C1. SPRES^C, $\Box$ MP, DW^C, OR^C and RMon^C are obtained as backtranslates from the originals SPRES, INC, DW, OR and RMon. Due to C0, each scheme X for $\triangleright$ has a *contraposed* (or converse-dual) X^c obtained by transforming $\varphi \triangleright \psi$ into $\neg\psi \triangleright \neg\varphi$. C0, C1, SPRES^C and $\Box$ MP are their own contraposed, and the contraposed to AND is OR. The stronger evidential conditional logic EC was proven complete in strongly centered Lewisean models (Raidl et al., 2021).[8]

To establish completeness, we need to prove three additional facts. First, for α a formula of $\mathcal{L}_\triangleright$ let's call $\alpha^{\circ\bullet}$ its *twin*. The following Lemma shows that α is provably equivalent to its twin:

Lemma 2 (Twin) *Let α in $\mathcal{L}_\triangleright$. Then $\vdash_C \alpha \leftrightarrow \alpha^{\circ\bullet}$.*

Proof. By induction on the complexity of the formula. It suffices to consider $\triangleright$ and $\Box$. Denote $\equiv$ provable equivalence in C. Let $\alpha = (\varphi \triangleright \psi)$. Assume $\varphi \equiv \varphi^{\circ\bullet}$ and $\psi \equiv \psi^{\circ\bullet}$ (IH). $(\varphi \triangleright \psi)^{\circ\bullet}$ resolves into $((\varphi^\circ > \psi^\circ) \land (\neg\psi^\circ > \neg\varphi^\circ))^\bullet$ which further resolves into the conjunction of $(\Box(\varphi^{\circ\bullet} \land \psi^{\circ\bullet}) \lor (\varphi^{\circ\bullet} \triangleright (\varphi^{\circ\bullet} \land \psi^{\circ\bullet})))$ and $(\Box(\neg\psi^{\circ\bullet} \land$

---

[8] EC = CE + ID + AND + CMon + MP + OR* + RM* + IA + C0 + D. $\Box$ is not needed and replaced by truth assumptions, e.g., MP [MI] instead of $\Box$ MP, and OR*, RM*, D are similarly obtained from OR^C, RMon^C and C1–2. For other axioms, see Raidl (2019b) and Rott (2020).

$\neg\varphi^{\circ\bullet}) \vee (\neg\psi^{\circ\bullet} \rhd (\neg\psi^{\circ\bullet} \wedge \neg\varphi^{\circ\bullet})))$. Using IH and substitution of equivalents (CE + N + K), this is equivalent to the conjunction of $(\Box(\varphi \wedge \psi) \vee (\varphi \rhd (\varphi \wedge \psi)))$ and $(\Box(\neg\varphi \wedge \neg\psi) \vee (\neg\psi \rhd (\neg\psi \wedge \neg\varphi)))$, where the last conditional can be replaced by $(\psi \vee \varphi) \rhd \psi$, due to C0 + CE. Due to the KD necessity, the mentioned conjunction is equivalent to the disjunction of $b = (\Box\varphi \wedge \Box\psi \wedge ((\psi \vee \varphi) \rhd \psi))$, $c = ((\varphi \rhd (\varphi \wedge \psi)) \wedge \Box\neg\varphi \wedge \Box\neg\psi)$, $d = ((\varphi \rhd (\varphi \wedge \psi)) \wedge ((\psi \vee \varphi) \rhd \psi))$. But $\varphi \rhd \psi$ implies d (ID + AND + OR). Conversely: each disjunct implies $\varphi \rhd \psi$: b implies $\varphi \rhd \psi$ by C2, C1 and $\Box$Cons. c: similar reasoning. d implies $\varphi \rhd \psi$ by C1.
Similarly for $\alpha = \Box \varphi$, using CE + N + K + $\Box$MP. □

Second, the following Lemma proves that any proof of α in C can be *simulated* in VN by proving its translate α°:

Lemma 3 (Simulation) *Let $\alpha \in \mathcal{L}_\rhd$. If $\vdash_C \alpha$ then $\vdash_{VN} \alpha^\circ$.*

Proof. Assume $\vdash_C \alpha$. We show $\vdash_{VN} \alpha^\circ$ by induction on the length of the proof. We freely use RW, INC and SPRES (Observation 1–2).

The proof is of length 1. Then α is an axiom of C. There are 15 possible cases, depending on whether α is an instance of PT, ID, AND, IA, C0, C1, C2, SPRES^C, $\Box$MP, K, $\Box$Cons, DW^C, OR^C, CMon, or RMon^C. For the simulation of ID, AND, CMon [CM], IA, C0 [C] see Raidl et al. (2021, Fact 11). For PT, see Raidl (2020b, Lemma 5). C1, $\Box$MP, K, Cons are easily simulated by RW, INC, AND + RW, and Cons.

C2. The translate is $(a \wedge b \wedge c) \to (d \wedge e)$, where $a = \neg(\top > \neg\varphi)$, $b = ((\psi \vee \varphi) > \psi)$, $c = (\neg\psi > \neg(\psi \vee \varphi))$, and $d = (\varphi > (\varphi \wedge \psi))$, $e = (\neg(\psi \wedge \varphi) > \neg\varphi)$. Assume $a \wedge b \wedge c$. From c we get $\neg\psi > \neg\varphi$ by RW, hence e by ID + OR + LLE. From b we get $\top > (\varphi \to \psi)$ by INC + RLE. With a this yields $\varphi > \psi$ by SPRES. Thus d by ID + AND.

SPRES^C. The translate is $(a \wedge b \wedge c) \to (d \wedge e)$, where $a = (\top > (\varphi \to \psi))$, $b = \neg(\top > \neg\varphi)$, $c = \neg(\top > \neg\neg\psi)$, and $d = (\varphi > \psi)$, $e = (\neg\psi > \neg\varphi)$. Assume $a \wedge b \wedge c$. $a \wedge b$ yields d by SPRES. From a we get $a' = (\top > (\neg\psi \to \neg\varphi))$ by RLE. But $a' \wedge c$ yields e by SPRES.

DW^C. The translate is CE-equivalent to $(a \wedge b \wedge c) \to (d \wedge e)$, where $a = (\varphi > (\varphi \wedge \psi))$, $b = ((\neg\varphi \vee \neg\psi) > \neg\varphi)$, $c = \neg(\top > \neg\neg\varphi)$, and $d = (\varphi > (\varphi \wedge (\psi \vee \chi)))$, $e = ((\neg\varphi \vee (\neg\psi \wedge \neg\chi)) > \neg\varphi)$. a implies d by RW. b implies $\Box((\neg\varphi \vee \neg\psi) \to \neg\varphi)$ by INC. Thus also $\Box((\neg\varphi \vee (\neg\psi \wedge \neg\chi)) \to \neg\varphi)$. But $c = \Diamond \neg\varphi$ and thus $\Diamond(\neg\varphi \vee (\neg\psi \wedge \neg\chi))$. Hence by SPRES, we obtain e.

Three Conditionals

OR^C. The translate is CE-equivalent to $(a_1 \wedge a_2 \wedge b_1 \wedge b_2 \wedge c_1) \to (d_1 \wedge d_2)$, where $a_1 = (\varphi > (\varphi \wedge \chi))$, $a_2 = ((\neg\varphi \vee \neg\chi) > \neg\varphi)$, $b_1 = (\psi > (\psi \wedge \chi))$, $b_2 = ((\neg\psi \vee \neg\chi) > \neg\psi)$, $c_1 = \neg(\top > (\varphi \vee \psi))$, $d_1 = ((\varphi \vee \psi) > ((\varphi \vee \psi) \wedge \chi))$, $d_2 = (((\neg\varphi \wedge \neg\psi) \vee \neg\chi) > (\neg\varphi \wedge \neg\psi))$. From a_1 we obtain $\varphi > \chi$ by RW, from b_1 we get $\psi > \chi$. Thus $(\varphi \vee \psi) > \chi$ by OR. Hence d_1 by ID + AND. Let's establish d_2: From a_1 we obtain $\Box(\varphi \to (\varphi \wedge \chi))$ by INC, hence $\Box(\varphi \to \chi)$. Similarly, from b_1 we get $\Box(\psi \to \chi)$. Thus $\Box((\varphi \vee \psi) \to \chi)$. Therefore $\Box(\neg\chi \to (\neg\varphi \wedge \neg\psi))$ and thus $e = \Box(((\neg\varphi \wedge \neg\psi) \vee \neg\chi) \to (\neg\varphi \wedge \neg\psi))$. From c_1 we obtain $\Diamond(\neg\varphi \wedge \neg\psi)$ and thus $f = \Diamond((\neg\varphi \wedge \neg\psi) \vee \neg\chi)$. But $e \wedge f$ imply d_2 by SPRES.

RMon^C. The translate is CE-equivalent to $(a_1 \wedge a_2 \wedge (\neg b_1 \vee \neg b_2) \wedge c_1) \to (d_1 \wedge d_2)$, where $a_1 = (\varphi > (\varphi \wedge \chi))$, $a_2 = ((\neg\varphi \vee \neg\chi) > \neg\varphi)$, $b_1 = (\varphi > (\varphi \wedge \neg\psi))$, $b_2 = ((\neg\varphi \vee \psi) > \neg\varphi)$, $c_1 = \neg(\top > \varphi)$, $d_1 = ((\varphi \wedge \psi) > (\varphi \wedge \psi \wedge \chi))$, $d_2 = ((\neg\varphi \vee \neg\psi \vee \neg\chi) > (\neg\varphi \vee \neg\psi))$. From a_2 we obtain $(\neg\varphi \vee \neg\chi) > (\neg\varphi \vee \neg\psi)$ by DW. But $\neg\psi > (\neg\varphi \vee \neg\psi)$ by ID + DW. Thus d_2 by OR + LLE. From a_1 we obtain $\varphi > \chi$ by RW and from $\neg b_1$ we get $\neg(\varphi > \neg\psi)$, using ID + AND. Thus $(\varphi \wedge \psi) > \chi$ by RMon. Hence d_1 by ID + AND. Suppose instead that b_1 and hence $\neg b_2$. Then SPRES yields $e_1 = \neg\Box((\neg\varphi \vee \psi) \to \neg\varphi)$ or $e_2 = \neg\Diamond(\neg\varphi \vee \psi)$. e_2 is excluded, since it would imply $\Box(\varphi \wedge \neg\psi)$, which contradicts $c_1 = \neg\Box\varphi$. Thus e_1. This implies successively $\neg\Box(\psi \to \neg\varphi)$, $\Diamond\neg(\psi \to \neg\varphi)$, $f = \Diamond(\varphi \wedge \psi)$. Yet a_1 also implies $\Box(\varphi \to (\varphi \wedge \chi))$ by INC and hence $g = \Box((\varphi \wedge \psi) \to (\varphi \wedge \psi \wedge \chi))$. But $g \wedge f$ yield d_1 by SPRES.

Suppose the proof of α in C is of length $n + 1$ and (IH) for all previous formulas β in the proof, until the n-th included, we have $\vdash_{VN} \beta^\circ$. If α is an axiom the reasoning is as before. Thus suppose α is obtained by one of the 4 rules of C from previous formulas. MoPo, LLE and RLE can be simulated in VN (Raidl, 2020b, Lemma 5). Suppose α is obtained by the necessitation rule N. Thus α° is of the form $\top > \varphi$ and by IH, we have $\vdash_{VN} \varphi$. Thus $\vdash_{VN} \top \to \varphi$. But $\top > \top$ by ID. Hence $\top > \varphi$ by RW. □

Third, the following Lemma proves that any proof of α in VN can be *back-simulated* in C by proving its backtranslate $\alpha^\bullet$:

Lemma 4 (Backsimulation) *Let $\alpha \in \mathcal{L}_>$. Then, if $\vdash_{VN} \alpha$ then $\vdash_C \alpha^\bullet$.*

Proof. Assume $\vdash_{VN} \alpha$. We show $\vdash_C \alpha^\bullet$ by induction on the proof length.
Length 1. Then α is an axiom of VN. (We use that $\Box$ is KD.) PT works as in Lemma 3. ID and Cons are simulated by ID and $\Box$Cons + $\Box$MP.

AND. The backtranslate is equivalent to $((a \vee b) \wedge (c \vee d)) \to (e \vee f)$, where $a = \Box(\varphi \wedge \psi)$, $b = (\varphi \triangleright (\varphi \wedge \psi))$, $c = \Box(\varphi \wedge \chi)$, $d = (\varphi \triangleright (\varphi \wedge \chi))$, and $e = \Box(\varphi \wedge \psi \wedge \chi)$, $f = (\varphi \triangleright (\varphi \wedge \psi \wedge \chi))$. If $a \wedge c$, we get e. If $a \wedge d$, we get $\Box \varphi$, and from d we get $\Box(\varphi \to (\varphi \wedge \chi))$ ($\Box$ MP), thus $\Box \chi$ (K). Hence e. Similarly if $b \wedge c$. And $b \wedge d$ implies f by AND + RLE.

DW. The backtranslate is $(a \vee b) \to (c \vee d)$, with $a = \Box(\varphi \wedge \psi)$, $b = (\varphi \triangleright (\varphi \wedge \psi))$, and $c = \Box(\varphi \wedge (\psi \vee \chi))$, $d = (\varphi \triangleright (\varphi \wedge (\psi \vee \chi)))$. a implies c. Suppose b. If $e = \Diamond \neg \varphi$, we get d (DW^C). If $\neg e = \Box \varphi$, and since b implies $\Box(\varphi \to (\varphi \wedge \psi))$ ($\Box$ MP), we get a, thus c.

OR. The backtranslate is $((a_1 \vee a_2) \wedge (b_1 \vee b_2)) \to (c_1 \vee c_2)$, with $a_1 = \Box(\varphi \wedge \chi)$, $a_2 = (\varphi \triangleright (\varphi \wedge \chi))$, $b_1 = \Box(\psi \wedge \chi)$, $b_2 = (\psi \triangleright (\varphi \wedge \chi))$, $c_1 = \Box((\varphi \vee \psi) \wedge \chi)$, $c_2 = ((\varphi \vee \psi) \triangleright ((\varphi \vee \psi) \wedge \chi))$. a_1 yields c_1. Similarly for b_1. Thus suppose $a_2 \wedge b_2$. From a_2 we get $\Box(\varphi \to (\varphi \wedge \chi))$ by $\Box$ MP, thus $\Box(\varphi \to \chi)$. Similarly from b_2 we get $\Box(\psi \to \chi)$. Hence $\Box((\varphi \vee \psi) \to \chi)$. Assume $d = \Box(\varphi \vee \psi)$. Then we obtain $\Box \chi$ by K and hence c_1. Assume $\neg d$, i.e., $\Diamond \neg(\varphi \vee \psi)$. Then c_2 by OR^D.

CMon. The backtranslate is equivalent to $((a_1 \vee a_2) \wedge (b_1 \vee b_2)) \to (c_1 \vee c_2)$, where $a_1 = \Box(\varphi \wedge \psi)$, $a_2 = (\varphi \triangleright (\varphi \wedge \psi))$, $b_1 = \Box(\varphi \wedge \chi)$, $b_2 = (\varphi \triangleright (\varphi \wedge \chi))$, $c_1 = \Box(\varphi \wedge \psi \wedge \chi)$, $c_2 = ((\varphi \wedge \psi) \triangleright (\varphi \wedge \psi \wedge \chi))$. Again a_2 implies $\Box(\varphi \to \psi)$ and b_2 implies $d = \Box(\varphi \to \chi)$. Thus: $a_1 \wedge b_1$ implies c_1. $a_1 \wedge b_2$ implies $\Box \varphi \wedge d$ and hence $\Box \chi$ (K), hence c_1. Similarly $a_2 \wedge b_1$ implies c_1. $a_2 \wedge b_2$ implies c_2 by CMon and LLE.

RMon. The backtranslate is equivalent to $(a_1 \wedge a_2 \wedge (b_1 \vee b_2)) \to (c_1 \vee c_2)$, where $a_1 = \neg \Box(\varphi \wedge \neg \psi)$, $a_2 = \neg(\varphi \triangleright (\varphi \wedge \neg \psi))$, $b_1 = \Box(\varphi \wedge \chi)$, $b_2 = (\varphi \triangleright (\varphi \wedge \chi))$, $c_1 = \Box(\varphi \wedge \psi \wedge \chi)$, $c_2 = ((\varphi \wedge \psi) \triangleright (\varphi \wedge \psi \wedge \chi))$. First suppose b_1. Thus $\Box \varphi$ and $\Box \chi$. If $\Box \psi$, we get c_1. Thus suppose $\neg \Box \psi$, i.e., $\Diamond \neg \psi$. By a_1 we have $\neg \Box \varphi$ or $\neg \Box \neg \psi$. Since the first is excluded, we get $\Diamond \psi$. Since we have $\Box \chi$, we also have $\Box((\varphi \wedge \psi) \to \chi)$ as well as $d_1 = \Box((\varphi \wedge \psi) \to (\varphi \wedge \psi \wedge \chi))$, since we have $\Box \varphi$ and $\Diamond \psi$, we also have $d_2 = \Diamond(\varphi \wedge \psi)$. Finally, since $\Diamond \neg \psi$, we also obtain $d_3 = \Diamond \neg(\varphi \wedge \psi \wedge \chi)$. But $d_1 \wedge d_2 \wedge d_3$ imply c_2 by SPRES^C. Now suppose $\neg b_1$. Thus b_2. If $\neg \Box \varphi$, then with $a_2 \wedge b_2$ this yields c_2 by RMon^C. Hence assume $\Box \varphi$. From b_2, we obtain $\Box(\varphi \to (\varphi \wedge \chi))$ by $\Box$ MP, hence $d_1 = \Box((\varphi \wedge \psi) \to (\varphi \wedge \psi \wedge \chi))$. From a_1, and since $\Box \varphi$, we get $\Diamond \psi$. Hence $d_2 = \Diamond(\varphi \wedge \psi)$. From $\neg b_1$, and since $\Box \varphi$, we get $\Diamond \neg \chi$. From this we get $d_3 = \Diamond \neg(\varphi \wedge \psi \wedge \chi)$. But $d_1 \wedge d_2 \wedge d_3$ implies c_2 (SPRES^C).

Length $n+1$. Suppose for all previous formulas β in the proof, until the n-th included, we have $\vdash_C \beta^\bullet$ (IH). If α is an axiom the reasoning is as before.

Suppose α is obtained by one of the 3 rules of VN. The simulation of MoPo works by MoPo.

Suppose α is obtained by LLE. Then there is a formula $\delta = (\varphi \leftrightarrow \psi)$ such that $\vdash_{VN} \varphi \leftrightarrow \psi$ with proof $\leq n$ and $\alpha^\bullet$ is of the form $(\Box(\varphi^\bullet \wedge \chi^\bullet) \vee (\varphi^\bullet \triangleright (\varphi^\bullet \wedge \chi^\bullet))) \to (\Box(\psi^\bullet \wedge \chi^\bullet) \vee (\psi^\bullet \triangleright (\psi^\bullet \wedge \chi^\bullet)))$. By IH we have $\vdash_C \varphi^\bullet \leftrightarrow \psi^\bullet$ and thus also $\vdash_C (\varphi^\bullet \wedge \chi^\bullet) \leftrightarrow (\psi^\bullet \wedge \chi^\bullet)$. Using standard reasoning for $\Box$, we obtain that $a = \Box(\varphi^\bullet \wedge \chi^\bullet) \to \Box(\psi^\bullet \wedge \chi^\bullet)$ is provable in C. Using LLE, $(\varphi^\bullet \triangleright (\varphi^\bullet \wedge \chi^\bullet)) \to (\psi^\bullet \triangleright (\varphi^\bullet \wedge \chi^\bullet))$ is provable in C, and by RLE, $(\psi^\bullet \triangleright (\varphi^\bullet \wedge \chi^\bullet)) \to (\psi^\bullet \triangleright (\psi^\bullet \wedge \chi^\bullet))$ is provable in C. Thus $b = (\varphi^\bullet \triangleright (\varphi^\bullet \wedge \chi^\bullet)) \to (\psi^\bullet \triangleright (\psi^\bullet \wedge \chi^\bullet))$ is provable in C. Since $a \wedge b$ are provable, we also have that $\alpha^\bullet$ is provable in C.

For α obtained by RLE, the reasoning is similar. □

From the previous results it easily follows that:

Theorem 2 C *is sound and complete for consistent Lewisean models in* $\mathcal{L}_\triangleright$.

Proof. Denote M the consistent Lewisean models. Suppose $\vdash_C \alpha$. Thus $\vdash_{VN} \alpha°$ (Lemma 3). Hence $M \vDash_> \alpha°$ (Theorem 1). Therefore $M \vDash_\triangleright \alpha$ (Lemma 1). Suppose $M \vDash_\triangleright \alpha$. Then $M \vDash_> \alpha°$ (Lemma 1). Thus $\vdash_{VN} \alpha°$ (Theorem 1). Therefore $\vdash_C \alpha°{}^\bullet$ (Lemma 4). Hence $\vdash_C \alpha$ (Lemma 2). □

5 Sufficient reason & difference making

It is easily seen that in $\mathcal{L}_\gg$ the modality $\Box^S \alpha$ defined by $f(w, W) \subseteq [\alpha]_\gg$ can be expressed by $\neg(\bot \gg \alpha)$, and $\Box^S \alpha$ defined by $f(w, [\neg\alpha]_\gg) \subseteq \emptyset$ can be expressed by $\neg \alpha \gg \bot$ (or by $\neg(\alpha \gg \alpha)$).

Definition 8 *The* difference-making logic S *is* CE + AND *augmented by:*

$(\varphi \gg \psi) \leftrightarrow ((\varphi \gg (\varphi \vee \psi)) \wedge \neg(\neg\varphi \gg (\neg\varphi \vee \psi)))$	S0
$\neg(\varphi \gg \top)$	S1
$(\varphi \gg (\varphi \vee \psi \vee \chi)) \to (\varphi \gg (\varphi \vee \psi))$	S2
$((\varphi \gg (\varphi \wedge \chi)) \wedge (\psi \gg (\psi \wedge \chi)) \wedge \neg \Box^S(\varphi \vee \psi)) \to ((\varphi \vee \psi) \gg ((\varphi \vee \psi) \wedge \chi))$	OR^S
$(\varphi \gg (\varphi \wedge \psi)) \wedge (\varphi \gg (\varphi \wedge \chi))) \to ((\varphi \wedge \psi) \gg (\varphi \wedge \psi \wedge \chi))$	CMon^r
$(\varphi \gg (\varphi \wedge \chi)) \wedge \neg(\varphi \gg (\varphi \wedge \neg\psi))) \to ((\varphi \wedge \psi) \gg (\varphi \wedge \psi \wedge \chi))$	RMon^r
$\neg \Box^S \bot$	$\Box^S$Cons
$\neg(\neg\varphi \gg (\neg\varphi \vee \psi)) \to \Box^S(\varphi \to \psi)$	INC^S
$(\Box^S \psi \wedge \Diamond^S \varphi) \to \neg(\neg\varphi \gg (\neg\varphi \vee \psi))$	PRES^S

S is a classical conditional logic.[9] But contrary to VN, it is not normal. Disjunctive weakening (DW) fails. Instead, we have the new axioms S0–S2, etc.

---
[9] Raidl (2020a) investigates the weaker logic $S_0 := CE + S0 + S1 + S2$ and extensions.

Eric Raidl

S0 is needed to prove the Twin Lemma. The remaining axioms are obtained as backtranslates of the originals modulo some additional modifications. ID itself fails. CMon^r and RMon^r are *relativizations* of the originals. The relativization consists in putting the antecedent of a conditional as an additional conjunct into the consequent of the conditional. OR^S is an augmented relativization, with an additional modal premiss.[10] $\gg$ may also be interpreted as Spohn's (2015) sufficient reason relation (Raidl, 2020a, Sections 7,9).

Observation 3 S *derives the following laws:*

$\boxdot^S \top$ $\boxdot^S$ N
$(\boxdot^S(\varphi \to \psi) \wedge \boxdot^S \varphi) \to \boxdot^S \psi$ $\boxdot^S$ K
$(\varphi \gg \psi) \to \boxdot^S(\varphi \to \psi)$ $\boxdot^S$ MP
$(\boxdot^S(\varphi \to \psi) \wedge \diamondsuit^S \varphi) \to (\boxdot^S(\varphi \wedge \psi) \vee (\varphi \gg (\varphi \wedge \psi)))$ SPRES'
$\boxdot^S \varphi \to \boxdot^S \varphi$ OI^S

Thus $\boxdot^S$ is a KD necessity and one can show the same for $\square^S$.

Observation 4 *In* CE + S2, *S0 is equivalent to the conjunction of:*

$(\varphi \gg \psi) \leftrightarrow ((\varphi \gg (\varphi \wedge \psi)) \wedge (\varphi \gg (\varphi \vee \psi)))$ Sa
$(\varphi \gg (\varphi \wedge \psi)) \leftrightarrow (\neg(\neg \varphi \gg (\neg \varphi \vee \psi)) \wedge (\varphi \gg \varphi))$ Sb

CE + S0 + S1 + S2 *implies:*

$(\varphi \gg (\varphi \wedge \psi \wedge \chi)) \to (\varphi \gg (\varphi \wedge \psi))$ Sc
$(\varphi \gg (\varphi \wedge \psi)) \to (\varphi \gg (\varphi \wedge (\psi \vee \chi)))$ DW^r

Thus S0 may be replaced by Sa + Sb (Raidl, 2020a, Lemma 9), and DW^r relativizes DW.

Lemma 5 $\vdash_S \neg(\neg\alpha \gg (\neg\alpha \vee \beta)) \leftrightarrow (\boxdot^S(\alpha \wedge \beta) \vee (\alpha \gg (\alpha \wedge \beta)))$.

Proof. $\alpha \gg (\alpha \wedge \beta)$ implies $\neg(\neg\alpha \gg (\neg\alpha \vee \beta))$ (Sb). And $\boxdot(\alpha \wedge \beta)$ implies $\boxdot \beta$ and $\boxdot \alpha$, thus $\diamondsuit \alpha$, hence $\neg(\neg\alpha \gg (\neg\alpha \vee \beta))$ (PRES^S). Conversely: $\neg(\neg\alpha \gg (\neg\alpha \vee \beta))$ implies $\boxdot(\alpha \to \beta)$ (INC^S). If $\neg(\alpha \gg (\alpha \wedge \beta))$ then $\neg(\alpha \gg \alpha)$ (Sb). Thus $\boxdot \alpha$ (INC^S). Hence $\boxdot \beta$. Therefore $\boxdot(\alpha \wedge \beta)$. □

Thus, it does not matter whether we use the backtranslations $\bullet_\gg$ or $\bullet'_\gg$.

Observation 5 S *derives*

$((\varphi \vee \psi) \gg \bot) \leftrightarrow ((\varphi \gg \bot) \wedge (\psi \gg \bot))$ MC
$(\varphi \gg (\psi \wedge \chi)) \to ((\varphi \gg \psi) \vee (\varphi \gg \chi))$ AND^S

---
[10]Thus the $\triangleright$-axioms SPRES^C, DW^C, OR^C, RMon^C are also augmented relativizations.

Proof. $\varphi \gg \bot$ and $\psi \gg \bot$ yield $\varphi \gg (\varphi \wedge \bot)$ and $\psi \gg (\psi \wedge \bot)$ by Sa. They also yield $\Box^S \neg \varphi, \Box^S \neg \psi$. Thus $\Box^S \neg \varphi, \Box^S \neg \psi$ (OI^S). Hence $\Box^S(\neg \varphi \wedge \neg \psi)$, thus $\Diamond^S(\neg \varphi \wedge \neg \psi)$, i.e., $\Diamond^S \neg (\varphi \vee \psi)$. Therefore $\Diamond^S \neg (\varphi \vee \psi)$ (OI^S). Hence $(\varphi \vee \psi) \gg ((\varphi \vee \psi) \wedge \bot)$ by OR^S. Thus $(\varphi \vee \psi) \gg \bot$. Conversely, $(\varphi \vee \psi) \gg \bot$ implies $(\varphi \vee \psi) \gg ((\varphi \vee \psi) \wedge \bot)$ by Sa. Thus $(\varphi \vee \psi) \gg ((\varphi \vee \psi) \wedge \varphi \wedge \bot)$. Hence $(\varphi \vee \psi) \gg ((\varphi \vee \psi) \wedge \varphi)$ by Sc. Thus $((\varphi \vee \psi) \wedge \varphi) \gg ((\varphi \vee \psi) \wedge \varphi \wedge \bot)$ by CMon^r. Hence $\varphi \gg \bot$.

$\varphi \gg (\psi \wedge \chi)$ implies $\varphi \gg (\varphi \wedge \psi \wedge \chi)$ (Sa) and thus $a = (\varphi \gg (\varphi \wedge \psi))$ and $b = (\varphi \gg (\varphi \wedge \chi))$ (Sc). We also get $c = (\varphi \gg (\varphi \vee (\psi \wedge \chi)))$ (S0). If we had, $\neg(\varphi \gg \psi)$ and $\neg(\varphi \gg \chi)$ we would have $\neg(\varphi \gg (\varphi \vee \psi))$ and $\neg(\varphi \gg (\varphi \vee \psi))$ (Sa). Thus $\Box(\neg \varphi \wedge \psi) \vee (\neg \varphi \gg (\neg \varphi \wedge \psi))$ and $\Box(\neg \varphi \wedge \chi) \vee (\neg \varphi \gg (\neg \varphi \wedge \chi))$ by Lemma 5. By K, $\Box$ MP and AND, we get $\Box(\neg \varphi \wedge \psi \wedge \chi) \vee (\neg \varphi \gg (\neg \varphi \wedge \psi \wedge \chi))$. By Lemma 5 $\neg (\varphi \gg (\varphi \vee (\psi \wedge \chi)))$, contradicting c. $\square$

The logic S can be compared to the logic of Rott (2019):[11] $\Box^S$ Cons is Rott's ($\gg 0$), AND^S is ($\gg 1$), Sa and Sb are ($\gg 2a, b$), INC^S is ($\gg 3$), PRES^S corresponds to ($\gg 4$), MC is ($\gg 5$) and encodes monotonicity and closure under conjunction of $\Box^S$, LLE + RLE are encoded in ($\gg 6$), and OR^S, CMon^r and RMon^r are ($\gg 7', 8c, 8$). From Observations 4–5 and Rott's Lemma 5.1 it follows that Rott's axiomatics is equivalent to our S, since the former derives AND (his d26), S1 (d14) and S2 (d25). Another equivalent axiomatization of S is given in Raidl (2020a, Table 3 last line).[12] The advantage of the present axiomatization is its comparability to the contraposition and dependency logic, due to AND and the choice of relativized companions.

We prove again the Twin equivalence and the two simulation Lemmas.

Lemma 6 (Twin) $\vdash_S \alpha \leftrightarrow \alpha^{\circ \bullet}$

Proof. Denote $\equiv$ provable equivalence in S. It suffices to check the conditional $\alpha = \varphi \gg \psi$. Assume $\varphi \equiv \varphi^{\circ \bullet}$ and $\psi \equiv \psi^{\circ \bullet}$ (IH). $(\varphi \gg \psi)^{\circ \bullet}$ resolves into $((\varphi^\circ > \psi^\circ) \wedge \neg(\neg \varphi^\circ > \psi^\circ))^\bullet$, which resolves into $(\Box^S(\varphi^{\circ \bullet} \wedge \psi^{\circ \bullet}) \vee (\varphi^{\circ \bullet} > (\varphi^{\circ \bullet} \wedge \psi^{\circ \bullet}))) \wedge \neg(\Box^S(\neg \varphi^{\circ \bullet} \wedge \psi^{\circ \bullet}) \vee (\neg \varphi^{\circ \bullet} > (\neg \varphi^{\circ \bullet} \wedge \psi^{\circ \bullet})))$. By Lemma 5 and CE, this is equivalent to $\neg(\neg \varphi^{\circ \bullet} \gg (\neg \varphi^{\circ \bullet} \vee \psi^{\circ \bullet})) \wedge (\varphi^{\circ \bullet} \gg (\varphi^{\circ \bullet} \vee \psi^{\circ \bullet}))$, which by IH and CE is equivalent to $\neg(\neg \varphi \gg (\neg \varphi \vee \psi)) \wedge (\varphi \gg (\varphi \vee \psi))$ which is equivalent to $\varphi \gg \psi$ (by S0). $\square$

Lemma 7 (Simulation)   Let $\alpha \in \mathcal{L}_\gg$. If $\vdash_S \alpha$ then $\vdash_{VN} \alpha^\circ$.

---

[11] Compare the comments in Raidl (2020a, p. 99 and Section 9).

[12] It is S_0 +CC^S + CA^S + CMon^S + CV^S + P^S, where P^S $\equiv \Box^S$ Cons and CA^S, CMon^S, CV^S are alternative companions to OR, CMon, RMon, rendering INC^S and PRES^S redundant.

Proof. Assume $\vdash_S \alpha$. We show $\vdash_{VN} \alpha°$ by induction on the length of the proof, using the fact that $\Box$ is a KD modality.

Length 1. Then α is an axiom of S. The simulation of PT, S0, S1, S2 was shown by Raidl (2020a, Theorem 8). AND can be simulated by AND + RW, PRES^S by PRES, INC^S by INC and $\Box^S$ Cons by Cons. For the remaining three axioms, I reason more directly.

OR^S. $\varphi > (\varphi \wedge \chi)$ and $\psi > (\psi \wedge \chi)$ imply $(\varphi \vee \psi) > ((\varphi \vee \psi) \wedge \chi)$ by RW + OR. $\neg(\varphi \vee \psi) > ((\varphi \vee \psi) \wedge \chi)$ would imply $\neg(\varphi \vee \psi) > \bot$ (ID + AND + RW), contradicting $\neg(\neg(\varphi \vee \psi) > \bot)$.

CMon^r. $\varphi > (\varphi \wedge \chi)$ and $\varphi > (\varphi \wedge \psi)$ imply $(\varphi \wedge \psi) > (\varphi \wedge \psi \wedge \chi)$ by ID + AND + RW + CMon. $\neg(\varphi \wedge \psi) > (\varphi \wedge \psi \wedge \chi)$ would imply $\neg(\varphi \wedge \psi) > \bot$, contradicting $\neg(\neg\varphi > (\varphi \wedge \chi))$.

RMon^r. $\varphi > (\varphi \wedge \chi)$ and $\neg(\varphi > (\varphi \wedge \neg\psi))$ imply $(\varphi \wedge \psi) > (\varphi \wedge \psi \wedge \chi)$ by RW + RMon. $\neg(\varphi \wedge \psi) > (\varphi \wedge \psi \wedge \chi)$ would imply $\neg(\varphi \wedge \psi) > \bot$, i.e., $\Box(\varphi \wedge \psi)$, contradicting $\Diamond \neg\varphi$ which follows from $\neg(\neg\varphi > (\varphi \wedge \chi))$. $\neg\varphi > (\varphi \wedge \neg\psi)$ would imply $\neg\varphi > \bot$, contradicting $\neg(\neg\varphi > (\varphi \wedge \chi))$.

Length $n + 1$: for MoPo, LLE and RLE see Raidl (2020a, Theorem 8). □

Lemma 8 (Backsimulation) *Let $\alpha \in \mathcal{L}_>$. If $\vdash_{VN} \alpha$ then $\vdash_S \alpha^\bullet$.*

Proof. By induction, using Observation 3.

Length 1. The backsimulation of PT, ID, AND, DW, and Cons are easy.

OR. $\Box^S(\varphi \wedge \chi)$ implies $\Box^S((\varphi \vee \psi) \wedge \chi)$. Similarly for $\Box^S(\psi \wedge \chi)$. $a = (\varphi \gg (\varphi \wedge \chi)) \wedge (\psi \gg (\psi \wedge \chi))$ with $\neg \Box^S(\varphi \vee \psi)$ implies $(\varphi \vee \psi) \gg ((\varphi \vee \psi) \wedge \chi)$ by OR^S. Assume $\Box^S(\varphi \vee \psi)$. Hence $\Box^S(\varphi \vee \psi)$ (OI^S). But a also implies $\Box^S((\varphi \vee \psi) \to \chi)$. Thus $\Box^S \chi$. Hence $\Box^S((\varphi \vee \psi) \wedge \chi)$.

CMon. $a_1 = \Box^S(\varphi \wedge \chi)$ and $b_1 = \Box^S(\varphi \wedge \psi)$ imply $c_1 = \Box^S(\varphi \wedge \psi \wedge \chi)$. a_1 and $b_2 = \varphi \gg (\varphi \wedge \psi)$, also imply c_1 by $\Box^S$ MP. Similarly for b_1 and $a_2 = \varphi \gg (\varphi \wedge \chi)$. $a_2 \wedge b_2$ imply $(\varphi \wedge \psi) \gg (\varphi \wedge \psi \wedge \chi)$ by CMon^r.

RMon. $\neg(\varphi \gg (\varphi \wedge \neg\psi))$ and $\varphi \gg (\varphi \wedge \chi)$ imply $(\varphi \wedge \psi) \gg (\varphi \wedge \psi \wedge \chi)$ by RMon^r. Suppose $\Box^S(\varphi \wedge \chi)$. Since $\neg \Box^S(\varphi \wedge \neg\psi)$ we obtain $\Diamond^S \psi$. Thus we get $\Box^S((\varphi \wedge \psi) \to (\varphi \wedge \psi \wedge \chi))$ and $\Diamond^S(\varphi \wedge \psi)$. Hence $\Box^S(\varphi \wedge \psi \wedge \chi) \vee ((\varphi \wedge \psi) \gg (\varphi \wedge \psi \wedge \chi))$ by SPRES$'$.

Length $n + 1$: for MoPo, LLE and RLE see Raidl (2020a, Theorem 8). □

By Theorem 1 and Lemma 1, 6, 7, 8, we obtain:

Theorem 3 S *is sound and complete for consistent Lewisean models in* $\mathcal{L}_\gg$.

6 Dependency conditional

In $\mathcal{L}_\succ$, $\square^D \alpha$ defined by $f(w, W) \subseteq [\alpha]_\succ$ is expressible by $\top \succ \alpha$ as in $\mathcal{L}_>$. Thus I drop the upper index ^{D}.

Definition 9 *The* dependency logic D *is* CE + AND + ID + OR^r + CMon^r + RMon^r + Cons *augmented by:*

$$((\varphi \succ (\varphi \wedge \psi)) \wedge (\varphi \succ (\varphi \vee \psi))) \to (\varphi \succ \psi) \qquad \text{D0}$$
$$(\varphi \succ \psi) \to (\neg \varphi \succ \neg \psi) \qquad \text{D1}$$
$$(\varphi \succ (\varphi \wedge \psi \wedge \chi)) \to (\varphi \succ (\varphi \wedge \psi)) \qquad \text{D2}$$

D is a logic for a weak counterfactual dependency (Lewis, 1973). It is a classical conditional logic but not a normal conditional logic, since DW fails. D0 is needed for the Twin Lemma. It is the '←'-part Sa$^\leftarrow$ of Sa. It allows to *derelativize* a conditional $\varphi \succ (\varphi \wedge \psi)$ to $\varphi \succ \psi$, when $\varphi \succ (\varphi \vee \psi)$. D1 follows by definition of $\succ$. D2 is obtained as backtranslate of DW and a weak substitute for it. Thus the difference to S is threefold. First, we have D0–D2 instead of S0–S2. D0 (=Sa$^\leftarrow$) and D2 (Sc) are derivable in S. Second, D contains ID, S does not. Third, the outer impossibility remains $\varphi \succ \bot$ and the inner necessity remains $\top \succ \varphi$. In this sense, D does better than C and S.

By D1, each of the other axioms X has an equivalent *dual* X^d, where $\wedge$ in the scope of a conditional is traded for $\vee$ and conversely. ID and D0 are their own duals. The dual to D2 is S2. The weak substitute DW^r for DW is derivable (from D2) and so are the following strengthenings of D0 and D1:

$$((\varphi \succ (\varphi \wedge \psi)) \wedge (\varphi \succ (\varphi \vee \psi))) \leftrightarrow (\varphi \succ \psi) \qquad \text{Sa} = \text{D0}^*$$
$$(\varphi \succ \psi) \leftrightarrow (\neg \varphi \succ \neg \psi) \qquad \text{D1}^*$$

Cons, AND, D2, D1, D0* and ID correspond to Rott's (2019) axioms ($\succ 0$, $1a, b, c, 2a, b$), and RLE + LLE, OR^r, CMon^r, RMon^r to ($\succ 6, 7', 8c, 8$).[13]

Establishing completeness works as before.

Lemma 9 (Twin) $\vdash_D \alpha \leftrightarrow \alpha^{\circ \bullet}$.

Proof. By induction, using $\circ, \bullet$, CE, D1* and D0*. □

Lemma 10 (Simulation) *Let* $\alpha \in \mathcal{L}_\succ$. *If* $\vdash_D \alpha$ *then* $\vdash_{VN} \alpha^\circ$.

Proof. By induction on the length of the proof. *Length* 1. PT as in Lemma 3. ID is simulated by ID, AND by AND + RW, Cons by Cons, D0 by RW, D1 by CE, D2 by RW + ID, OR^r by OR + RW + ID, CMon^r by CMon + RW +

---
[13] Rott's ($\succ$ 3-4) and ($\succ$ 5) are derivable.

ID, RMon^r by RMon + RW + ID. *Length* $n+1$: MoPo, LLE and RLE work as in Lemma 3. □

Lemma 11 (Backsimulation) *Let* $\alpha \in \mathcal{L}_>$. *If* $\vdash_{VN} \alpha$ *then* $\vdash_D \alpha^\bullet$.

Proof. By induction on the proof length. *Length* 1. The backsimulation of PT, ID, AND, DW, and Cons are clear. OR is backsimulated by OR^r, CMon by CMon^r, and RMon by RMon^r. *Length* $n+1$ as in Lemma 3. □

By Theorem 1 and Lemma 1, 9, 10, 11, we get

Theorem 4 D *is sound and complete for consistent Lewisean models in* $\mathcal{L}_\succ$.

7 Conclusion

In consistent Lewisean models one easily proves the following validities and invalidities of the usual Lewisean principles:

	CE	RW	ID	AND	OR	CMon	RMon
≫	√	×	×	√	×	×	×
≻	√	×	√	√	×	×	×
▷	√	×	√	√	√	√	×

The validities follow from our results. ID is invalid for ≫, due to S1, and counter-examples to prove the other invalidities are easily constructed.[14] Thus ▷ scores best when it comes to the Lewisean axioms.

Overall, the definable conditionals ▷, ≫, ≻ have a classical, but non-normal conditional logic, since Right Weakening fails, although AND holds. In the context of an unnecessary antecedent, ≻ implies ≫. The essential differences are these: First, ID is valid for ≻ and ▷, but invalid for ≫. Second, whereas OR and CMon are valid for ▷, they are invalid for ≫ and ≻. Third, whereas the inner modality □ is definable by ≫ and by ≻, it is not by ▷ alone. In all three cases, the companion axioms to the basic conditional axioms are mainly relativized versions of the originals, some with additional modal assumptions for ≫ and ▷. In brief, although the ground axiomatics for ▷ looks bad, it does better than the axiomatics for ≫ or ≻, by the standards of the Lewisean account for conditionals.

---

[14]For the invalidity of CMon for ≫ consider three worlds: w_0 is abc, w_1 is $a\neg bc$, w_2 is $\neg a \neg b \neg c$. Seen from w_0, order them according to their indicies. Thus in w_0: $a \gg b$ (since $a > b$ and $\neg(\neg a > b)$), $a \gg c$ (since $a > c$ and $\neg(\neg a > c)$), but $\neg((a \wedge b) \gg c)$ (since $(\neg a \vee \neg b) > c$). The same example proves invalidity of CMon for ≻ and invalidity of RMon for ≫ and ≻. For the invalidity of RMon for ▷, remove w_0 and add $w_3 = ab\neg c$.

References

Booth, R., & Chandler, J. (2020). On strengthening the logic of iterated belief revision: Proper ordinal interval operators. *Artificial Intelligence, 285,* 103289.

Chellas, B. F. (1975). Basic conditional logic. *Journal of Philosophical Logic, 4*(2), 133–153.

Crupi, V., & Iacona, A. (2020). The evidential conditional. *Erkenntnis.* doi: https://doi.org/10.1007/s10670-020-00332-2

Gomes, G. (2019). Meaning-preserving contraposition of conditionals. *Journal of Pragmatics, 1*(152), 46–60.

Lewis, D. (1971). Completeness and decidability of three logics of counterfactual conditionals. *Theoria, 37*(1), 74–85.

Lewis, D. (1973). Causation. *Journal of Philosophy, 70*(17), 556–567.

Raidl, E. (2019a). Completeness for counter-doxa conditionals – using ranking semantics. *The Review of Symbolic Logic, 12*(4), 861–891.

Raidl, E. (2019b). *Quick Completeness for the Evidential Conditional.* http://philsci-archive.pitt.edu/16664/

Raidl, E. (2020a). Definable conditionals. *Topoi, 40,* 87–105.

Raidl, E. (2020b). Strengthened conditionals. In B. Liao & Y. Wáng (Eds.), *Context, Conflict and Reasoning* (pp. 139–155). Singapore: Springer.

Raidl, E., Iacona, A., & Crupi, V. (2021). The logic of the evidential conditional. *Review of Symbolic Logic.*
doi: https://doi.org/10.1017/S1755020321000071

Rott, H. (1986). Ifs, though and because. *Erkenntnis, 25*(3), 345–370.

Rott, H. (2019). Difference-making conditionals and the Relevant Ramsey Test. *The Review of Symbolic Logic.*
doi: https://doi.org/10.1017/S1755020319000674

Rott, H. (2020). *Notes on contraposing conditionals.* http://philsci-archive.pitt.edu/17092/

Spohn, W. (2015). Conditionals: A unifying ranking-theoretic perspective. *Philosophers' Imprint, 15*(1), 1–30.

Stalnaker, R. C. (1968). A theory of conditionals. In N. Rescher (Ed.), *Studies in Logical Theory* (pp. 98–112). Oxford: Blackwell.

Eric Raidl
University of Tübingen
Excellence Cluster "Machine Learning: New Perspectives for Science"
Germany
E-mail: `eric.raidl@uni-tuebingen.de`

A Logic of Affordances

SEBASTIAN SEQUOIAH-GRAYSON[1]

Abstract: I motivate and develop a formal account of inference rules as specifications of environmental affordance-types such that they afford us opportunities for successful epistemic action across abstract environments. The formal account is given by a weak substructural logic, motivated by a proceeding philosophical discussion.

Keywords: affordances, inference rules, omniscience, substructural logic, types

1 Introduction

Actions are things that we do, perform, or execute. What we are able to do successfully will depend typically on two things. Firstly, what we are able to do successfully will depend on our skillset or acumen. Secondly, what we are able to do successfully will depend on environmental affordances—those opportunities for action that our environment affords us (Bermudez, 1998; Gibson, 1966; Rowlands, 1997).

For example, suppose that the action in question is my playing Bach's *Cello Suite no. 1* on the cello. For me to be able to perform this action successfully at a given point in time, then instances of the two conditions above must obtain. Firstly, I must know how to play Cello Suite no. 1 on cello in the first place, and secondly, there needs to be an actual cello around. If either of these conditions are not met, then I shall not be able to perform the action in question successfully.

The example above (as with examples of affordances traditionally), targets the physical environment. The proposal for which I shall argue here is that we may think of certain abstract environments, namely those populated by logic-mathematical entities and structures, as *bona fide* environments such

---

[1] I would like to thank Igor Sedlar and Andrew Tedder for their invaluable feedback and discussions, as well as all of those at SEGA in Bayreuth, July 18-20, 2019 for their insightful comments. The comments from an anonymous referee have been invaluable, and I am in their debt. I would also like to thank the editors of *The Logica Yearbook 2020* for their patience!

that they may contain affordances in a manner that is not totally disanalogous to that which is described above. In particular, I want to motivate and develop a formal account of *inference rules as specifications of environmental affordances such that they afford us opportunities for successful epistemic action across abstract environments.*

2 Metaphysical prerequisites — Platonism

We assume a robust logico-mathematical platonism (see Williamson, 2002). The abstract environment therein is populated by logically existent objects that are necessarily non-concrete hence abstract. The truth-makers for statements referring to them are mind-independent, but this is not to say that there is no role for the mind to play when it comes to navigating or perceiving this environment. Indeed, perception of the logico-mathematical environment (LME hereafter) appears to be uniquely mental insofar as it is via mental actions, acts of the mind, by which we traverse it.

Like the physical environment, the LME affords us multiple opportunities for action. These actions will be *epistemic actions*. We will explore epistemic actions in some detail in Section 5 below. For now, we can understand an epistemic action to be any action that is precipitated by a desire to relieve an epistemic deficit. Inferences are canonical examples. Again like the physical environment, what we are able to do successfully in the LME will depend typically on two things. Firstly, what we are able to do successfully will depend on our skillset or acumen. Secondly, what we are able to do successfully will depend on environmental affordances—those opportunities for action that our environment affords us.

In the LME, it is our logical acumen, that is our relevant mental skillset, that will make the difference. Similarly, the opportunities for action that this environment will afford us will depend on the abstract artefacts that populate the proper part of the LME to which we are attending. It is a mistake to think that attending to one part of the LME is to attend to all of it. In a slogan:

Slogan 1 a priori *knowability is not knowability for free.*[2]

Whatever the informational architecture of the LME might be, it is what it is necessarily. Hence our actions cannot change it. Rather, our actions in the LME are actions of the mind or understanding such that they may, all things

---

[2] By way of a suitably dramatic example, consider Frege's initial failure to notice the inconsistency resulting from his unrestricted comprehension axiom (Basic Law V).

A Logic of Affordances

going well, take us to new logico-mathematical facts. The perception of such logico-mathematical facts will involve new understandings, or new mental states, since it is through our faculties of understanding that such facts are perceived. In another slogan:

Slogan 2 *Our actions in the LME cannot change that environment, but they can change us.*

In the following section, we build on our understanding of the LME and develop the proposal that inference rules are specifications of environmental affordances such that they afford us opportunities for successful epistemic action across abstract environments in more detail.

3 Development

We will begin this development with a third slogan:

Slogan 3 *Inference rules may be understood as specifications of affordances across the LME, with respect to target propositions.*

Take as an example disjunction ($\vee$) and consider the rule of *disjunctive syllogism* (DS). What DS tells us is that if we "have" $\phi \vee \psi$, then if we "have" $\neg \phi$ then we "may get" ψ, and if we "have" $\neg \psi$ then we "may get" ϕ.

The scare quotes are to flag the following. By "have $\phi \vee \psi$", we mean that we are attending to the proper part of the LME that is $\phi \vee \psi$. By "attending to" we mean that $\phi \vee \psi$ is the explicit target proposition of some propositional attitude or other. Such an attitude may be entirely non-committal, such as *entertaining*, *supposing*, or *assuming*, or involve non-factive assent, such as *belief*, or factive assent, such as *knowledge*. By "may get ψ" if we have $\neg \phi$, we mean a bundle of two things. The first thing is that DS is an inference rule with which we are competent to a reasonable degree. By reasonable degree we do not mean the exclusion of error, any more than we would mean the exclusion of error by stating that we can play Bach's *Cello Suite no. 1*. The second thing is that it does not follow from it being true that we may get ψ that it is true that we *have gotten* ψ. In order to get ψ from $\phi \vee \psi$ and $\neg \phi$ with DS, we need to perform the inference in question, to execute the *epistemic action*. In a fourth slogan:

Slogan 4 *When performing epistemic actions, we avail ourselves of the relevant environmental affordance.*

Again, epistemic actions are those actions whose purpose is to relieve us of some epistemic deficit or other. Just as observations are canonical epistemic actions in the physical environment, inferences are canonical epistemic actions in the logico-mathematical one.

Environmental artefacts present multiple affordances typically, and this is no less true for the abstract artefacts populating the LME than it is for the concrete objects populating the physical one. What DS tells us is that a disjunction presents to us a pair of affordances. In more detail, what DS tells us in particular, is that the artefact $\phi \vee \psi$ is the type of thing that might afford us ϕ on the one hand, or ψ on the other, but it does not do so unconditionally! Rather, DS tells us that $\phi \vee \psi$ is of the type $\neg\phi \to \psi$, and the type $\neg\psi \to \phi$. Using standard type-theoretic notation, we may say that DS tells us that $\phi \vee \psi : \neg\phi \to \psi$ and $\phi \vee \psi : \neg\psi \to \phi$. Moreover, DS will specify affordance types when the target propositions are negations. Here too, the affordances are not unconditional. Rather, DS tells us that $\neg\phi : (\phi \vee \psi) \to \psi$, and $\neg\psi : (\phi \vee \psi) \to \phi$ and so on. Hence the following slight adjustment to Slogan 3 above:

Slogan 5 *Inference rules specify affordance* types *in the LME, with respect to target propositions.*

It is important to recognise that the specification of an affordance type is not the same thing as our having availed ourselves of the affordance, or *actualised* it. The affordances above are conditional on our targeting the antecedent, or attending to the proper part of the LME populated by the antecedent, and *then* combining it with the conditional type in the appropriate manner. Just what it is to which "combine in appropriate manner" amounts will depend on the properties (for example, *identity and logical form*) of the target propositions and affordance types in question. This is not all that surprising on reflection. In the physical environment, the properties of objects impose constraints on the types of actions to which they are amenable. In the LME, the properties of abstract objects will impose constraints on the epistemic actions to which they are amenable also. We will look at this in detail in Section 5 below.

For now, note that with the example of DS above, the target proposition does not itself provide the antecedent of the relevant conditional types. Although we will see that this is true for many target proposition/inference rule pairings, it is not true in general. Consider *conjunction elimination* (CE). What CE tells us is that we may have a conjunct on its own, conditional on the conjunction of which the conjunct is a part being the target

A Logic of Affordances

proposition. That is, from $\phi \wedge \psi$ being the target proposition, which is to say the part of the LME to which we are attending, CE specifies that we are afforded ϕ and afforded ψ. The affordance of which we avail ourselves in practice will depend on the nature of the epistemic action that we perform. In affordance-type terms, CE is telling us that $\phi \wedge \psi$ is of multiple types. In particular, that $\phi \wedge \psi : (\phi \wedge \psi) \to \phi$ and $\phi \wedge \psi : (\phi \wedge \psi) \to \psi$. In contrast to DS, with CE the antecedent of the conditional type is *given to us* by the target proposition. Again, crucially, this does not mean that we are afforded the relevant conjunct directly. Rather, we must *act* on the target proposition in the way specified by the relevant conditional affordance type.

Before we say more about the epistemic actions at work, we need to identify a third and final affordance-type mechanism. With this done, we can complete an affordance-type taxonomy of inference rules in the following section. As an example of this third affordance-type mechanism, consider conditional introduction/*the rule of conditional proof* (CP). CP specifies an affordance type with respect to a target proposition, however, the nature of the affordance is *underspecified*. Unlike DS and CE above, the affordance type that is specified in practice will depend on the epistemic actions that one performs. With CP, the conditional affordance type is arrived at by a chain of epistemic actions that begin with the assumption that we have the antecedent as an affordance completely. This is an assumption that we are attending to a proper part of the LME via a factive mental state.

CP begins with a *hypothetical affordance* via a target proposition, ϕ say, which is then used to reach, via epistemic actions, another target proposition, ψ say. The result of this is the conditional affordance $\phi \to \psi$. Unlike the case with DS and CE, ψ may be anything at all. What matters with CP is that ψ was gotten to via a chain of epistemic actions that begin with ϕ. We might think of CP usefully as a method by which we may *discover affordance types*. Put another way, our epistemic deficit might be with regard to the topography of the LME itself, and not merely with regard to target proposition considered *in situ*. CP is a way for us to discover affordance types via discoveries about said topography.

Having made a case for the claim that inference rules are specifications of environmental affordance types, in the following section we will test this claim against a system of natural deduction.

4 Inference rules and affordance types

Consider a system of natural deduction, N_1 (Smith, 2012). Here we have a intro/elim rule pair for implication, conjunction, negation, and disjunction. We will start with the intro/elim pair for *implication*.

The introduction rule for implication in N_1 is simply the rule of conditional proof (CP) above. Since we have dealt with it already, we will move directly to *conditional elimination*. The elimination rule for the conditional in N_1 is simply *modus ponens* (MP). The affordance type specified by MP with regard to a target proposition ϕ, is $\phi : (\phi \to \psi) \to \psi$. With regard to $\phi \to \psi$, the affordance type specified by MP is *identified by the target proposition completely*. That is, $\phi \to \psi$ just *is* an affordance type.

The affordance type specified *conjunction introduction* (CIn)[3] with respect to ϕ is as follows — $\phi : \psi \to (\phi \wedge \psi)$. The important thing here is to note that a conjunction is a mere *aggregation* action, as opposed to the *combinatorial* type actions underpinning MP say. We say more about the distinction between aggregation and combinatorial actions in the following section. The elimination rule for conjunction in N_1 has been dealt with in the section above, so we will move directly the intro/elim pair for *negation*.

The affordance type specified by *negation introduction* (NI) operates similarly to that of conditional elimination/conditional proof (CP) described in the preceding section above. Like CP, NI begins with an assumption that we have reached the relevant part of the LME, ϕ say, which is then used to reach, via epistemic actions, a pair of target propositions, one of which is the negation of the other. This pair will then return the negation of the hypothetical affordance. That is, $\phi : (\psi \wedge \neg\psi) \to \neg\phi$. *Negation elimination* (NE) is typed similarly, as $\neg\phi : (\psi \wedge \neg\psi) \to \phi$, again with the specification that $(\psi \wedge \neg\psi)$ is arrived at via a sequence of epistemic actions from the hypothetical affordance $\neg\phi$. More generally, we might think of pairs of inconsistent propositions as *universal affordances* such that they afford the opportunity to attend to any target proposition in the LME whatsoever.

The affordance type specified by *disjunction introduction* (DI) on the basis of the target proposition ϕ is one where the target proposition itself provides the antecedent of the relevant conditional type. That is, $\phi : \phi \to (\phi \vee \psi)$. The affordance type specified by *disjunction elimination* (DE) on the basis of the target proposition $\phi \vee \psi$ is complicated in comparison. Slightly differently to CP and NI, DE proceeds on the basis of *two distinct hypothetical affordances*,

---

[3] We write "CIn" instead of the intuitive "CI" here in order to avoid confusion with the structural rule of Weak Commutation in Section 6 below.

with one of each corresponding to one of each of the disjuncts of the target proposition $\phi \vee \psi$ itself. If from each of these hypothetical affordances ϕ and ψ, the same target proposition γ can be reached via a series of epistemic actions, then γ is reached definitively from the original target disjunction $\phi \vee \psi$. Hence DE types disjunction targets as $\phi \vee \psi : ((\phi \rightarrow \gamma) \wedge (\psi \rightarrow \gamma)) \rightarrow \gamma$.

Although the claim that inference rules are specifications of environmental affordance types is still hardly uncontroversial, I hope that enough has been demonstrated so far for it to be plausible at least. With this hope in mind, in the following section we will discuss epistemic actions on their own terms and in detail.

5 Epistemic actions

Epistemic actions have been doing some heavy lifting so far, so something substantial had best be said about them. As noted above, an epistemic action is any action performed in order to alleviate an epistemic deficit. This much is uncontroversial.

Although we do not pretend to anything like a complete taxonomic breakdown of such actions and deficits, we can say the following. If you suffer from an epistemic deficit such that the epistemic route to alleviating that deficit is *a posteriori*, then performing the relevant observation—a canonical example of an epistemic action if there is such a thing—might likely relieve the relevant deficit itself.[4]

We find ourselves in such deficits on account of our failing to be omniscient. If we were omniscient, then epistemic actions of the above sort would not alleviate our epistemic deficits about empirical matters of fact for the simple reason that we would not be in any such deficits in the first place.

If a route to alleviating an epistemic deficit is *a posteriori*, then we will be acting in the concrete physical domain by definition. Many routes to alleviating epistemic deficits are *a priori*, and the epistemic actions that take us along such routes will not be empirically directed actions such as observations or announcements. *a priori* routes are travelled by acts of reason, acts of the mind. These latter epistemic actions are acts of the mind on its

---

[4] For a trite but useful example, if you do not know how much money is in your wallet, and you know that you suffer from this deficit and you want to alleviate it, then looking in your wallet should, barring exceptional accident, do the trick. Of course plenty of epistemic deficits whose route to relief is *a posteriori* might not lend themselves to observations for resolution in practice, if only because they are in the past. In such cases *testimony/announcement* actions will be playing a crucial role. I shall not discuss such cases here.

own states, and it is via such inferential epistemic actions that we traverse the abstract environment of logico-mathematical structures.

"Inference" is very broad, but it captures the important fact that we can and do recognise that we suffer from epistemic deficits from which recovery requires us to think about things. Thinking is an act of the mind, but no less an action for this. We find ourselves in the relevant epistemic deficits here on account of our not being *logically omniscient*. If we *were* logically omniscient, then reasoning-based epistemic actions could never alleviate the relevant epistemic deficits for the simple reason that we would not be in any such deficits in the first place.

To reemphasise the running theme so far, inference *rules* specify the epistemic actions afforded to us by the parts of the LME to which we are attending. The part of the LME to which we are attending is the part that constitutes the target proposition of some propositional attitude of ours. In keeping with the terminology popular in theoretical computer science (if not mainstream philosophy), we will call *any* mental state of an agent underpinning such propositional attitudes "epistemic".

Although we do not pretend to anything like a complete taxonomic breakdown of epistemic actions of this latter psychological sort, we can make some headway. One such psychological epistemic action is something that we might, tentatively, call *aggregation*. Consider the case where you bear a propositional attitude, call it A, towards some proposition ϕ, written $A\phi$. In this case you are in a state of mind, you bear some particular mental state m, that is directed towards, or is about, ϕ itself, or takes ϕ as its object. In this case we say that ϕ is the part of the LME to which we are attending.

Suppose that you bear another instance of this same propositional attitude type towards some other proposition ψ, hence $A\psi$. In this case you bear some mental state m' that is directed towards ψ. In this case ψ is the part of the LME to which we are attending. It does not follow from these facts alone that $m = m'$. This is just to say that bearing a propositional attitude towards a proposition and bearing another instance of that same propositional attitude towards another proposition in no way entails that you bear a single instance of that propositional attitude to both of these propositions taken together. This is to say no more than that a careful account of propositional attitudes will not understand them to be closed under conjunction. From $A\phi$ and $A\psi$ it does not *follow* that $A(\phi \wedge \psi)$ for the simple reason that you might not have borne a mental state m'' that takes the conjunction of ϕ and ψ as its object. That one has attended to the part of the LME populated by ϕ, and that one has attended to the part of the LME populated by ψ, does not imply

A Logic of Affordances

that one has attended to the part of the LME populated by $\phi \wedge \psi$. To get there requires labour on our part—an *aggregative* epistemic action.

In spite of $A(\phi \wedge \psi)$ not being a mere logical consequence of $A\phi$ and $A\psi$, $A(\psi \wedge \phi)$ is still an attitude that you might achieve on the basis of $A\psi$ and $A\phi$, *along with* some mental effort on your part. Again, the mental effort will comprise an epistemic action of the aforementioned psychological sort, what we are calling *aggregation*. Aggregation actions are those that we perform in order to bring together within the scope of a single instance of a propositional attitude those propositions that were previously within the scope of distinct instances of propositional attitudes. Although there is no restriction in principle that all attitudinal instances in such cases be of the same type—I might through aggregations come to know that I both believe that ϕ and desire that ψ say, in practice we will consider aggregation actions that operate on instances of epistemic attitudes of the same type only.

Aggregation is labour, and like any act requiring labour on our part, it is prone to error. Again, we are not logically omniscient, and neither are what we might call *maximally psychologically introspective*. Our mental states are not transparent to us, and neither do we possess infallible memories of facts in general, of which our previously transparent mental states are a proper subset. Hence:

Slogan 6 *We can get lost in the LME, and have accidents too.*

A different psychological epistemic action is that which we will call *combination*. The result of a successful aggregation action is a mental state that is no greater than the sum of the parts of mental states that the action aggregated. Combination actions, by contrast, are *generative* actions *on* the contents of mental states such that the results of such actions *are* greater than the sum of their parts. This is a fine distinction. Consider the following example. You ask a new logic student to consider or entertain (we take *entertaining* to be a non-committal propositional attitude) $\neg \psi$ and $\phi \rightarrow \psi$. In this case both $\neg \psi$ and $\phi \rightarrow \psi$ are aggregated within the scope of a single attitude of the students. You then ask the student what, if anything, follows from this pair of propositions. The mere aggregation is insufficient for the student to answer correctly, that is for the student to bear the belief or knowledge attitude towards the claim the $\neg \phi$ (or any other logical consequence beyond aggregation for that matter).

Anyone who has taught introductory logic to a large cohort knows just how counterintuitive *modus tollens* is to many students on their first encounter.

In spite of having aggregated the premises and being in a mental state m that is directed towards the conjunction of the relevant pair of propositions, this aggregation alone is insufficient for the student to give a novel answer, to move to a mental state m' such that m' is directed towards $\neg\psi$. When confronted with such a pair of propositions, students answer often with "I don't know". In order for the student to move from m to m', they need to *combine* the propositions borne by m and m', and combine them in the right way.

Before we look more closely at combination actions themselves, especially at how we might say something philosophically robust with regard to what comprises combining the contents of mental states in "the right way", we will say something about the relationship between aggregation actions and combinations actions.

The first thing is that aggregation is a necessary condition on combination. This does not seem too controversial. If I am going to *combine* the propositions towards which a pair of my mental states are directed, I need to aggregate this pair before any such combination action can be performed. Aggregation actions are not sufficient for combination actions however, as the *modus tollens* example above demonstrates.

Aggregation actions can stand as necessary conditions on combination actions in a second manner that is distinct to that described above. Suppose that you bear some attitude A towards $(\phi \wedge \psi) \rightarrow \gamma$, and suppose also that you are looking for a way to perform *modus ponens* in order to discharge the consequent γ. In this case you will need to perform an aggregation action on ϕ and ψ in order to get $\phi \wedge \psi$ so that the conjunction may be used as a second premise in order to be able to perform the relevant consequent discharge. In other words, you might need to perform an aggregation action in order to form the very thing that will be one of the components in a combination action. In yet *other* words, *you might need to avail yourself of several affordances in order to get to where you want to be*.

The next thing that we note about our action pair is that there is a sense in which aggregation actions can never fail. They can fail to be realised, but when realised they are never illegitimate. This is not to say that they can never be *false*. Some aggregation actions will result in aggregations such as $\phi \wedge \neg\phi$, which are always false. The point is rather that the aggregation action itself is not illegitimate. Even in the case of explicit contradictions, we aggregate such things for the purposes of demonstrating explosion, or when we are explaining contradictions themselves. If the attitude under which such aggregation is taking place is one that comprises assent, such as belief, then the resulting aggregation will result in a mental state that is in error, but this error *depends* on the aggregation having taken place.

In contrast to aggregation actions, combination actions *can* be illegitimate. They can fail outrightly in the sense that a combination may be attempted that simply cannot achieve its goal. Suppose that I am trying to perform *modus ponens* on $\phi \to \psi$ and γ. Any attempt to do so will be an outright failure. An attempt to combine $\phi \to \psi$ and γ will not result in anything at all. Importantly, we should not be tempted into thinking that it will result in $(\phi \to \psi) \wedge \gamma$. The latter is the result of an aggregation action, not a combination action. The aggregation action must have been performed successfully in order for the (doomed, tragically) combination action to be attempted in the first place. There is simply no mental state m that is an accurately developed state on the basis of an illegitimate combination action. By analogy, they are akin to functions given an input that they do not accept, or programs being fed data of the wrong sort. There is simply no output at all, *because of a misidentification of affordance types with respect to inference rules*. Aggregation actions are un-typed instances of CI, whereas combination actions of the sort that we are discussing presently are typed instances of CE! With regard to the target proposition $\phi \to \psi$, what CI tells us is that $\phi \to \psi : \gamma \to ((\phi \to \psi) \wedge \gamma)$. By contrast, CE tells us that $\phi \to \psi : \phi \to \psi$.

There is a strongly normative flavour to this story, because in actual practice I might reason badly and possess false beliefs about the veracity of my combination actions and their resulting mental states. I might have false beliefs about just what actions are afforded to me by my local environment. Good. As stated at the beginning of the paragraph above, we are trying to say something philosophically robust about what it means for combination actions to be performed in the right way.

Whether a combination action is legitimate and successful or not will depend on both the logical form of the proposition towards which one's attitude is directed, as well as the inference rule that one is attempting to apply. That is, *their success will depend on one's availing one's self of the correct affordance given the target proposition that is the object of one's propositional attitude.*

6 Modelling combination actions/affordance actualisation

We want a formal model theoretic structure that allows a natural interpretation in terms of attitudinal states and psychological combination actions of the sort introduced in Section 5 above. To this end we introduce a frame $\mathbf{F}: \langle S, \bullet, \sqsubseteq \rangle$, where S is a set of information states $x, y, z \ldots$, $\bullet$ is a binary composition operator on members of S, and $\sqsubseteq$ is a partial order on S. A model $\mathbf{M}: \langle \mathbf{F}, \Vdash \rangle$, where $\Vdash$ is a relation between members of S and propositions $A, B, C \ldots$

We need to give our model a robust attitudinal interpretation. To this end, we take the domain of S to be a set of attitudinal states of an agent α. Although we think that the proposal below is general enough to apply to attitudinal states of any type, in practice we will limit our discussion to attitudes involving assent, such as doxastic or epistemic attitudes.

In this case we may read $x \Vdash A$ as α *knows/believes that A*; or equivalently, α *is attending to the part of the LME populated by* A. Importantly, we place a restriction on $\mathbf{M}$ such that we understand $x \Vdash A$ to mean that A is the *only* part of the LME constituting α's state x.

$x \bullet y$ is understood as the combination of α's attitudinal states x and y by α themselves, as an explicit psychological action.

$x \sqsubseteq y$ indicates informational-relevance in general, and explicit attitudinal relevance in particular. So if we are interpreting S as a set of epistemic states, $x \sqsubseteq y$ will be read as *x is epistemically relevant to y*, and $x \bullet y \sqsubseteq z$ as *the act of combining x and y is epistemically relevant to z*. We may also read $x \sqsubseteq y$ as *the part of the LME that is the target of* α's *epistemic state x is contained in the part of the logico-mathematical environment that is the target of* α's *epistemic state y*.

We can say the following about attitudinal relevance/LME containment. Firstly, any attitudinal state will be relevant to itself, hence $\forall x\, x \sqsubseteq x$. Similarly, any part of the LME will be a part of itself. Secondly, the relevance of an epistemic action $x \sqsubseteq y$ to some epistemic state y will depend on the logical form of the propositions towards which the attitudes x, y, and z are directed. For example, if $x \Vdash A \to B$ and $y \Vdash A$ and $z \Vdash B$, then $x \bullet y \sqsubseteq z$, but $y \not\sqsubseteq z$. The latter is the case because A is not informationally relevant to B on its own, not without further context which is given in this case by x. Similarly, we may say that A is not a proper part of the LME that is populated by B, at least not without further environmental context.

What we need is some way of specifying constraints on our action operation $\bullet$ such that these constraints will preserve the epistemic relevance of the corresponding psychological action. Put another way, we want a way of

A Logic of Affordances

specifying properties of • so that these properties guarantee that the result of performing the successful psychological epistemic action is preserved. Put yet a third way, we want some way of guaranteeing progress across the logic-mathematical environment.

We can specify such constraints (and *ipso facto* permissions) with the familiar structural rules of substructural logic. Via (Restall, 2000, p. 250) we list the frame conditions for the most common structural rules (B) Associativity, (B^c) Converse Associativity, (B') Twisted Associativity, (C) Commutation, (CI) Weak Commutation, (W) Contraction, (WI) Weak Contraction, (M) Mingle, (K) Weakening, and (K') Commuted Weakening below (reading $\Rightarrow$, $\wedge$, and $\vee$ as "if then", "and", and "or" in the metalanguage, respectively):

$$\exists u((x \bullet y \sqsubseteq u) \wedge (u \bullet z \sqsubseteq w)) \Rightarrow \exists u((y \bullet z \sqsubseteq u) \wedge (x \bullet u \sqsubseteq w)) \quad (B)$$

$$\exists u((y \bullet z \sqsubseteq u) \wedge (x \bullet u \sqsubseteq w)) \Rightarrow \exists u((x \bullet y \sqsubseteq u) \wedge (u \bullet z \sqsubseteq w)) \quad (B^c)$$

$$\exists u((y \bullet x \sqsubseteq u) \wedge (u \bullet z \sqsubseteq w)) \Rightarrow \exists u((y \bullet z \sqsubseteq u) \wedge (x \bullet u \sqsubseteq w)) \quad (B')$$

$$\exists u((x \bullet z \sqsubseteq u) \wedge (u \bullet y \sqsubseteq w)) \Rightarrow \exists u((x \bullet y \sqsubseteq u) \wedge (u \bullet z \sqsubseteq w)) \quad (C)$$

$$(x \bullet y \sqsubseteq z) \Rightarrow (y \bullet x \sqsubseteq z) \quad (CI)$$

$$(x \bullet y \sqsubseteq z) \Rightarrow \exists w((x \bullet y \sqsubseteq w) \wedge (w \bullet y \sqsubseteq z)) \quad (W)$$

$$x \bullet x \sqsubseteq x \quad (WI)$$

$$(x \bullet x \sqsubseteq y) \Rightarrow (x \sqsubseteq z \vee y \sqsubseteq z) \quad (M)$$

$$(x \bullet y \sqsubseteq z) \Rightarrow x \sqsubseteq z \quad (K)$$

$$(y \bullet x \sqsubseteq z) \Rightarrow x \sqsubseteq z \quad (K')$$

We give the evaluation conditions for two instances of our combination action operator •. The first where a conditional is being combined with a potential input, the second where conditionals themselves are being combined. This is just to say that the first is where *affordances are being actualised*, and the second is where *affordances are being composed*. Because we are talking about the properties of dynamic (epistemic) actions here, and not merely the results of the same, we must treat affordance-type operations intensionally:

$$x \Vdash A \rightarrow B \text{ iff } \forall x \forall y : x \bullet y \sqsubseteq z, \text{ if } y \Vdash A \text{ then } z \Vdash B \quad (1)$$

$$x \Vdash A \rightarrow B \text{ iff } \forall x \forall y : x \bullet y \sqsubseteq z, \text{ if } y \Vdash B \rightarrow C \text{ then } z \Vdash A \rightarrow C \quad (2)$$

Claim 1 *Combination actions of type (1), affordance actualising actions, are guaranteed to have their success preserved by (CI), but destroyed by all other structural rules.*

Sebastian Sequoiah-Grayson

Claim 2 *Combination actions of type* (2), *affordance composing actions, are guaranteed to have their success preserved by* (B), (B^c), *but destroyed by all other structural rules.*

We start by justifying Claim 1. Consider (CI). Given that the action here is combining a conditional with its antecedent, the success of this action is order-invariant, so (CI) holds. The logical form of the propositions being combined in *affordance actualisation* type actions forces the discharge of the consequent. There is only one way that things can go, so to speak.

We should reemphasise the restriction on **M** outlined above, such that we understand $x \Vdash A$ to mean that A is the *only* part of the LME constituting α's state x. In the more abstract terms of our model *qua* model, we take *satisfaction* to be a primitive notion, there to be no points/information states other than those mentioned, and that they satisfy or support exactly and only the formulas that they are stated to satisfy.[5]

To see why it is that combination actions of type (1) have their success destroyed by the other structural rules, we start with (B). Suppose that $x \Vdash A \to B, y \Vdash A, u \Vdash B, w \Vdash C$, and $z \Vdash B \to C$. In this case the antecedent of (B) is satisfied whist the consequent is false. The consequent is false on account of its left hand conjunct being false. The is no u such that it is the result of $y \bullet z$. Why is this? Recall the restriction on M above—that we understand $x \Vdash A$ to mean that A is the *only* part of the LME constituting α's state x. our models here are not general substructural models, but rather those in which certain states are *identified* with certain formulas. Syntactically speaking, attempting to discharge the consequent from $B \to C$ by combining it with A is doomed. Note that the argument against (B) depends on the argument for (CI) above. If (CI) were not acceptable then the antecedent of (B) would not be satisfied, on account of $u \bullet z \not\sqsubseteq w$ in this case.[6]

Now consider (B^c), and suppose that $y \Vdash A, z \Vdash A \to B, u \Vdash b, w \Vdash C$, and $x \Vdash B \to C$. In this case the antecedent of (B^C) will be satisfied whilst the consequent is false, again on account of the consequent's left hand conjunct being false for reasons similar to those concerning the left hand conjunct of the consequent of (B).

Now consider (B'), and suppose that $y \Vdash A, x \Vdash A \to B, u \Vdash B, z \Vdash B \to C$, and $w \Vdash c$. Here the antecedent of (B') will be satisfied whilst its

---

[5]I am indebted both to Igor Sedlár and to an anonymous referee for making me be clearer on this point than I would have been had I been left to my own devices.
[6]I am indebted to the same anonymous referee for making me be clearer here.

consequent will be false on account of its consequent's left hand conjunct being false for reasons analogous to those given above.

Now consider (C), and suppose that $x \Vdash A, z \Vdash A \to B, u \Vdash B, y \Vdash B \to C$, and $w \Vdash C$. In this case the antecedent of (C) will be satisfied, but the left hand conjunct of its consequent will be false, hence its consequent will be false.

Now consider (W), and suppose that $x \Vdash A \to B, y \Vdash A$, and $z \Vdash B$. In this case there is no way to satisfy both of the consequent's conjuncts. We can satisfy the left hand conjunct with $w \Vdash B$, in which case the right hand conjunct will be false. Alternatively, we could satisfy the right hand conjunct with $w \Vdash A \to B$, but now the left hand conjunct will be false.

Now consider (WI). (WI) is not a conditional. It states that any combination of an information state with itself develops into that same information state. In our epistemic attitudinal gloss, the combination action of an epistemic state with itself is epistemically relevant to that same state. This is false no matter what we choose for x. Speaking syntactically, combining (as opposed to aggregating) any proposition with itself is a doomed attempt at epistemic advancement.

Now consider (M), and suppose that $x \Vdash A \to B, y \Vdash A$, and $z \Vdash B$. In this case the antecedent of (M) will be satisfied but the consequent is false. We have it that $x \bullet y \sqsubseteq z$, however it is neither the case that $x \sqsubseteq z$, nor $y \sqsubseteq z$. With regard to the former, $A \to B$ is not non-contextually epistemically relevant to B (we need A as further context). With regard to the latter, A is not non-contextually epistemically relevant to B, as here too we need further context.

Now consider (K), and suppose that $x \Vdash A \to B, y \Vdash A$, and $z \Vdash B$. In this case the antecedent of (K) will be satisfied but the consequent will be false. The consequent states that $x \sqsubseteq z$, but this is not the case as speaking syntactically $A \to B$ is not non-contextually epistemically relevant to B. Any attempt to discharge B from $A \to B$ without A is doomed. The reasoning with regard to (K') is identical to that surrounding (K).

We now justify Claim 2. Consider (B). There is no way of satisfying the antecedent that will make the consequent false. As an illustrative exercise, suppose that $x \Vdash A \to B, y \Vdash B \to C, u \Vdash A \to C, z \Vdash C \to D$, and $w \Vdash A \to D$.

The reasoning with regard to (B^C) is identical. This is not surprising, since conditionals can be thought of as functions, and function composition is associative, and association is meant often in its biconditional form, which is just the conjunction of (B) and (B^C).

Now consider (B'), and suppose that $y \Vdash A \to B, x \Vdash B \to C, u \Vdash A \to C, z \Vdash C \to D$, and $A \to D$. In this case the antecedent will be satisfied but the consequent will be false. The consequent is false on account of both of its conjuncts being false. Speaking syntactically, attempting to compose either $A \to B$ with $C \to D$, or $B \to C$ with $A \to C$ is doomed.

Now consider (C), and suppose that $x \Vdash A \to B, z \Vdash B \to C, u \Vdash A \to C, y \Vdash C \to D$, and $w \Vdash A \to D$. In this case the antecedent will be satisfied by the left hand conjunct of the consequent false for reasons similar to those surrounding (B') above.

Now consider (CI), and suppose that $x \Vdash A \to B, y \Vdash B \to C$, and $z \Vdash A \to C$. Here the antecedent will be true but the consequent will be false. Speaking syntactically, trying to feed the consequent of $B \to C$ to the antecedent of $A \to B$ is the wrong order for cutting out the middle or joining proposition.

Now consider (W), and suppose that $x \Vdash A \to B, y \Vdash B \to A$, and $z \Vdash A \to A$. In this case the consequent will be false as there is no w that will satisfy both conjuncts. Either $w \Vdash A \to A$ in which case the right hand conjunct is false, or $w \Vdash A \to B$ in which case the left hand conjunct is false.

Now consider (WI). The reasoning here is identical to that surrounding (WI) for Claim 1 above.

Now consider (M), and suppose that $x \Vdash A \to B, y \Vdash B \to C$, and $A \to C$. Here the antecedent of (M) will be satisfied whilst its consequent will be false. Although we do have it that $x \bullet y \sqsubseteq z$, we have it neither that $x \sqsubseteq z$, nor that $y \sqsubseteq z$. With regard to the former, it is not the case that $A \to B$ is non-contextually epistemically relevant to $A \to C$. With regard to the latter, it is not the case that $B \to C$ is non-contextually epistemically relevant to $A \to C$.

Now consider (K), and suppose that $x \Vdash A \to B, y \Vdash B \to C$, and $z \Vdash A \to C$. In this case the antecedent will be satisfied but the consequent false. x is not informationally relevant to z all on its own, that is non-contextually. Speaking syntactically, trying to get $A \to C$ from $A \to B$ without any other informational artefacts or actions is doomed. The reasoning with regard to (K') is identical to that surrounding (K).

This completes our proposal for a logic of affordances. We have seen that combination actions of types corresponding to *affordance actualising actions*, have their success preserved by (CI), but destroyed all other structural rules. We have seen also that combination actions of types corresponding to *affordance composing actions*, have their success preserved by (B), (B^c), but destroyed by all other structural rules.

7 Conclusion

We have covered a lot of ground above, and I do not pretend to anything like confidence that I will have convinced all readers. My hope is that the above is worked out sufficiently for it to strike most as plausible, and worthy of further pursuit.

I have made a case for the abstract LME affording opportunities for action in ways not dissimilar entirely from the the ways in which the concrete physical environment affords the same. I have motivated, or tried to motivate at least, an understanding of inference rules as specifications of affordance types, such that they afford us opportunities for successful action across abstract environments. I have proposed that such actions be understood properly as epistemic actions of a psychological sort, and I have put forward and argued for a weak substructural logic as a plausible model for said affordance types and their related epistemic actions (or more strictly, a unique weak substructural logic for discrete affordance-actions).

The frame conditions given in (1) and (2) in Section 6 above correspond to the serial and parallel composition of information channels in channel theory. Given the by now, and increasingly, well known correspondence between channel theory and the ternary frame semantics of relevance and substructural logics (Mares, 1996; Restall, 1996), this is not itself a surprise. It is less of a surprise again when one considers the recent applications of substructural logics to epistemic problems (Aucher, 2014; Sedlár, 2015). Indeed the modelling of channel-theoretic phenomena in frame semantics terms has led to a recent revival of interest (Tedder, 2017). What might be a surprise is that a channel-theoretic interpretation of the subject matter at hand is both straightforward and natural. The insight motivating channel-theory in the first place is that one part of our environment may carry information to, or about, another part of it. Recall also the initial promise of situation theory to contribute to an analysis of hyperintensional phenomena and mathematical knowledge (Barwise & Perry, 1983). A robust platonism of the sort proposed above allows for a sensible mapping of this insight from the concrete environment over to the abstract environment of logico-mathematical objects. In a final slogan:

Slogan 7 *Affordances are information channels.*

References

Aucher, G. (2014). Dynamic epistemic logic as a substructural logic. In A. Baltag & S. Smets (Eds.), *Johan van Benthem on Logic and Information Dynamics* (pp. 855–880). Cham: Springer International Publishing.

Barwise, J., & Perry, J. (1983). *Situations and Attitudes*. Cambridge, MA: MIT Press.

Bermudez, J. L. (1998). *The Paradox of Self-Consciousness*. Cambridge, MA: MIT Press.

Gibson, J. J. (1966). *The Senses Considered as Perceptual Systems*. Boston: Houghton Mifflin.

Mares, E. D. (1996). Relevant logic and the theory of information. *Synthese, 109*(3), 345–360.

Restall, G. (1996). Information flow and relevant logics. In J. Seligman & D. Westerstahl (Eds.), *Logic, Language and Computation* (pp. 463–477). Standford: CSLI Publications.

Restall, G. (2000). *An Introduction to Substructural Logics*. Boston: Routledge.

Rowlands, M. (1997). Teleological semantics. *Mind, 106*(422), 279–303.

Sedlár, I. (2015). Substructural epistemic logics. *Journal of Applied Non-Classical Logics, 25*(3), 256–285.

Smith, N. J. J. (2012). *Logic: The Laws of Truth*. Princeton University Press.

Tedder, A. (2017). Channel composition and ternary relation semantics. *IfCoLog Journal of Logics and Their Applications, 4*(3), 731–753.

Williamson, T. (2002). Necessary existents. *Royal Institute of Philosophy Supplement, 51*, 233–251.

Sebastian Sequoiah-Grayson
University of Sydney, Department of Philosophy
Australia
E-mail: sequoiah@gmail.com

Imperative Bilateralism

KAI TANTER[1]

Abstract: This paper provides a proof-theoretic account of imperative logical consequence by generalising Greg Restall's multiple conclusion bilateralism for declarative logic. According to imperative bilateralism, a sequent $\Gamma \vdash \Delta$ is valid iff jointly commanding all the imperatives $\Phi \in \Gamma$ and prohibiting all the imperatives $\Psi \in \Delta$ clashes. This account has three main virtues: (1) it provides a proof-theoretic account of imperatives; (2) it does not rely on the controversial notion of imperative inference; and (3) it is neutral regarding cognitivism about imperatives.

Keywords: imperatives, bilateralism, logical consequence, inferentialism

1 Introduction

This paper provides a proof-theoretic account of imperative logical consequence by generalising Greg Restall's multiple conclusion bilateralism for declarative logic. According to imperative bilateralism, a sequent $\Gamma \vdash \Delta$ is valid iff jointly commanding all the imperatives $\Phi \in \Gamma$ and prohibiting all the imperatives $\Psi \in \Delta$ clashes. In the following Section 2, a trilemma for imperative logic is introduced, along with the contemporary debate between cognitivists and non-cognitivists about imperatives. A virtue of imperative bilateralism is that it allows one to remain neutral about this debate. Next, in Section 3, bilateralism is introduced, first for declaratives, second for imperatives, and third for a mixed language containing both imperatives and declaratives. A second virtue of this theory is that, in spite of its proof-theoretic orientation, it does not rely on the controversial notion of imperative inferences. A third, is that it provides a general definition of logical consequence and a sequent calculus in which the meaning of logical connectives is neutral regarding sentence type.

---

[1] I would like to thank David Ripley and Lloyd Humberstone for their comments on this paper, as well as Greg Restall and Shawn Standefer for their help and comments on earlier versions of this work. This work is funded by an Australian Government Research Training Program Scholarship.

2 Introducing imperatives

Imperative logic faces the following trilemma:

Claim 1 (Relations)   *Imperatives stand in logical relations.*

Claim 2 (Truth Apt)   *Imperatives are not truth-apt.*

Claim 3 (Relata)   *The relata of logical relations must be truth apt.*[2]

One cannot consistently affirm all three claims, yet each seems well motivated. Consider Claim 1 (Relations). Imperatives stand in logical relations such as incompatibility (1a and 1b) and equivalence (2a and 2b).

Example 1 (Incompatibilities and Equivalences)

1. (a) Both buy jam and don't buy marmalade $[\phi \wedge \neg \psi]$
   (b) Either don't buy jam or do buy marmalade $[\neg \phi \vee \psi]$

2. (a) Neither buy quinoa nor soy falafel mix $[\neg(\phi \vee \psi)]$
   (b) Don't buy quinoa falafel mix and also not the soy $[\neg \phi \wedge \neg \psi]$

There also appear to be valid arguments involving imperatives. Consider the following example from Peter Vranas (2011). Say someone is sitting an exam. They read instructions 1, 2, and 3 below. They then notice that the third follows from the first two.

Example 2 (Exam)

1. Answer exactly three out of the six questions;

2. Do not answer both questions 3 and 5;

3. Answer at least one even-numbered question (Vranas, 2011, p.369).

For a mixed declarative and imperative argument, consider:

Example 3 (Umbrella)

1. If it's raining, then bring your umbrella

2. It's raining.

---

[2] I have taken and slightly modified this presentation from Hannah Clark-Younger (2012).

3. Therefore: bring your umbrella

Claim 2 (Truth Apt) also has intuitive motivations, as well as critics. For the motivation, one appears to make a mistake in calling imperative sentences true or false. Consider:

Example 4 ('True' and 'False')

3. (a) Wipe the bench
   (b) #That's true.

4. (a) Do your homework
   (b) #That's false.

Truth and falsity simply don't seem applicable to imperatives (Charlow, 2018; Parsons, 2013; Portner, 2004). Cognitivists about imperatives argue, however, that this move is too hasty, partly because truth is central to both standard theories of semantics and logical consequence. In response, various cognitivist theories of imperatives have been proposed in which the proponents attempt to explain away this apparent non-truth aptness (Kaufmann, 2012; Lewis, 1970, 1979).

Before continuing, let us explicitly distinguish between sentences, speech acts and contents, as it allows us to be clearer about neutrality regarding cognitivism. Declarative sentences are paradigmatically used in speech acts of asserting and denying and are normally taken to express truth apt propositions. Similarly, imperative sentences are paradigmatically used to command and prohibit and express some kind of content, truth apt or not. For example, if imperatives express sets of actions that comply with the imperative, as in Fine (2018), then we can naturally say that one commands and prohibits actions, analogous to asserting and denying propositions. In contrast, if the content of imperatives is propositional (Kaufmann, 2012), some rephrasing would be required to avoid the unidiomatic "commanding and prohibiting propositions". All that's intended is that imperatives express some kind of content and that the following is neutral regarding this content. A further terminological note is that one often speaks of asserting or denying some declarative A, even though, perhaps, strictly speaking one only uses the sentence to assert or deny a proposition. In the same way, we will speak of commanding or prohibiting some imperative Φ, even though, strictly speaking, one commands or prohibits the content expressed by using the sentence.

3 Bilateralism

Most theories of imperative logical consequence generalise standard truth-conditional approaches to content and logical relations formalised using model theory. Such theories either treat imperatives as truth apt or apply a different predicate analogous to truth. For example, Josh Parsons (2013) and Hannah Clark-Younger (2014) apply the notion of *compliance* to imperatives, with valid pure imperative arguments preserving compliance. In contrast, proof-theoretic approaches remain unexplored.[3] Much proof-theory for declarative logics is motivated by inferentialism, according to which meaning is understood in terms of inferential relations rather than truth and reference (Peregrin, 2014; Steinberger & Murzi, 2017). Adopting inferentialism allows one to remain neutral about Claim 3 (Relata). For if inference is taken as basic, truth need not be built into the notion of logical relations. Neutrality regarding Claim 3 (Relata) also allows the inferentialist to remain neutral regarding Claim 2 (Truth Apt), because unlike someone who already endorses Claim 1 and Claim 3, the inferentialist's hand is not forced on Claim 2. They are open to either endorse Claim 2 and reject Claim 3 or vice versa.

This line of reasoning, however, runs into a problem. This is because there is an ongoing debate about whether there are genuine imperative inferences, in the sense of 'inference' as an activity (Clark-Younger, 2012; Hansen, 2008; Vranas, 2010; Williams, 1963). Inferentialists would seem to only avoid one debate at the cost of having to stake a substantial position in the other. A way to avoid both debates is to adopt an interpretation of proof theory that takes some notion other than inference *per se* as basic. An example is Greg Restall's bilateralist interpretation of the classical multiple conclusion sequent calculus (2005). Restall's bilateralism is outlined below in Section 3.1 before showing how it can be generalised to imperatives in Section 3.2.

3.1 Declarative bilateralism

We briefly summarise Restall's bilateralism (2005), in order to ground the approach taken to imperatives in the next section. To represent declaratives in our metalanguage lower case Latin p and q are used for atomics, upper case,

---

[3] See Fox (2012) for one of the few proof-theoretic approaches to imperatives, and Vranas (2019) and Fine (2018) for proof systems for their respective model-theoretic definitions of imperative consequence.

A and B, for arbitrary declaratives, and upper case X and Y for multisets of declaratives. X, A is shorthand for $X + [A]$.[4]

Declarative logical consequence will be defined for the standard (declarative) propositional language $\mathcal{L}_D$.

Definition 1 ($\mathcal{L}_D$) *$\mathcal{L}_D$ is the language whose vocabulary is made up of denumerably many atomic declaratives $p_1, p_2, p_3, \ldots$; the one-place connective $\neg$ and the two-place connectives $\wedge$, $\vee$ and $\supset$; and whose sentences are all and only those generated recursively from the following rule: all atomic declaratives p are sentences and if A and B are sentences then so are $\neg A$, $A \wedge B$, $A \vee B$ and $A \supset B$.*

Instead of interpreting proofs in terms of inference, according to Restall's declarative bilateralism, proofs show that there is a clash or incoherence in a position that both asserts the premises and denies the conclusions.

Definition 2 (Declarative Positions and Clash) *If X and Y are both multisets of declaratives, then $X : Y$ is the declarative position that asserts all $A \in X$ and denies all $B \in Y$. $X \vdash Y$ expresses the claim that the position $X : Y$ clashes.*

For example, $A \vee B \vdash A, B$ records that the position which both asserts the disjunction $A \vee B$ and denies both disjuncts A and B clashes.

Clash is used to define declarative logical consequence.

Definition 3 (Declarative Consequence) *Y is a declarative consequence of X iff the declarative position $X : Y$ clashes.*

This provides a natural reading of the following classical multiple conclusion sequent calculus. Each rule says that if each position above the line clashes then so does the position below the line. The rules can also be read contrapositively, saying that if the position below the line does not clash, then neither does at least one of the positions above the line.

$$\frac{}{A \vdash A}\text{Id} \qquad \frac{X \vdash A, Y \quad X, A \vdash Y}{X \vdash Y}\text{Cut}$$

$$\frac{X \vdash Y}{X, A \vdash Y}\text{KL} \qquad \frac{X \vdash Y}{X \vdash A, Y}\text{KR} \qquad \frac{X, A, A \vdash Y}{X, A \vdash Y}\text{WL} \qquad \frac{X \vdash A, A, Y}{X \vdash A, Y}\text{WR}$$

---

[4] $[A]$ is a singleton multiset, whose only member is A. $X + Y$ is the sum of X and Y, where for any x that occurs n times in X and m time in Y, x occurs $n + m$ times in $X + Y$.

$$\frac{X, A \vdash Y}{X, A \wedge B \vdash Y} \wedge L_1 \quad \frac{X, B \vdash Y}{X, A \wedge B \vdash Y} \wedge L_2 \quad \frac{X \vdash A, Y \quad X \vdash B, Y}{X \vdash A \wedge B, Y} \wedge R$$

$$\frac{X, A \vdash Y \quad X, B \vdash Y}{X, A \vee B \vdash Y} \vee L \quad \frac{X \vdash A, Y}{X \vdash A \vee B, Y} \vee R_1 \quad \frac{X \vdash B, Y}{X \vdash A \vee B, Y} \vee R_2$$

$$\frac{X \vdash A, Y \quad X, B \vdash Y}{X, A \supset B \vdash Y} \supset L \quad \frac{X, A \vdash B, Y}{X \vdash A \supset B, Y} \supset R$$

$$\frac{X \vdash A, Y}{X, \neg A \vdash Y} \neg L \quad \frac{X, A \vdash Y}{X \vdash \neg A, Y} \neg R$$

As examples of how to read the rules: Id records that assertion and denial are incompatible in the sense that asserting and denying the same thing always clashes. Id is unrestricted, in the sense that it applies to all sentences rather than just atomics. However, if it were restricted to atomics, then identity sequents could be derived for sentences of arbitrary logical complexity; ¬L says that if a position that denies A clashes, then so does one that swaps A's denial for the assertion of it's negation $\neg A$. Read contrapositively, ∨L says that if it is coherent to assert a disjunction, then, in the same context, it is coherent to assert at least one of the disjuncts. The same rules will be applied to imperatives in the following section where they will be given a more detailed interpretation.

Restall's bilateralism has two main advantages as an account for generalising to imperatives. First, the central notion of clash makes no reference to truth and instead takes incompatibilities between assertions and denials as basic. Thus, in generalising to imperatives the question of truth doesn't come up. Second, although it is nominally an inferentialist theory, it is the norms governing clash, rather than inference as a cognitive process, that are front and centre. This suits imperatives, because incompatibilities between imperatives are uncontroversial regardless of whether there are genuine imperative inferences. Whether Restall's theory is *really* a form of inferentialism is open to debate and largely hinges on how liberal we're being about what counts as inferentialism. *If* for a theory to count as inferentialist it needs to treat the activity of inference as central in explaining meaning, then the clash-based interpretation of the sequent calculus in this paper is certainly not inferentialist. However, many inferentialists take the relevant notion of inference to be inferential rules, instead of the activity. This is the approach of Jaroslav Peregrin (2014) who, like Restall, also focuses on the constraining nature of rules of language use—a rule of inference from 'A' to 'B' does not require that we infer 'B' whenever we know 'A' and instead prohibits rejecting 'B' when we affirm 'A'. If the rules of the multiple conclusion sequent calculus count as inference rules, then their clash-based reading could count as a form of inferentialism like Peregrin's. I will not argue for the antecedent here. Even if this is not accepted, Restall's theory is an adjacent form of normative pragmatism, in the sense that meaning is explained in terms of norms of language use.

3.2 Imperative bilateralism

Restall's bilateralism is now generalised to imperatives, drawing on the virtues for doing so identified above. This section only discusses imperatives, rather than declaratives also, and uses lower case Greek ξ and ρ for atomic imperatives, upper case Φ and Ψ for arbitrary imperatives, and upper case Γ and Δ for multisets of imperatives. In the following section, the system will be extended to accommodate both imperatives and declaratives together.

We define an imperative language $\mathcal{L}_\mathrm{I}$, analogous to the declarative language $\mathcal{L}_\mathrm{D}$.

Definition 4 $\mathcal{L}_\mathrm{I}$ *is the language whose vocabulary is made up of denumerably many atomic imperatives $\xi_1, \xi_2, \xi_3, \ldots$; the one-place connective $\neg$ and the two-place connectives $\wedge$, $\vee$ and $\supset$; and whose sentences are all and only those generated recursively from the following rule: all atomic imperatives ξ are sentences, and if Φ and Ψ are sentences then so are $\neg \Phi$, $\Phi \wedge \Psi$, $\Phi \vee \Psi$ and $\Phi \supset \Psi$.*

In the declarative case, assertions "rule in" propositions or ways the world could be, whereas denials rule these out. In contrast, uses of imperatives shift the practical commitments of agents, an idea shared by cognitivists and non-cognitivists alike (Charlow, 2014; Kaufmann, 2012; Lewis, 1979; Mastop, 2011; Parsons, 2013; Portner, 2004). What's needed for an imperative bilateralism are speech acts that rule in and out actions for agents, shifting their practical commitments. Commands rule in actions for someone by requiring their performance, whereas prohibitions rule out actions for someone by forbidding their performance. Commands and prohibitions also clash just as assertions and denials do. For example, the respective command and prohibition of 2a and 2b from Section 2. Note that, if one thinks that imperatives express propositions, then commands and prohibitions will rule actions in and out by, e.g., making true deontic modals or requiring that propositions expressing the fulfilment conditions of the imperative are made true.

We first define the notion of an *imperative position*.

Definition 5 (Imperative Positions and Clash)   *If Γ and Δ are both multisets of imperatives, then $\Gamma : \Delta$ is the* imperative position *that commands all $\Phi \in \Gamma$ and prohibits all $\Psi \in \Delta$. $\Gamma \vdash \Delta$ expresses the claim that $\Gamma : \Delta$ clashes.*

Think of these positions as representing a simple "Overseer-Underling" type situation, where Overseer gives Underling commands and prohibitions but not the reverse. This gives us a natural notion of imperative logical consequence.

Definition 6 (Imperative Consequence)   *Σ is an imperative consequence of Γ iff the imperative position $\Gamma : \Sigma$ clashes.*

We can apply this definition to give an intuitive interpretation of the same classical multiple conclusion sequent calculus rules as for declaratives. As with assertions and denials in the declarative case, commands and prohibitions are both, in a sense, mutually exclusive and exhaustive. The sense in which they are mutually exclusive is represented by the identity axiom Id. Id says that a position that commands and prohibits the same thing clashes. It is of course possible to command and prohibit the same thing, but not coherently.

$$\frac{}{\Phi \vdash \Phi} \text{ Id} \qquad \frac{\Gamma \vdash \Phi, \Sigma \quad \Gamma, \Phi \vdash \Sigma}{\Gamma \vdash \Sigma} \text{ Cut}$$

The sense in which commands and prohibitions are exhaustive is represented by the Cut rule. Cut, read contrapositively, tells us that commanding and prohibiting are exhaustive in the sense that whenever an imperative position does not clash, then neither does its extension with either commanding or prohibiting any arbitrary Φ. This does not mean that every position in fact either commands or prohibits Φ.

$$\frac{\Gamma \vdash \Sigma}{\Gamma, \Phi \vdash \Sigma} \text{ KL} \qquad \frac{\Gamma \vdash \Sigma}{\Gamma \vdash \Phi, \Sigma} \text{ KR} \qquad \frac{\Gamma, \Phi, \Phi \vdash \Sigma}{\Gamma, \Phi \vdash \Sigma} \text{ WL} \qquad \frac{\Gamma \vdash \Phi, \Phi, \Sigma}{\Gamma \vdash \Phi, \Sigma} \text{ WR}$$

The other structural rules of weakening K and contraction W fall out of the intended reading of clash. Weakening holds because once a position clashes then adding further commands or prohibitions to it will not remove the clash. Contraction records that commanding or prohibiting once is equivalent to doing so multiple times. If a position that commands or prohibits Φ, Φ clashes, then it will still clash if it commands or prohibits Φ just once. This makes sense of readings according to which someone commanding 'Buy a bottle of milk' and then saying so again later is reinforcing the first command with the second rather than telling them to buy two bottles of milk.

The operational rules receive readings very similar to in the declarative case, but replacing assertions and denials with commands and prohibitions. Conjunction and disjunction are primarily connected respectively to commands and prohibitions.

$$\frac{\Gamma, \Phi \vdash \Sigma}{\Gamma, \Phi \wedge \Psi \vdash \Sigma} \wedge L_1 \quad \frac{\Gamma, \Psi \vdash \Sigma}{\Gamma, \Phi \wedge \Psi \vdash \Sigma} \wedge L_2 \quad \frac{\Gamma \vdash \Phi, \Sigma \quad \Gamma \vdash \Psi, \Sigma}{\Gamma \vdash \Phi \wedge \Psi, \Sigma} \wedge R$$

$$\frac{\Gamma, \Phi \vdash \Sigma \quad \Gamma, \Psi \vdash \Sigma}{\Gamma, \Phi \vee \Psi \vdash \Sigma} \vee L \quad \frac{\Gamma \vdash \Phi, \Sigma}{\Gamma \vdash \Phi \vee \Psi, \Sigma} \vee R_1 \quad \frac{\Gamma \vdash \Psi, \Sigma}{\Gamma \vdash \Phi \vee \Psi, \Sigma} \vee R_2$$

Read contrapositively, the $\wedge L$ rules say that if there is no clash in commanding $\Phi \wedge \Psi$, then neither is there in commanding both Φ and also Ψ. Hence, commanding $\Phi \wedge \Psi$ rules out prohibiting either conjunct. However, prohibiting $\Phi \wedge \Psi$ does not require prohibiting both of Φ and Ψ. Rather, what $\wedge R$ records, read contrapositively, is that if there is no clash in prohibiting $\Phi \wedge \Psi$, then it does not clash to prohibit at least one of Φ or Ψ in the same context. Prohibiting $\Phi \wedge \Psi$ only clashes with commanding both conjuncts. Dually, prohibitions of disjunctions clash with commanding each disjunct (from $\vee R_1$ and $\vee R_2$). Commands of disjunctions importantly do not command either disjunct in particular. Instead, a command of $\Phi \vee \Psi$ rules out the prohibition of both disjuncts together (from $\vee L$).

Imperative bilateralism provides a good grasp on the meaning of imperative negation. A negated imperative $\neg \Phi$ should be read as 'Don't Φ' rather than 'It is not the case that Φ'.

$$\frac{\Gamma \vdash \Phi, \Sigma}{\Gamma, \neg \Phi \vdash \Sigma} \neg L \quad \frac{\Gamma, \Phi \vdash \Sigma}{\Gamma \vdash \neg \Phi, \Sigma} \neg R$$

$\neg L$ tells us that if a position that prohibits Φ clashes then so does one that swaps the prohibition of Φ for a command for $\neg \Phi$. Read contrapositively, if there is no clash in a position that commands $\neg \Phi$ then neither is there in prohibiting Φ in the same context. $\neg L$ allows us to derive $\Phi, \neg \Phi \vdash$, meaning that commands of imperatives clash with commands of their negations. Similarly, $\neg R$ has the result that $\vdash \Phi, \neg \Phi$ meaning that prohibitions of imperatives clash with prohibitions of their negations. This is because $\neg R$ says that if commanding Φ clashes then prohibiting ¬Φ also clashes, and contrapositively, that if prohibiting ¬Φ doesn't clash then neither does commanding Φ. In essence, the two rules make equivalent the force of commanding an imperative and prohibiting its negation and vice versa. Together with the rules for conjunction, disjunction and contraction, the two negation rules allow for the following two derivations:

$$\frac{\frac{\frac{\frac{\Phi \vdash \Phi}{\Phi, \neg \Phi \vdash} \neg L}{\Phi \wedge \neg \Phi, \neg \Phi \vdash} \wedge L}{\frac{\Phi \wedge \neg \Phi, \Phi \wedge \neg \Phi \vdash}{\Phi \wedge \neg \Phi \vdash} WL} \wedge L}$$

$$\frac{\frac{\frac{\frac{\Phi \vdash \Phi}{\vdash \neg \Phi, \Phi} \neg R}{\vdash \Phi \vee \neg \Phi, \Phi} \vee R}{\frac{\vdash \Phi \vee \neg \Phi, \Phi \vee \neg \Phi}{\vdash \Phi \vee \neg \Phi} WR} \vee R}$$

Imperative LNC   Imperative LEM

Φ ∧ ¬Φ and Φ ∨ ¬Φ are examples of imperative contradictions and tautologies. An imperative contradiction is an imperative whose command always clashes. One can, of course, command Φ ∧ ¬Φ, just not coherently. Similarly, imperative tautologies are not imperatives that are "always commanded", but rather whose prohibition always clashes. This is directly analogous to the declarative case, where it is always incoherent to deny $A \vee \neg A$ without it being the case that every declarative position does in fact assert $A \vee \neg A$.

$$\frac{\Gamma \vdash \Phi, \Sigma \quad \Gamma, \Psi \vdash \Sigma}{\Gamma, \Phi \supset \Psi \vdash \Sigma} \supset L \qquad \frac{\Gamma, \Phi \vdash \Psi, Y}{\Gamma \vdash \Phi \supset \Psi, \Sigma} \supset R$$

English, at least, lacks conditionals with imperative antecedents. For example, 'If go to the beach, put on sunscreen' is ungrammatical. If one wished, one could exclude such conditionals from the syntax of $\mathcal{L}_I$ because they have no intuitive interpretation.[5] If these conditionals are kept in the language, ⊃R tells us that if a position that commands Φ and prohibits Ψ clashes, then so does one that prohibits Φ ⊃ Ψ in the same context. Read contrapositively, ⊃L says that if a position that commands Φ ⊃ Ψ does not clash, then neither does at least one of the positions that prohibit Φ or command Ψ in the same context.[6]

---

[5] I am here remaining neutral on whether the lack of imperative antecedents is a mere quirk of grammar or an indication of deeper features of imperatives and conditionals.

[6] The syntax of $\mathcal{L}_I$ treats imperatives as syntactic primitives, as in Fine (2018), Fox (2012) and Mastop (2011). This differs from approaches that start with a declarative language whose vocabulary is extended with an imperative forming operator !. The operator ! takes a declarative p and forms an imperative !p, where the declarative p will often state the fulfillment conditions

Imperative Bilateralism

The two advantages of Restall's bilateralism identified in the previous section carry over naturally to imperative bilateralism. First, as in the declarative case, the notion of clash between commands and prohibitions doesn't draw on notions of truth and falsity. Because of this, imperative bilateralism is able to remain neutral regarding Claim 3 (Relata); hence, also Claim 2 (Truth Apt) and debates about cognitivism. Second, the central notion of clash involves discursive norms, rather than inference as an activity, and therefore does not require taking a stand on the debate about imperative inference.

We finish with two example derivations, using the incompatible and equivalent sentences from Example 1.

$$
\cfrac{\cfrac{\cfrac{\phi \vdash \phi}{\phi \wedge \neg\psi \vdash \phi} \wedge L}{\phi \wedge \neg\psi, \neg\phi \vdash} \neg L \quad \cfrac{\cfrac{\cfrac{\psi \vdash \psi}{\psi, \neg\psi \vdash} \neg L}{\psi, \phi \wedge \neg\psi \vdash} \wedge L}{}}{\neg\phi \vee \psi, \phi \wedge \neg\psi \vdash} \vee L
$$

In the above derivation read ϕ as 'Buy jam' and ψ as 'Buy marmalade'. What results is a sequent telling us that it is incoherent to jointly assert Example 1a 'Both buy marmalade and don't buy marmalade' and Example 1b 'Either don't buy jam or do buy marmalade'. In the derivation below read ξ as 'Buy quinoa falafel mix' and ρ as 'Buy soy falafel mix'. The derivation shows that it is incoherent to jointly assert 2a 'Don't buy quinoa falafel mix and also not the soy' and deny 2b 'Neither buy quinoa nor soy falafel mix'.

$$
\cfrac{\cfrac{\cfrac{\cfrac{\xi \vdash \xi}{\xi, \neg\xi \vdash} \neg L}{\xi, \neg\xi \wedge \neg\rho \vdash} \wedge L \quad \cfrac{\cfrac{\rho \vdash \rho}{\rho, \neg\rho \vdash} \neg L}{\rho, \neg\xi \wedge \neg\rho \vdash} \wedge L}{\xi \vee \rho, \neg\xi \wedge \neg\rho \vdash} \vee L}{\neg\xi \wedge \neg\rho \vdash \neg(\xi \vee \rho)} \neg R
$$

---

of the imperative (Charlow, 2018; Parsons, 2013; Vranas, 2019). For example, from 'The door is shut' could be formed 'Make it the case that the door is shut'. This approach allows for a syntactic difference between sentences where the imperative operator takes wide versus narrow scope, such as $!(p \vee q)$ and $!p \vee !q$, and opens up the question of whether there should be a semantic difference in such cases. For example, is there a semantic difference between 'Make it the case that either the door is shut or the window is shut' and 'Either make it the case that the door is shut or make it the case that the window is shut'. For better or worse these sorts of questions do not arise in $\mathcal{L}_I$.

3.3 Mixed inference rules

So far, Restall's bilateralism has been generalised to a purely imperative logic, but not to one that includes both imperatives and declaratives. It would, however, be desirable to have a general account of logical consequence covering declaratives and imperatives. For it seems that logical consequence (or logical relations) is a general notion and this is required to explain mixed arguments, such as Example 3 Umbrella, that feature both imperatives and declaratives.

To be clear about what kind of sentences we are talking about, say that a sentence is *purely declarative* or *purely imperative* if all its subformulas are, respectively, declaratives or imperatives. If a sentence is neither purely declarative nor imperative we say that it is *mixed*. We keep the previous convention of using standard Latin to represent sentences that are purely declarative and Greek for ones that are purely imperative. For example, Φ may be of arbitrary logical complexity but is purely imperative. To represent sentences in general we use lower case Latin maths sans serif: p and q for atomics; upper case A and B for arbitrary sentences; and upper case X and Y for multisets of sentences. For example, A is a sentence of arbitrary complexity that may be mixed, purely declarative or purely imperative.

General logical consequence will be defined for a mixed language $\mathcal{L}_\text{M}$.

Definition 7 $\mathcal{L}_\text{M}$ *is the language whose vocabulary is made up of denumerably many atomic declaratives* $p_1, p_2, p_3, \ldots$; *denumerably many atomic imperatives* $\xi_1, \xi_2, \xi_3, \ldots$; *the one-place connective* $\neg$ *and the two-place connectives* $\wedge$, $\vee$ *and* $\supset$; *and whose sentences are all and only those generated recursively from the following rule: all atomic declaratives* p *and atomic imperatives* ξ *are sentences, and if* A *and* B *are sentences then so are* $\neg$A, A $\wedge$ B, A $\vee$ B *and* A $\supset$ B.

Note that $\mathcal{L}_\text{M} \neq \mathcal{L}_\text{D} \cup \mathcal{L}_\text{I}$. $\mathcal{L}_\text{D}$ and $\mathcal{L}_\text{I}$ consist respectively of purely declarative and purely imperative sentences. Whereas, while $\mathcal{L}_\text{M}$ is built up from purely declarative and purely imperative atomics, it contains mixed mooded sentences such as the conditional $p \supset \phi$.[7]

The notion of mixed positions and mixed inference rules will now be introduced. In the previous section, the declarative nature of assertion and denial was abstracted from, to thinking of them as forms of ruling-in and

---

[7] In additional to lacking conditionals with imperative antecedents, sentential negation, in English at least, appears not to scope over mixed-mooded sentences. For example, '# It is not the case that, go to the party and I'll see you later' and '# Don't, go to the party and I'll see you later'. See Boisvert (1999) for an analysis of mixed-mooded sentences, arguing that many possible combinations are absent for pragmatic rather than semantic or syntactic reasons.

ruling-out, and that commanding and prohibiting were their respective imperative forms. Assertions and commands are both instances of ruling-in, with denials and with prohibitions instances of ruling-out. However, to allow for uses of mixed-mooded sentences, such as conditionals with declarative antecedents and imperative consequents, then some rulings-in and rulings-out will themselves be mixed.

Definition 8 (Mixed Positions and Clash) *If* X *and* Y *are both multisets of sentences, then* $X : Y$ *is the* mixed position *that rules in all* $A \in X$ *and rules out all* $B \in Y$. $X \vdash Y$ *expresses the claim that the position* $X : Y$ *clashes.*

General logical consequence can now be defined as expected.

Definition 9 (General Consequence) Y *is a consequence of* X *iff the mixed position* $X : Y$ *clashes.*

We can represent structural rules and those for logical connectives in a sentence-type neutral way.

$$\frac{}{A \vdash A} \text{Id} \qquad \frac{X \vdash A, Y \quad X, A \vdash Y}{X \vdash Y} \text{Cut}$$

$$\frac{X \vdash Y}{X, A \vdash Y} KL \qquad \frac{X \vdash Y}{X \vdash A, Y} KR \qquad \frac{X, A, A \vdash Y}{X, A \vdash Y} WL \qquad \frac{X \vdash A, A, Y}{X \vdash A, Y} WR$$

$$\frac{X \vdash A, Y}{X, \neg A \vdash Y} \neg L \qquad \frac{X, A \vdash Y}{X \vdash \neg A, Y} \neg R$$

$$\frac{X \vdash A, Y \quad X, B \vdash Y}{X, A \supset B \vdash Y} \supset L \qquad \frac{X, A \vdash B, Y}{X \vdash A \supset B, Y} \supset R$$

$$\frac{X, A \vdash Y}{X, A \wedge B \vdash Y} \wedge L_1 \qquad \frac{X, B \vdash Y}{X, A \wedge B \vdash Y} \wedge L_2 \qquad \frac{X \vdash A, Y \quad X \vdash B, Y}{X \vdash A \wedge B, Y} \wedge R$$

$$\frac{X, A \vdash Y \quad X, B \vdash Y}{X, A \vee B \vdash Y} \vee L \qquad \frac{X \vdash A, Y}{X \vdash A \vee B, Y} \vee R_1 \qquad \frac{X \vdash B, Y}{X \vdash A \vee B, Y} \vee R_2$$

Interpret these rules as before but in terms of ruling-in and ruling-out. Keep in mind that some ruling-ins will be either assertions or commands but that some will be themselves mixed (and ditto for ruling-outs). The above system has two main upshots. The first is that logical connectives have the same meanings regardless of the sentence types involved. This avoids the consequence that a connective, e.g., 'and', means different things in different sentence types, and allows for a simple way to form "mixed mooded" sentences. The second is that it allows for an account of logical consequence defined in terms of clash that is also completely general regarding sentence type. We then have a valid derivation for Example 3 Umbrella:

$$\frac{\text{It's raining} \vdash \text{It's raining} \qquad \text{Bring your umbrella} \vdash \text{Bring your umbrella}}{\text{If it's raining then bring your umbrella, It's raining} \vdash \text{Bring your umbrella}} \supset\text{L}$$

As an example of a more complex derivation, consider an argument taken from Clark-Younger (2014, p. 3). Suppose, before a philosophy exam, your friend advises you 'If you have a choice, don't answer the Kant question'. In the exam, the instructions say to choose between a question on Hume and a question on Kant. The derivation below shows that ruling in your friend's advice and the exam instructions makes ruling out 'Answer the Hume question' incoherent. Read p as 'You have a choice', ϕ as 'Answer the Kant question', and ψ as 'Answer the Hume question'.

$$\cfrac{p \vdash p \qquad \cfrac{\cfrac{\phi \vdash \phi}{\phi, \neg\phi \vdash} \neg\text{L}}{\phi, p \supset \neg\phi, p \vdash} \supset\text{L} \qquad \cfrac{\cfrac{\psi \vdash \psi}{\psi, p \supset \neg\phi \vdash \psi} \text{KL}}{\psi, p \supset \neg\phi, p \vdash \psi} \text{KL}}{p \vee \psi, p \supset \neg\phi, p \vdash \psi} \text{VL}$$

In the above the material conditional has been used. Some may have concerns about the material conditional, specific to modelling conditionals with declarative antecedents and imperative consequents. In model-theoretic approaches that generalise truth and falsity for declaratives to satisfaction and violation for imperatives, if the logic is two-valued and the conditional material, then a conditional imperative with a false antecedent counts as satisfied. Some have argued, *pace* Dummett (1978, pp. 8–9), that this is an unintuitive result and that conditional imperatives require a three-valued logic (Sosa, 1967; Vranas, 2008). The material conditional is adopted here for simplicity's sake and I make no commitment to it as accurately modelling natural language conditionals. I suspect that imperatives are not a special case and that the reasons that would count against the material conditional for imperatives apply equally well to declaratives and vice versa. Arguing this case is, however, left to future work.

4 Conclusion

This paper has outlined a novel theory of imperative, and general, logical consequence by generalising Restall's declarative bilateralism to one that includes imperatives. Imperative consequence is defined in terms of clash between commands and prohibitions, just as declarative consequence is defined in terms of clash between assertions and denials. As an account of imperatives, it has three main virtues: (1) it provides a proof-theoretic account of imperatives; (2) it does not rely on the controversial notion of imperative inference; and (3) it is neutral regarding cognitivism about imperatives. Commands and assertions are both instances of ruling-in; prohibitions and denials instances of ruling-out. General logical consequence is then defined in terms of clash between ruling-ins and ruling-outs. This has the virtue of (4) being a general account of logical consequence that covers declaratives and imperatives together; and (5) having rules for logical connectives that are neutral regarding sentence type and which allow connectives to have the same meaning regardless of sentence type.

References

Boisvert, D. R. (1999). *Pragmatics and semantics of mixed sentential mood sentences* (Unpublished master's thesis). University of Florida.

Charlow, N. (2014). Logic and semantics for imperatives. *Journal of Philosophical Logic, 43*(4), 617–664.

Charlow, N. (2018). Clause-type, force, and normative judgment in the semantics. In D. Fogal, D. W. Harris, & M. Moss (Eds.), (pp. 223–33). Oxford University Press.

Clark-Younger, H. (2012). Is imperative inference impossible? In J. Maclaurin (Ed.), *Rationis Defensor* (pp. 275–292). Springer.

Clark-Younger, H. (2014). Imperatives and the more generalised Tarski thesis. *Thought: A Journal of Philosophy, 3*(4), 314–320.

Dummett, M. A. E. (1978). *Truth and Other Enigmas*. Harvard University Press.

Fine, K. (2018). Compliance and command I—categorical imperatives. *Review of Symbolic Logic, 11*(4), 609–633.

Fox, C. (2012). Imperatives: A judgemental analysis. *Studia Logica, 100*(4), 879–905.

Hansen, J. (2008). Imperative logic and its problems. In D. Gabbay, J. Horty, X. Parent, R. van der Meyden, & L. van der Torre (Eds.), *Handbook of Deontic Logic and Normative Systems* (pp. 137–192). College Publications.

Kaufmann, M. (2012). *Interpreting imperatives*. Springer.

Lewis, D. K. (1970). General semantics. *Synthese, 22*(1-2), 18–67.

Lewis, D. K. (1979). A problem about permission. In E. Saarinen, R. Hilpinen, I. Niiniluoto, & M. Provence (Eds.), *Essays in Honour of Jaakko Hintikka on the Occasion of His Fiftieth Birthday on January 12, 1979* (pp. 163–175). Reidel.

Mastop, R. (2011). Imperatives as semantic primitives. *Linguistics and Philosophy, 34*(4), 305–340.

Parsons, J. (2013). Command and consequence. *Philosophical Studies, 164*(1), 61–92.

Peregrin, J. (2014). *Inferentialism: Why Rules Matter*. Springer.

Portner, P. (2004). The semantics of imperatives within a theory of clause types. In *Proceedings of semantics and linguistic theory* (Vol. 14, pp. 235–252).

Restall, G. (2005). Multiple conclusions. In *Logic, methodology and philosophy of science: Proceedings of the twelfth international congress* (pp. 189–205).

Sosa, E. (1967). The semantics of imperatives. *American Philosophical Quarterly, 4*(1), 57–64.

Steinberger, F., & Murzi, J. (2017). Inferentialism. In B. Hale, C. Wright, & A. Miller (Eds.), *Blackwell Companion to Philosophy of Language* (pp. 197–224). Wiley Blackwell.

Vranas, P. (2008). New foundations for imperative logic I: Logical connectives, consistency, and quantifiers. *Noûs, 42*(4), 529–572.

Vranas, P. (2010). In defense of imperative inference. *Journal of Philosophical Logic, 39*(1), 59–71.

Vranas, P. (2011). New foundations for imperative logic: Pure imperative inference. *Mind, 120*(478), 369–446.

Vranas, P. (2019). New foundations for imperative logic IV: Natural deduction. *Journal of Applied Logics, 6*(3), 431-446.

Williams, B. A. O. (1963). Imperative inference. *Analysis, 23*(Supplement 1), 30–36.

Kai Tanter
Monash University, Philosophy Department
Australia
E-mail: `Kai.Tanter@monash.edu`

Presupposition, Admittance and Karttunen Calculus

YOAD WINTER[1]

Abstract: Classic works define presuppositions of a sentence S as conclusions that follow from both S and its negation. Other studies focus on the necessary conditions for admitting S as true or false, assuming that those conditions converge with S's presuppositions. Here we study this assumption in three systems: asymmetric Kleene truth tables, Heim's admittance-based theory, and a new propositional calculus inspired by Karttunen's entailment-based approach. Common versions of the Kleene and Heim systems are known to be semantically congruent, and we show that they identify presuppositions with admittance conditions. By contrast, it is proved that the proposed *Karttunen calculus* distinguishes the two notions. This aspect of the Karttunen calculus avoids the "proviso problem" for the Kleene/Heim approaches: the generation of presuppositions that appear to be too weak.

Keywords: presupposition, admittance, propositional logic, Kleene truth tables, three valued logic, proviso problem

1 Introduction

Presuppositions may disappear when the expression that triggers them is embedded in a complex sentence. For instance, the term *"the king of France"* famously presupposes that France is a monarchy, but the sentence "if France has a king, *the king of France* must be living at the Élysée Palace" does not. In such cases, we say that the presupposition "France is a monarchy" does not *project*. Karttunen (1973, 1974) analyzed presupposition projection and the lack thereof using rules that draw on entailment relations between logical forms. Peters (1979) suggested to emulate Karttunen's proposals using a truth-functional analysis that employs an asymmetric version of the Strong

---

[1] Special thanks to Matthew Mandelkern for many remarks and discussions. For their remarks, the author is also grateful to Jakub Dotlačil, Danny Fox, Rick Nouwen and Philippe Schlenker, as well as to audiences at Institut Jean Nicod and Hebrew University. Work on this paper was partially funded by the European Research Council (ERC) under the European Union's Horizon 2020 research and innovation programme (grant agreement No 742204).

Kleene tables. In this three-valued semantics, a *presupposition* of a sentence S is classically defined as a proposition that follows from both S and its negation (van Fraassen, 1971). The Kleene-Peters analysis has been opposed to the "dynamic" approach in (Heim, 1983; Stalnaker, 1978), which defines a presupposition of a sentence S as a proposition that is entailed by all contexts that *admit* S, i.e., make S true or false.

This paper first shows that the Heim-Stalnaker account derives the same consequence relation as the Kleene-Peters system. In both systems, a close relation is rendered between classic presuppositions and admittance conditions: a proposition is a logically *strongest* presupposition of a sentence S if and only if it is a *weakest* admittance condition of S. This property has linguistically undesirable ramifications, known as the *proviso problem* (Winter, 2019 and references therein). For example, in the sentence "if Sue is busy, her spouse is away", the Kleene-Peters and Heim-Stalnaker analyses expect an unintuitive presupposition: "Sue is married if she is busy". We propose a solution of the proviso problem that generalizes Karttunen's rules into a so-called *Karttunen calculus*. This calculus derives the same admittance conditions as in the Kleene/Heim system. However, the presuppositions that are derived in the Karttunen calculus may be stronger than those admittance conditions, which allows the system to avoid the proviso problem.

The Karttunen calculus is similar to the Kleene-Peters system in relying on *left-determinant* values of binary operators for defining presupposition projection. Like the Heim-Stalnaker system, it uses local contexts for satisfying presuppositions. However, unlike the Kleene/Heim systems, the calculus relies on entailment between propositional formulas as in Karttunen's work, rather than on implication or set inclusion between their denotations. Contrary to Peters' claims, this makes context a non-redundant element of the Karttunen calculus, indeed of Karttunen's (1974) original proposal.

Section 2 introduces the notions of *left-determinant value* and *projection calculus* and illustrates their use for presenting the Kleene-Peters tables. Section 3 shows that the Heim-Stalnaker semantics leads to the same equivalence and entailment relations as those tables. Section 4 shows that the Kleene/Heim system conflates strongest presuppositions with weakest admittance conditions. It is conjectured that this conflation is inadequate for describing natural language and leads to the proviso problem. Section 5 introduces the *Karttunen calculus* and shows that it distinguishes a sentence's strongest presuppositions from its, possibly weaker, weakest admittance conditions, thus avoiding the proviso problem. Section 6 concludes. For proofs and further technical notes see (Winter, 2021).

2 Kleene truth tables and projection calculi

The Strong Kleene truth tables are one of the earliest logical treatments of presupposition. While these tables are symmetric, presupposition projection is often not (Mandelkern, Zehr, Romoli, & Schwarz, 2020). In view of this fact, Peters (1979) proposed the tables in Figure 1, where '1', '0' and '∗' stand for *true, false* and *undefined*, respectively. These trivalent *Kleene-Peters* (KP) tables asymmetrically extend the standard bivalent tables. A bivalent conjunction (disjunction/implication) is false (true/true) when the lefthand operand is false (true/false, respectively). This property is preserved in the KP truth tables, also when the righthand operand is undefined. However, when the lefthand operand is undefined, the result is undefined with no respect to the value of the righthand operand.

α	$\neg\alpha$		$\alpha \wedge \beta$	0	1	∗		$\alpha \vee \beta$	0	1	∗		$\alpha \to \beta$	0	1	∗
0	1		0	0	0	0		0	0	1	∗		0	1	1	1
1	0		1	0	1	∗		1	1	1	1		1	0	1	∗
∗	∗		∗	∗	∗	∗		∗	∗	∗	∗		∗	∗	∗	∗

Figure 1: The Kleene-Peters (KP) truth tables

Presuppositional and assertive elements of English sentences are analyzed as *bivalent*, and are expressed using a standard *propositional language*: a closure of a non-empty set of constants C under the propositional operators $\neg, \wedge, \vee$ and $\to$. When the constants in C are arbitrary we assume that they are assigned a bivalent interpretation, and refer to the propositional language over C as 'L_2'. English sentences are analyzed as simple *trivalent* propositions, which are represented as pairs of formulas from L_2: a presuppositional part and an assertive content. Such pairs are denoted $(\alpha{:}\beta)$ and are interpreted in $\{0, 1, *\}$ using Blamey's (1986) *transplication* operator, which is defined below:

Definition 1 (transplication) *For any bivalent interpretation $[\![\cdot]\!]^{bi}$ of L_2, we extend the interpretation $[\![\cdot]\!]^{bi}$ of L_2 into an interpretation of $L_2 \times L_2$ by defining, for any $\alpha, \beta \in L_2$:*

$$[\![(\alpha{:}\beta)]\!]^{bi} = \begin{cases} [\![\beta]\!]^{bi} & [\![\alpha]\!]^{bi} = 1 \\ * & [\![\alpha]\!]^{bi} = 0 \end{cases}$$

Complex trivalent formulas are obtained using Definition 2 below:

Definition 2 (L_3) *Given a propositional language L_2 over arbitrary constants, the language L_3 is a propositional language over $L_2 \times L_2$.*

One way to analyze presupposition projection is by defining the trivalent denotation of complex L_3 formulas for any bivalent interpretation of L_2. Definition 3 below uses the KP truth tables to extend the trivalent interpretation of $L_2 \times L_2$ in Definition 1 into a trivalent interpretation of L_3:

Definition 3 (KP interpretation of L_3) *Let $[\![\cdot]\!]^{bi}$ be a bivalent interpretation of L_2, which is extended to $L_2 \times L_2$ as in Definition 1. For any formula $\kappa \in L_3$, the KP interpretation of κ is denoted $[\![\kappa]\!]^{KP}$ and is defined as follows:*

$$[\![\kappa]\!]^{KP} = \begin{cases} [\![\kappa]\!]^{bi} & \kappa \in L_2 \times L_2 \\ [\![\neg]\!]^{KP}([\![\varphi]\!]^{KP}) & \kappa = \neg\varphi \\ [\![\text{op}]\!]^{KP}([\![\varphi]\!]^{KP}, [\![\psi]\!]^{KP}) & \kappa = \varphi\,\text{op}\,\psi \end{cases}$$

where $[\![\varphi]\!]^{KP}$ and $[\![\psi]\!]^{KP}$ are inductively defined, and negation and the binary operator 'op' are interpreted using the KP tables (Fig. 1)

It is useful to note that the corresponding equivalence relation ($\stackrel{KP}{\equiv}$) over L_3 satisfies the following standard equivalences, for any $\varphi, \psi \in L_3$:

Fact 1 $\quad \varphi \vee \psi \stackrel{KP}{\equiv} \neg((\neg\varphi) \wedge \neg\psi) \qquad \varphi \to \psi \stackrel{KP}{\equiv} (\neg\varphi) \vee \psi$

The following example illustrates how KP semantics is used for analyzing presupposition projection:

Example 1 Sentences S1 and S2 below are represented as L_3 formulas:

S1 = *if Sue is married her spouse is away* = $(\top:\alpha_1) \to (\beta:\gamma)$
S2 = *if Sue is busy her spouse is away*   = $(\top:\alpha_2) \to (\beta:\gamma)$

where α_1="Sue is married", α_2="Sue is busy", β="Sue has a spouse", and γ="Sue has a spouse who is away" are bivalent propositions. We now observe the following KP equivalence:

$(\top:\alpha) \to (\beta:\gamma) \equiv (\alpha \to \beta : \alpha \to \gamma)$

While $\alpha_1 \to \beta$ is tautological, $\alpha_2 \to \beta$ is not. Thus, according to the KP semantics, the presupposition of sentence S1 is expected to be patently true, in agreement with linguistic judgements, where S1 shows no presupposition. By contrast, S2 is analyzed as presupposing "if Sue is busy, she has a spouse", which is weaker than the presupposition that ordinary speakers report ("Sue has a spouse"). This incongruence between theory and speaker judgements illustrates the *proviso problem* (Karttunen, 1973, p. 188; Geurts, 1996).

Presupposition, Admittance and Karttunen Calculus

An alternative way of analyzing presupposition projection is by rewriting any L_3 formula κ into a formula $(\kappa_1 : \kappa_2)$ in $L_2 \times L_2$, where the bivalent formula κ_1 is viewed as κ's strongest presupposition and κ_2 is viewed as κ's assertive context. We refer to this technique as a *projection calculus*.

The *Weak Kleene* (WK) tables let a propositional formula be interpreted as '$*$' if any of its sub-formulas is interpreted as '$*$'. Thus, the WK tables trivially "project" all presuppositions of κ's sub-formulas by letting κ_1 be their conjunction. This is modelled by the following projection calculus:

Definition 4 (WK calculus) *For any formula κ in L_3, let $WK(\kappa)$ be the formula in $L_2 \times L_2$ that is inductively defined as follows:*

$$WK(\kappa) = \begin{cases} \kappa & \kappa = (\kappa_1 : \kappa_2) \\ (\varphi_1 : \neg\varphi_2) & \kappa = \neg\varphi \\ (\varphi_1 \wedge \psi_1 : \varphi_2 \, op^{bi} \, \psi_2) & \kappa = \varphi \, op \, \psi \end{cases}$$

where: - op^{bi} is the bivalent propositional operator corresponding to op
- $\kappa_1, \kappa_2 \in L_2$, and inductively: $(\varphi_1 : \varphi_2) = WK(\varphi)$ and $(\psi_1 : \psi_2) = WK(\psi)$

A similar rewriting technique describes the KP tables (Figure 1). We first assign a unary operator 'LDV_{op}' (*left determinant value*) to any bivalent binary operator op. This is defined below:

Definition 5 (left determinant value) *For any binary operator op, the corresponding unary operator specifying the* left determinant value(s) *of op is defined as follows for any $\alpha \in L_2$:*

$LDV_{op}(\alpha) = (\alpha \, op \perp \leftrightarrow \alpha \, op \top)$.

Thus, we have: $LDV_\wedge(\alpha) = LDV_\rightarrow(\alpha) \equiv \neg\alpha$ and $LDV_\vee(\alpha) \equiv \alpha$.

Using the LDV operator, we define the *KP calculus* as follows:

Definition 6 (KP calculus) *For any formula κ in L_3, let $KP(\kappa)$ be the formula in $L_2 \times L_2$ that is inductively defined as follows:*

$$KP(\kappa) = \begin{cases} \kappa & \kappa = (\kappa_1 : \kappa_2) \\ (\varphi_1 : \neg\varphi_2) & \kappa = \neg\varphi \\ WK((\varphi_1 : \varphi_2) \, op \, ((\psi_1 \vee LDV_{op}(\varphi_2)) : \psi_2)) & \kappa = \varphi \, op \, \psi \end{cases}$$

where $\kappa_1, \kappa_2 \in L_2$, and inductively: $(\varphi_1 : \varphi_2) = KP(\varphi)$ and $(\psi_1 : \psi_2) = KP(\psi)$

By Definitions 4, 5 and 6 we have:
$$KP(\varphi \wedge \psi) \equiv (\varphi_1 \wedge (\psi_1 \vee \neg \varphi_2) : \varphi_2 \wedge \psi_2)$$
$$KP(\varphi \vee \psi) \equiv (\varphi_1 \wedge (\psi_1 \vee \varphi_2) : \varphi_2 \vee \psi_2)$$
$$KP(\varphi \rightarrow \psi) \equiv (\varphi_1 \wedge (\psi_1 \vee \neg \varphi_2) : \varphi_2 \rightarrow \psi_2)$$

Definition 6 of the KP calculus is *sound* with respect to KP interpretations:

Fact 2 *For any formula $\kappa \in L_3$ and KP interpretation:* $[\![KP(\kappa)]\!]^{KP} = [\![\kappa]\!]^{KP}$.

Example 2 $KP((\top:\alpha) \rightarrow (\beta:\gamma)) \equiv (\top \wedge (\beta \vee \neg \alpha) : \alpha \rightarrow \gamma)$, which is equivalent to $(\alpha \rightarrow \beta : \alpha \rightarrow \gamma)$ as in Example 1.

3 Heim-Stalnaker semantics

Heim (1983) analyzes presupposition projection in terms of a sentence's admittance by a given context. Following Stalnaker (1978), Heim defines a context as a set of possible worlds, which *admits* a sentence S if it is contained in the set of possible worlds where S's presuppositions hold. A sentence S is analyzed using a pair $\langle A, B \rangle$, where A and B are the sets of possible worlds denoted by S's presupposition and S's assertive content, respectively. Such pairs are used to update the context. Propositional connectives modify the updates induced by their operand(s). This view of presupposition projection seems quite different from the Kleene tables in traditional three-valued logic. However, following Peters (1979), this section shows that in terms of the entailment and equivalence relations they describe over formulas in L_3, the *Heim-Stalnaker* (HS) semantics and the KP truth tables are congruent.

3.1 Heim-Stalnaker semantics – language and interpretation

When representing a sentence's semantic import as its *context change potential* (CCP), it is convenient to use the following propositional language:

Definition 7 $L_{CCP} \stackrel{def}{=} L_2 \cup \{\chi[\kappa] : \chi \in L_{CCP} \text{ and } \kappa \in L_3\}$

Thus, any L_{CCP} formula is made of a context formula in L_2 and a (possibly empty) sequence of formulas in L_3.

Example 3 Given $C, \alpha, \beta, \gamma \in L_2$, the following are all L_{CCP} formulas:
$C, \ C[(\alpha:\beta)], \ (C[(\alpha:\beta)])[(\top:\gamma) \vee (\alpha:\beta)]$.

Adding disjunction to Heim's system, we get the following canonical semantics of L_{CCP} (Nouwen, Brasoveanu, van Eijck, & Visser, 2016; Rothschild, 2011):

Definition 8 (HS interpretation of L_{CCP}) *Let $W \neq \emptyset$ be an arbitrary set of possible worlds, and let $[\![\cdot]\!]_W^M$ (in short: '$[\![\cdot]\!]^M$') be a modal interpretation of L_2, which assigns any constant $p \in L_2$ a set $[\![p]\!] \subseteq W$, and interprets any complex L_2 formula using the set-theoretical operators corresponding to the propositional connectives. An HS interpretation over W is a function $[\![\cdot]\!]_W^{HS}$ (in short: '$[\![\cdot]\!]^{HS}$' or '$[\![\cdot]\!]$') from L_{CCP} to $\wp(W) \cup \{*\}$ that inductively extends $[\![\cdot]\!]_W^M$ to any formula $\chi \in L_{CCP}$. This is defined as follows:*

- *For any $\chi \in L_2$, we define:* $[\![\chi]\!] = [\![\chi]\!]^M$.

- *For any $\chi = \mu[\kappa] \in L_{CCP} \setminus L_2$:*

(a) *If $[\![\mu]\!] = *$, we define:* $[\![\mu[\kappa]]\!] = *$.

(b) *If $[\![\mu]\!] \neq *$ and $\kappa = (\kappa_1 : \kappa_2) \in L_2 \times L_2$, we define:*

$$[\![\mu[(\kappa_1:\kappa_2)]]\!] = \begin{cases} [\![\mu]\!] \cap [\![\kappa_2]\!] & \text{if } [\![\mu]\!] \subseteq [\![\kappa_1]\!] \\ * & \text{otherwise} \end{cases}$$

(c) *If $[\![\mu]\!] \neq *$ and $\kappa \in L_3 \setminus (L_2 \times L_2)$, we define inductively:*

$$[\![\mu[\neg\varphi]]\!] = \begin{cases} [\![\mu]\!] \setminus [\![\mu[\varphi]]\!] & \text{if } [\![\mu[\varphi]]\!] \neq * \\ * & \text{otherwise} \end{cases}$$

$$[\![\mu[\varphi \wedge \psi]]\!] = [\![(\mu[\varphi])[\psi]]\!]$$

$$[\![\mu[\varphi \vee \psi]]\!] = \begin{cases} [\![\mu[\varphi]]\!] \cup [\![(\mu[\neg\varphi])[\psi]]\!] & \text{if } [\![\mu[\varphi]]\!] \neq * \\ & \text{and } [\![(\mu[\neg\varphi])[\psi]]\!] \neq * \\ * & \text{otherwise} \end{cases}$$

$$[\![\mu[\varphi \to \psi]]\!] = \begin{cases} [\![\mu[\neg\varphi]]\!] \cup [\![(\mu[\varphi])[\psi]]\!] & \text{if } [\![\mu[\neg\varphi]]\!] \neq * \\ & \text{and } [\![(\mu[\varphi])[\psi]]\!] \neq * \\ * & \text{otherwise} \end{cases}$$

Similarly to KP connectives (Fact 1), the corresponding *HS equivalence* relation ($\stackrel{HS}{\equiv}$) over L_{CCP} satisfies, for any $\chi \in L_{CCP}$ and $\varphi, \psi \in L_3$:

Fact 3 $\chi[\varphi \vee \psi] \stackrel{HS}{\equiv} \chi[\neg((\neg\varphi) \wedge \neg\psi)]$ $\chi[\varphi \to \psi] \stackrel{HS}{\equiv} \chi[(\neg\varphi) \vee \psi]$

For the *proof* of Fact 3 see (Winter, 2021, Appendix A).

In HS semantics, the analysis of presupposition projection in sentences S1 and S2 from Example 1 goes as follows:

Example 4 Sentences S1 and S2 below are represented as L_{CCP} formulas:
$$S1 = \textit{if Sue is married her spouse is away} = C[(\top:\alpha_1)\to(\beta:\gamma)]$$
$$S2 = \textit{if Sue is busy her spouse is away} = C[(\top:\alpha_2)\to(\beta:\gamma)]$$
where C is arbitrary, and $\alpha_1, \alpha_2, \beta$ and γ are as in Example 1. We observe that under HS interpretations:
$$C[(\top:\alpha)\to(\beta:\gamma)] \equiv C[(\alpha\to\beta:\alpha\to\gamma)]$$
Thus, for any context C and interpretation $[\![\cdot]\!]^{HS}$, the formula $\kappa = (\top:\alpha)\to(\beta:\gamma)$ is well-defined relative to C (i.e., has a non-'$*$' interpretation) *iff* the set $[\![C]\!]^M$ is contained in $[\![\alpha\to\beta]\!]^M$. Accordingly, and similarly to KP semantics (Example 1), the proposition $\alpha\to\beta$ is viewed as κ's presupposition.

3.2 HS semantics and KP semantics

Any HS interpretation over a set of possible worlds $W \neq \emptyset$ has a modal interpretation of L_2 at its basis. Such a modal interpretation corresponds with a family F of bivalent interpretations of L_2 that is indexed by W. Thus, a modal interpretation of L_2 gives rise to a family of KP interpretations of L_3. In this section we show that any HS interpretation can be represented as such a family of KP interpretations. First, for any family of bivalent interpretations of L_2 we define an alternative semantics of L_{CCP} that directly employs the KP semantics of L_3. Definition 9 specifies this *KP-based interpretation*:

Definition 9 *Given a set $W \neq \emptyset$, let $F = [\![\cdot]\!]_i^{bi}|_{i\in W}$ be a family of bivalent interpretations of L_2. For any $i \in W$, let $[\![\cdot]\!]_i^{KP}$ be the KP interpretation of L_3 corresponding to $[\![\cdot]\!]_i^{bi}$. A KP-based interpretation of L_{CCP} relative to F is a function $[\![\cdot]\!]_W^{KP}$ from L_{CCP} to $\wp(W) \cup \{*\}$ that is inductively defined as follows for any $\chi \in L_{CCP}$:*

If $\chi = \alpha$, s.t. $\alpha \in L_2$:

$$[\![\alpha]\!]_W^{KP} = \{i \in W : [\![\alpha]\!]_i^{bi} = 1\}$$

If $\chi = \mu[\kappa]$, s.t. $\mu \in L_{CCP}$ and $\kappa \in L_3$:

$$[\![\mu[\kappa]]\!]_W^{KP} = \begin{cases} [\![\mu]\!]_W^{KP} \cap \{i \in W : [\![\kappa]\!]_i^{KP} = 1\} & [\![\mu]\!]_W^{KP} \neq * \text{ and} \\ & [\![\mu]\!]_W^{KP} \subseteq \{i \in W : [\![\kappa]\!]_i^{KP} \neq *\} \\ * & \text{otherwise} \end{cases}$$

In the standard Definition 8 of HS interpretations, a complex formula $\mu[\kappa]$ is interpreted by updating the context μ inductively using sub-formulas of κ. By contrast, in KP-based interpretations, $\mu[\kappa]$ is interpreted using the KP semantics of κ in L_3, through the given family F of bivalent interpretations. Theorem 1 shows that this way of interpreting L_{CCP} using the KP tables covers all HS interpretations. As summarized in (Winter, 2021, Appendix B), this theorem makes the same point as the main property proved by Peters (1979).

Theorem 1 *Let $[\![\cdot]\!]_W^{HS}$ be an HS interpretation of L_{CCP} for some $W \neq \emptyset$. For any $i \in W$, let $[\![\cdot]\!]_i^{bi}$ be the bivalent interpretation of L_2 s.t. for any $\alpha \in L_2$: $[\![\alpha]\!]_i^{bi} = 1$ iff $i \in [\![\alpha]\!]^{HS}$. Let $[\![\cdot]\!]_W^{KP}$ be the KP-based interpretation of L_{CCP} relative to the family $F = [\![\cdot]\!]_i^{bi}|_{i \in W}$. Then for any $\chi \in L_{CCP}$ we have:*

$$[\![\chi]\!]_W^{HS} = [\![\chi]\!]_W^{KP}.$$

The *proof* of Theorem 1 in (Winter, 2021, Appendix C) is by induction on the structure of χ for the subset of L_{CCP} involving only negation and conjunction. This proof is directly applicable to disjunction and implication due to the standard Facts 1 and 3 under KP and HS interpretations.

Using Theorem 1, we now show that HS semantics is congruent with KP semantics in two senses. First, we show the soundness of a so-called *HS calculus*, which uses the KP calculus to rewrite any formula $\chi \in L_{CCP}$ as a maximally simple formula in L_{CCP}. Second, we use the HS calculus to show that entailment and equivalence relations over L_3 that are naturally induced by the HS semantics are identical to those induced by KP semantics.

3.3 HS calculus

Relying on Theorem 1, we first show that the KP calculus can be used to simplify any L_{CCP} formula while preserving its HS semantics:

Corollary 1 *For any $\chi \in L_{CCP}$ and $\kappa \in L_3$: $\chi[\kappa] \stackrel{HS}{\equiv} \chi[KP(\kappa)]$.*

For the *proof* see (Winter, 2021, Appendix D).

On the basis of Corollary 1, we show that the KP calculus extends into a sound method of rewriting L_{CCP} formulas into equivalent, maximally simple CCP formulas. Rewriting in this *HS calculus* is defined below:

Definition 10 (HS calculus) *Let $min(L_{CCP})$ be the following set of minimal L_{CCP} formulas:*

$$min(L_{CCP}) = L_2 \cup \{ C[(\kappa_1:\kappa_2)] : C, \kappa_1, \kappa_2 \in L_2 \}.$$

For any formula χ in L_{CCP}, we define $HS(\chi)$ as the formula in $min(L_{CCP})$

that is inductively defined as follows:
   For any $\chi = C \in L_2$:
      $HS(C) = C$.
   For any $\chi = C[\kappa] \in L_{CCP}$ where $C \in L_2$ and $\kappa \in L_3$:
      $HS(C[\kappa]) = C[KP(\kappa)]$.
   For any $\chi = (\mu[\varphi])[\kappa]$ where $\mu \in L_{CCP}$ and $\varphi, \kappa \in L_3$, inductively:
      $HS((\mu[\varphi])[\kappa]) = HS(\mu[\varphi \wedge \kappa])$.

This calculus maps any simple L_2 formula in L_{CCP} to itself. Any formula $C[\kappa]$ where $C \in L_2$ is mapped to $C[KP(\kappa)]$, where $KP(\kappa)$ is the $L_2 \times L_2$ formula obtained from κ in the KP calculus. More complex L_{CCP} formulas are of the form $(\ldots((C[\varphi_1])[\varphi_2])\ldots)[\varphi_n]$, where $\varphi_1, \varphi_2, \ldots, \varphi_n \in L_3$. Such formulas are "flattened" to the form $C[\varphi_1 \wedge \varphi_2 \wedge \ldots \wedge \varphi_n]$, which is inductively simplified using the KP calculus.

Example 5 $HS(C[(\varphi_1:\varphi_2)])[(\psi_1:\psi_2)]) = HS(C[(\varphi_1:\varphi_2)\wedge(\psi_1:\psi_2)])$
$= C[KP((\varphi_1:\varphi_2)\wedge(\psi_1:\psi_2))] = C[(\varphi_1\wedge(\psi_1\vee\neg\varphi_2):\varphi_2\wedge\psi_2)]$

This HS calculus is *sound* with respect to HS interpretations of L_{CCP} formulas:

Corollary 2 *For any formula $\chi \in L_{CCP}$ and HS interpretation:*
$$[\![HS(\chi)]\!]^{HS} = [\![\chi]\!]^{HS}.$$

The *proof* for formulas $C[\kappa]$ where $C \in L_2$ follows from Corollary 1. For other formulas of the form $\mu[\kappa]$, Corollary 2 is proved inductively, relying on its proof for μ. See (Winter, 2021, Appendix E) for details.

3.4 KP/HS entailment and KP/HS equivalence

KP interpretations naturally specify an equivalence relation ($\stackrel{KP}{=}$) over L_3. As for *entailment* over L_3, we standardly employ the following "Tarskian" definition in trivalent semantics (van Fraassen, 1971):

Definition 11 (trivalent entailment) *Let $\mathcal{C}$ be a class of trivalent interpretations mapping a language L to $\{0, 1, *\}$. For any two formulas $\varphi, \psi \in L$, we denote $\varphi \stackrel{\mathcal{C}}{\Rightarrow} \psi$ if for every interpretation $[\![\cdot]\!] \in \mathcal{C}$:*
   if $[\![\varphi]\!] = 1$ then $[\![\psi]\!] = 1$.

When $\mathcal{C}$ in Definition 11 is the class of KP interpretations of L_3, we obtain a relation of *KP-entailment* ($\stackrel{KP}{\Rightarrow}$). We note that Definition 11 distinguishes bidirectional entailment from equivalence: trivalent propositions may agree on the interpretations that make them *true* without necessarily agreeing on the interpretations that make them *false*.

HS semantics is defined over the language L_{CCP}, hence specifies an equivalence relation ($\stackrel{HS}{\equiv}$) over that language. To allow comparing HS and KP semantics, this relation is extended to an equivalence relation over L_3:

Definition 12 (HS equivalence over L_3)   *For any two formulas $\varphi, \psi \in L_3$, we denote $\varphi \stackrel{HS}{\equiv} \psi$ iff for every $\chi \in L_{CCP}$: $\chi[\varphi] \stackrel{HS}{\equiv} \chi[\psi]$.*

For any two L_3 formulas φ and ψ, we also define *HS entailment*, by requiring that whenever φ leaves a context intact, so does ψ. Formally:

Definition 13 (HS entailment over L_3)   *For any two formulas $\varphi, \psi \in L_3$, we denote $\varphi \stackrel{HS}{\Rightarrow} \psi$ iff for every $\chi \in L_{CCP}$ and HS interpretation:*

if $[\![\chi[\varphi]]\!]^{HS} = [\![\chi]\!]^{HS}$ then $[\![\chi[\psi]]\!]^{HS} = [\![\chi]\!]^{HS}$.

The claim below follows from the soundness of HS calculus (Corollary 2):

Corollary 3   *For any two formulas $\varphi, \psi \in L_3$:*

   (i) $\varphi \stackrel{KP}{\equiv} \psi$ iff $\varphi \stackrel{HS}{\equiv} \psi$ (ii) $\varphi \stackrel{KP}{\Rightarrow} \psi$ iff $\varphi \stackrel{HS}{\Rightarrow} \psi$

The *proof* of Corollary 3 is in (Winter, 2021, Appendix F).

The KP/HS entailment relation is *monotonic*, in the following sense:

Fact 4   *For all $\varphi, \psi, \kappa \in L_3$: if $\varphi \stackrel{KP}{\Rightarrow} \psi$ then $\kappa \wedge \varphi \stackrel{KP}{\Rightarrow} \psi$.*

This fact is related to the proviso problem, discussed in the following section.

4 Admittance vs. presupposition

In the HS semantics of L_{CCP}, *admittance* of a proposition by a context is defined as follows:

Definition 14 (HS-admittance)   *We say that a context $C \in L_2$ HS-admits a formula $\kappa \in L_3$ if $[\![C[\kappa]]\!]^{HS} \neq *$ for all HS interpretations $[\![\cdot]\!]^{HS}$.*

A parallel notion is defined over L_3 using the KP semantics. We first introduce the following general notation:

Notation. Given a projection calculus Ω mapping L_3 to $L_2 \times L_2$, for any formula $\kappa \in L_3$ we denote:

$$\Omega(\kappa) = (\alpha_\kappa^\Omega : \beta_\kappa^\Omega), \text{ where } \alpha_\kappa^\Omega, \beta_\kappa^\Omega \in L_2.$$

In KP semantics we define admittance by first observing the following fact:

Fact 5 *For any $C \in L_2$ and $\kappa \in L_3$: C HS-admits κ iff $\alpha_{(\top:C) \wedge \kappa}^{KP} \equiv \top$.*

Thus, C admits κ in HS semantics *iff* the KP calculus rewrites the conjunction $(\top : C) \wedge \kappa$ into a pair $(\alpha : \beta)$ where α is a tautology. By soundness of KP calculus, this means that no KP interpretation makes the formula $(\top : C) \wedge \kappa$ *undefined* ('*'). When this condition holds we say that C *KP-admits* κ.

Presuppositions are standardly defined using entailment:

Definition 15 (presupposition) *Given an entailment relation $\stackrel{C}{\Rightarrow}$ over L_3, we say that $\kappa \in L_3$ C-presupposes $\beta \in L_2$ if $\kappa \stackrel{C}{\Rightarrow} (\top : \beta)$ and $\neg\kappa \stackrel{C}{\Rightarrow} (\top : \beta)$.*

Due to the convergence of the entailment relations in KP and HS semantics (Corollary 3), KP-presupposition and HS-presupposition converge as well.

Furthermore, in KP/HS semantics, the logically *weakest admitting context* and *strongest presupposition* converge for any formula $\kappa \in L_3$. This is shown by the following theorem:

Theorem 2 *For $\kappa \in L_3$, let C be a weakest formula in L_2 that KP-admits κ, and let β be a strongest KP-presupposition of κ in L_2. Then $C \equiv \beta \equiv \alpha_\kappa^{KP}$.*

Standardly, we here say that $\alpha \in L_2$ is a weakest (strongest) formula in L_2 with a property Π if any $\alpha' \in L_2$ that has the property Π and satisfies $\alpha \Rightarrow \alpha'$ (respectively: $\alpha' \Rightarrow \alpha$) satisfies $\alpha' \equiv \alpha$. The *proof* of Theorem 2 is in (Winter, 2021, Appendix G).

Theorem 2 is closely related to the *proviso problem* for KP/HS semantics (Example 1). To highlight this, we propose the following empirical conjecture about English, which stands in opposition to Theorem 2:

Conjecture 1 *There exists an English sentence S that is admitted by a context C such that C is logically weaker than any strongest presupposition of S.*

Example 6 Sentences S3 and S4 below are represented as L_3 formulas:

S3 = *if Sue visited Dan, his beard annoyed her* = $(\top : \alpha) \to (\beta : \gamma)$

S4 = *if Sue visited Dan, he had grown a beard before she arrived*
= $(\top : \alpha) \to (\top : \beta')$

Where α="Sue visited Dan", β= "Dan had a beard", β'= "Dan had grown

a beard before Sue arrived" and γ= "Dan had a beard that annoyed Sue".
Substantiating Conjecture 1, we make the following empirical claims:

(a) Sentence *S3* presupposes that Dan had a beard.

(b) The conjunction *S4 and S3* does not presuppose that Dan had a beard. Furthermore, *S4 and S3* has no non-tautological presupposition.

Claim (b) is consistent with the expectation of KP/HS-semantics that the weakest admittance condition of S3 is S5 below, which is entailed by S4:

S5 = *if Sue visited Dan, he had a beard* = $(\top:\alpha) \to (\top:\beta)$

However, claim (a) is inconsistent with the expectation of KP/HS-semantics that S5 is also the strongest presupposition of S3.

5 The Karttunen calculus

Conjecture 1 as illustrated in Example 6 suggests that Theorem 2 is problematic for using KP/HS semantics as a model of presupposition projection in English. To solve this problem, we propose an alternative projection calculus called the *Karttunen* (K) *calculus*. Like the KP calculus, the K-calculus maps any L_3 formula to a formula in $L_2 \times L_2$. However, unlike the KP calculus, the K-calculus does not emerge from any straightforward trivalent semantics. Rather, as in (Karttunen, 1973, 1974), the K-calculus takes *entailment* between bivalent formulas (or "logical forms") as the key to admitting a sentence by way of satisfying its presuppositions.

At the basis of the mechanism lie two assumptions: (i) a context $C \in L_2$ *admits* a simple L_3 formula $(\kappa_1 : \kappa_2)$ iff C entails κ_1 in bivalent logic; (ii) in binary constructions φ op ψ, the assertive content of φ updates the context of ψ's evaluation using the LDV operator. The reliance on entailment in (i) prevents a direct interpretation of L_3 according to the K-calculus. Rather, L_3 formulas need to first be transformed into formulas in $L_2 \times L_2$ before they can be semantically interpreted. This representational analysis of presupposition projection follows Karttunen's reliance on logical forms, but it squarely aligns with the truth-functional practice of involving left-determinant values as the key to presupposition projection and admittance, as in Kleene-Peters semantics (Winter, 2019). Unlike HS semantics, where contexts are arguably redundant due to the operational equivalence with KP semantics (see Peters 1979 and Corollary 3 above), the K-calculus uses L_2 formulas non-redundantly for recording *local contexts*. These local contexts are not denotations like sets of possible worlds as in HS semantics but L_2 formulas (or "logical forms") as in (Karttunen, 1974).

Formally, the K-calculus maps any L_3 formula to a formula in $L_2 \times L_2$ using a bivalent *context* $C \in L_2$, which is assumed to be tautological in the base case. This is specified in Definition 16 below:

Definition 16 (K-calculus) *For any formula $C[\kappa]$ in L_{CCP} where $C \in L_2$, let $K(C[\kappa])$ be the formula in L_3 that is inductively defined as follows:*
If $\kappa = (\kappa_1:\kappa_2) \in L_2 \times L_2$:
$$K(C[(\kappa_1:\kappa_2)]) = \begin{cases} (\top:\kappa_2) & C \Rightarrow \kappa_1 \\ (\kappa_1:\kappa_2) & \textit{otherwise} \end{cases}$$
If $\kappa \in L_3 \setminus (L_2 \times L_2)$:
$$K(C[\kappa]) = \begin{cases} (\varphi_1:\neg\varphi_2) & \kappa = \neg\varphi \\ WK(\,(\varphi_1:\varphi_2)\,\text{op}\,K((C \wedge \varphi_1 \wedge \neg\text{LDV}_{op}(\varphi_2))[\psi])\,) & \kappa = \varphi\,\text{op}\,\psi \end{cases}$$
where inductively: $(\varphi_1:\varphi_2) = K(C[\varphi])$

For any $\kappa \in L_3$ *(without any given C), we abbreviate:*
$K(\kappa) = K(\top[\kappa])$.

By definition of the WK calculus and the LDV_{op} operator we now have:
$K(C[\varphi \wedge \psi]) = WK((\varphi_1:\varphi_2) \wedge K((C \wedge \varphi_1 \wedge \varphi_2)[\psi]))$
$K(C[\varphi \vee \psi]) = WK((\varphi_1:\varphi_2) \vee K((C \wedge \varphi_1 \wedge \neg\varphi_2)[\psi]))$
$K(C[\varphi \rightarrow \psi]) = WK((\varphi_1:\varphi_2) \rightarrow K((C \wedge \varphi_1 \wedge \varphi_2)[\psi]))$

According to Definition 16, in binary constructions both the presuppositional content and the (negation of) the assertive content of the lefthand operand are accommodated into the context of the righthand operand. This is useful in sentences like *if Sue stopped smoking, then Dan knows that Sue stopped smoking*. This sentence inherits the presupposition of the antecedent ("Sue used to smoke"), but not the presupposition of the consequent ("Sue used to smoke and doesn't smoke now"). According to the K-calculus, this happens due to the accommodation of the whole antecedent (both presupposition and assertive content) into the context of the consequent. This local context of the consequent entails its presupposition, hence that presupposition is not projected.

Let us now consider the application of the K-calculus to the analysis of sentences S3 and S5 from Example 6:

Example 7 For S3 and S5 from Example 6, we denote, respectively:
$\eta = (\top:\alpha) \rightarrow (\beta:\gamma)$, and $\theta = (\top:\alpha) \rightarrow (\top:\beta')$.

Since $K(\top[(\top{:}\alpha)] = (\top{:}\alpha))$, and since $\alpha \not\Rightarrow \beta$, we conclude:
$K(\eta) = K((\top{:}\alpha) \to (\beta{:}\gamma)) = K(\top[(\top{:}\alpha) \to (\beta{:}\gamma)])$
$\quad = \mathit{WK}((\top{:}\alpha) \to K((\top \wedge \top \wedge \alpha)[(\beta{:}\gamma)])) = \mathit{WK}((\top{:}\alpha) \to (\beta{:}\gamma)) = (\beta{:}\alpha \to \gamma)$
$K(\theta \wedge \eta) = \ldots = \mathit{WK}((\top{:}\alpha \to \beta') \wedge K((\alpha \to \beta')[(\top{:}\alpha) \to (\beta{:}\gamma)]))$
$\quad = \mathit{WK}((\top{:}\alpha \to \beta') \wedge \mathit{WK}((\top{:}\alpha) \to K(((\alpha \to \beta') \wedge \alpha)[(\beta{:}\gamma)])))$, since $\beta' \Rightarrow \beta$:
$\quad = \mathit{WK}((\top{:}\alpha \to \beta') \wedge \mathit{WK}((\top{:}\alpha) \to (\top{:}\gamma))) = \ldots = (\top{:}\alpha \to (\beta' \wedge \gamma))$

Unlike the KP/HS semantics, these derivations are consistent with claims (a) and (b) in Example 6. They show that β is the strongest K-presupposition of η, but the bivalent proposition $\alpha \to \beta'$ ($=K(\theta)$'s assertive content) K-admits η although it does not logically entail that presupposition. Thus, K-admittance and KP/HS-admittance converge in this case, although the K-presupposition is stronger than its KP/HS correlate.

More generally, we claim that weakest admittance conditions in the K-calculus are the same as in the KP/HS-calculus, for all L_3 formulas. By contrast, presuppositions in the K-calculus are at least as strong as those of the KP/HS-calculus, but they may also be properly stronger as in Example 7. For this comparison between calculi, we first define the necessary semantic notions in the K-calculus. Definition 17 below *K-interprets* any $\kappa \in L_3$ by rewriting it into $K(\kappa)$ — an $L_2 \times L_2$ formula interpreted by transplication under any bivalent interpretation (Definition 1):

Definition 17 (K-interpretation of L_3) *Let $[\![\cdot]\!]^{bi}$ be a bivalent interpretation of L_2, and let κ be an L_3 formula. The **Karttunen (K) interpretation** of κ is defined by $[\![\kappa]\!]^K = [\![K(\kappa)]\!]^{bi}$.*

Using K-interpretations, we define *K-equivalence* ($\stackrel{K}{\equiv}$), *K-entailment* ($\stackrel{K}{\Rightarrow}$) and *K-presupposition*, similarly to KP semantics. It is useful to note that similarly to KP/HS semantics (Facts 1 and 3), K-interpretations satisfy the following standard equivalences, for any $\varphi, \psi \in L_3$:

Fact 6 $\varphi \vee \psi \stackrel{K}{\equiv} \neg((\neg\varphi) \wedge \neg\psi)$ $\varphi \to \psi \stackrel{K}{\equiv} (\neg\varphi) \vee \psi$

The *proof* in (Winter, 2021, Appendix H) simply applies the K-calculus.

Unlike entailment in KP/HS semantics (Fact 4), K-entailment is *not monotonic*. This is illustrated by Example 7, where η K-entails $(\top{:}\beta)$ but $\theta \wedge \eta$ does not.

K-admittance of $\kappa \in L_3$ by a context $C \in L_2$ is defined, similarly to KP-admittance, as follows:

Definition 18 (K-admittance) *We say that a context $C \in L_2$ K-admits a formula $\kappa \in L_3$ if $\alpha^k_{(\top{:}C) \wedge \kappa} \equiv \top$.*

267

By Definition 17, this boils down to requiring that no K-interpretation assigns the formula $(\top : C) \wedge \kappa$ an *undefined* value ('$*$').

We now observe the following general fact about the K-calculus:

Theorem 3 *For any $\kappa \in L_3$, let C be a weakest formula in L_2 that K-admits κ, and let α be a strongest K-presupposition of κ in L_2. Then we have:*

$$\alpha \equiv \alpha_\kappa^K,\ \alpha_\kappa^K \Rightarrow C,\ \text{and}\ C \equiv \alpha_\kappa^{KP}.$$

In words: the strongest *K-presupposition* of κ is directly obtained in the K-calculus as α_κ^K. This K-presupposition entails any weakest context that *K-admits* κ, although it is not necessarily entailed by it (see Example 7). Rather, any weakest context that admits κ in the K-calculus is equivalent to any weakest context that <u>KP</u>-admits κ. See (Winter, 2021, Appendix I) for a *proof* of Theorem 3.

6 Conclusions

The Kleene-Peters (KP) and the Heim-Stalnaker (HS) systems are at the basis of many on-going attempts to describe presupposition projection. The proviso problem threatens these attempts. Following Peters (1979), this paper has argued that the KP system and the HS systems are logically congruent. However, against Peters' claim that his system adequately mimics the proposal in (Karttunen, 1974), we have developed the so-called K-calculus. This system maintains Karttunen's aim of avoiding the proviso problem by distinguishing presuppositions from admittance conditions. Thus, the K-calculus is also distinguished from the KP/HS systems. The proviso problem for these proposals is argued to result from these systems' conflation of *strongest presuppositions* with *weakest admittance conditions*. Both systems rely on a truth-functional account, where the semantic value of a sentence's presupposition is fully determined by the base language's bivalent interpretation. By contrast, the K-calculus relies, following Karttunen, on bivalent *entailment* as the basis for presupposition projection. This system is conjectured to be empirically more adequate than the KP/HS semantics. Notwithstanding, similarly to the KP tables, the K-calculus relies on *left determinant values*, and like the HS semantics, it uses local contexts operationally in its account of presupposition projection. Furthermore, the admittance conditions that the K-calculus derives are the same as in those two systems.

References

Blamey, S. (1986). Partial logic. In D. Gabbay & F. Guenthner (Eds.), *Handbook of Philosophical Logic* (Vol. 3, pp. 1–70). Dordrecht: D. Reidel Publishing Company.

Geurts, B. (1996). Local satisfaction guaranteed: A presupposition theory and its problems. *Linguistics and Philosophy*, *19*(3), 259–294.

Heim, I. (1983). On the projection problem for presuppositions. In D. P. Flickinger (Ed.), *West Coast Conference on Formal Linguistics (WCCFL)* (Vol. 2, pp. 114–125). Stanford, CA: CSLI Publications.

Karttunen, L. (1973). Presuppositions of compound sentences. *Linguistic Inquiry*, *4*(2), 169–193.

Karttunen, L. (1974). Presupposition and linguistic context. *Theoretical Linguistics*, *1*(1-3), 181–194.

Mandelkern, M., Zehr, J., Romoli, J., & Schwarz, F. (2020). We've discovered that projection across conjunction is asymmetric (and it is!). *Linguistics and Philosophy*, *43*, 473–514.

Nouwen, R., Brasoveanu, A., van Eijck, J., & Visser, A. (2016). Dynamic Semantics. In E. N. Zalta (Ed.), *The Stanford Encyclopedia of Philosophy*.

Peters, S. (1979). A truth-conditional formulation of Karttunen's account of presupposition. *Synthese*, *40*(2), 301–316.

Rothschild, D. (2011). Explaining presupposition projection with dynamic semantics. *Semantics and Pragmatics*, *4*, 1–43.

Stalnaker, R. C. (1978). Assertion. In P. Cole (Ed.), *Pragmatics* (pp. 315–322). New York: Academic Press. (Volume 9 of *Syntax and Semantics*)

van Fraassen, B. C. (1971). *Formal Semantics and Logic*. New York: Macmillan.

Winter, Y. (2019). On presupposition projection with trivalent connectives. In K. Blake, F. Davis, K. Lamp, & J. Rhyne (Eds.), *Semantics and Linguistic Theory* (Vol. 29, pp. 582–608).

Winter, Y. (2021). Presupposition, Admittance and Karttunen Calculus (unpublished version of the present paper, with appendices). (https://doi.org/10.24416/UU01-C3WPSV)

Yoad Winter
Utrecht University, Utrecht Institute of Linguistics OTS
The Netherlands
E-mail: y.winter@uu.nl

www.ingramcontent.com/pod-product-compliance
Lightning Source LLC
Chambersburg PA
CBHW070727160426
43192CB00009B/1349